Progress in Mathematics
Volume 237

Petr P. Kulish
Nenad Manojlovic
Henning Samtleben
Editors

Infinite Dimensional Algebras and Quantum Integrable Systems

Birkhäuser Verlag
Basel · Boston · Berlin

Authors:

Petr P. Kulish
St. Petersburg Department of
Steklov Mathematical Institute
Russian Academy of Sciences
Fontaka 27
191011 St. Petersburg
Russia
e-mail: kulish@pdmi.ras.ru

Nenad Manojlovich
Departamento de Matemática
Faculdade de Ciências e Tecnologia
Universidade do Algarve
Campus de Gambelas
8005-139 Faro
Portugal
e-mail: nmanoj@ualg.pt

Henning Samtleben
IInd Institute for Theoretical Physics
University of Hamburg
Luruper Chaussee 149
22761 Hamburg
Germany
e-mail: henning.samtleben@desy.de

2000 Mathematics Subject Classification 14H15, 14H70, 17B37, 17B55, 17B67, 17B68, 17B69, 17B80, 17B81, 20G10, 32F60, 32G15, 32G34, 33C70, 33C80, 35Q58, 37J35, 37K10, 46E20, 58B20, 58F07, 81R10, 81R50, 81U15, 81U20, 81T10, 81T40, 82B20, 82B23

A CIP catalogue record for this book is available from the Library of Congress, Washington D.C., USA

Bibliographic information published by Die Deutsche Bibliothek
Die Deutsche Bibliothek lists this publication in the Deutsche Nationalbibliografie; detailed bibliographic data is available in the Internet at <http://dnb.ddb.de>.

ISBN 3-7643-7215-X Birkhäuser Verlag, Basel – Boston – Berlin

Part of Springer Science+Business Media
Printed on acid-free paper produced of chlorine-free pulp. TCF ∞
Printed in Germany
ISBN-10: 3-7643-7215-X
ISBN-13: 978-3-7643-7215-6

9 8 7 6 5 4 3 2 1

www.birkhauser.ch

Contents

Preface vii

E. Frenkel
Gaudin Model and Opers 1

O.A. Castro-Alvaredo and A. Fring
Integrable Models with Unstable Particles 59

V.G. Kac and M. Wakimoto
Quantum Reduction in the Twisted Case 89

A. Gerasimov, S. Kharchev and D. Lebedev
Representation Theory and Quantum Integrability 133

H.E. Boos, V.E. Korepin and F.A. Smirnov
Connecting Lattice and Relativistic Models
via Conformal Field Theory 157

Kanehisa Takasaki
Elliptic Spectral Parameter and Infinite-Dimensional
Grassmann Variety 175

Takashi Takebe
Trigonometric Degeneration and Orbifold
Wess-Zumino-Witten Model. II 205

L.A. Takhtajan and Lee-Peng Teo
Weil-Petersson Geometry of the Universal Teichmüller Space 225

V. Tarasov
Duality for Knizhinik-Zamolodchikov and Dynamical Equations,
and Hypergeometric Integrals 235

Preface

The workshop "Infinite dimensional algebras and quantum integrable systems" was held in July 2003 at the University of Algarve, Faro, Portugal, as a satellite workshop of the XIV International Congress on Mathematical Physics. Recent developments in the theory of infinite dimensional algebras and their applications to quantum integrable systems were reviewed in invited lectures and a number of contributions from the participants. This volume presents the invited lectures of the workshop.

V. Kac and M. Wakimoto describe the representation theory of twisted vertex algebras obtained by quantum Hamiltonian reduction from affine superalgebras. They present a unified representation theory of twisted superconformal algebras. In particular this leads to unified free field realizations and determinant formulas. Examples include the Ramond type sectors and twisted sectors of the $N = 1, 2, 3, 4$ and the big $N = 4$ superconformal algebras.

E. Frenkel reviews relations between the Gaudin model and opers. He introduces the Gaudin algebra to a Lie algebra $\mathfrak{g}$ as a commutative subalgebra of $U(\mathfrak{g})^{\otimes N}$ that contains in particular the Hamiltonians of the Gaudin model. The spectrum of this algebra can be identified with the space of opers associated to the Langlands dual Lie group ${}^LG$ to $\mathfrak{g}$. Eventually, that allows to relate solutions of the Bethe Ansatz equations to Miura opers and further to the flag varieties associated to ${}^LG$.

L. Takhtajan and Lee-Peng Teo give a brief summary of recent work on geometrical structures on the universal Teichmüller space $T(1)$. They define a Weil-Petersson metric on $T(1)$ by Hilbert space inner products on tangent spaces, compute its Riemann curvature tensor, and show that $T(1)$ is a Kähler-Einstein manifold with negative constant Ricci curvature.

Several lectures are devoted to the applications to quantum integrable models, conformal field theory, and in particular the Knizhnik-Zamolodchikov equations. A. Gerasimov, S. Kharchev and D. Lebedev describe various constructions in the representation theory of classical and quantum groups that are inspired by the Quantum Inverse Scattering Method. Using the separation of variable method in the modern group-theoretical framework, they review recent results on the analytic continuation of Gelfand-Zetlin theory to infinite-dimensional representations of $U(\mathfrak{gl}_n)$ and present the generalization to the quantum groups $U_q(\mathfrak{gl}_n)$. They further demonstrate the applications to quantum integrable systems of Toda type.

H. Boos, V. Korepin and F. Smirnov present new results on correlation functions of the quantum group invariant XXZ-model. These results are based on the relation previously found by Jimbo and Miwa between XXZ correlators and solutions of the q-deformed Knizhnik-Zamolodchikov equations on level -4. These solutions are further related to level 0 solutions; the new formulae suggest the decomposition of general matrix elements with respect to states of the infrared CFT.

Takashi Takabe in his lecture discusses the trigonometric Wess-Zumino-Witten (WZW) model. Based on the result that the trigonometric WZW model is factorized into the orbifold WZW models, he shows that it arises as degeneration of the twisted WZW model on elliptic curves. This is natural as the elliptic r-matrix describing the elliptic Knizhnik-Zamolodchikov equations likewise degenerates to the trigonometric r-matrix. The rigorous proof requires careful algebro-geometric arguments.

V. Tarasov reviews the generalization of the Knizhnik-Zamolodchikov equations to the system of so-called differential dynamical equations. Both systems have a complete set of hypergeometric solutions. It is shown how the known $(\mathfrak{gl}_k, \mathfrak{gl}_n)$ dualities between the two systems of differential equations lead to nontrivial relations between hypergeometric integrals of different dimensions. Extensions to trigonometric and difference versions of the Knizhnik-Zamolodchikov and dynamical equations are briefly discussed.

Recent progress in the theory of classical integrable systems is reported by Kanehisa Takasaki. He analyzes new classes of integrable partial differential equations admitting a zero-curvature representation on algebraic curves of arbitrary genus. He first reviews how conventional soliton equations are treated in the Grassmannian perspective, considering as example the nonlinear Schrödinger hierarchy in great detail. Subsequently, recent results on the elliptic analogues of these systems are presented.

Finally, O. Castro-Alvaredo and A. Fring present a lecture on two-dimensional quantum field theories with unstable particles. They review the main facts on analytic scattering theory of factorizable integrable models before presenting a new bootstrap principle that allows to include unstable particles in the spectrum. They describe the underlying Lie algebraic structure and the construction of an S-matrix like object characterizing the scattering between unstable particles.

We gratefully acknowledge the financial support provided by the Centre for Mathematics and its Applications (CEMAT) of the Instituto Superior Técnico, the Luso-American Foundation and the Portuguese Foundation for Science and Technology, project POCTI/33858/MAT/2000. We wish to express our gratitude to José Ferreira Pereira Ferraz, Vice-Rector of the University of Algarve, and António Ferreira dos Santos, CEMAT and Department of Mathematics of Instituto Superior Técnico, for their support. Finally, we would like to thank all the participants for creating an excellent atmosphere of the workshop, and especially the contributors of this volume for writing a wonderful set of lecture notes.

P.P. Kulish, N. Manojlović, H. Samtleben

Progress in Mathematics, Vol. 237, 1–58

Gaudin Model and Opers

Edward Frenkel

Abstract. This is a review of our previous works [FFR, F1, F3] (some of them joint with B. Feigin and N. Reshetikhin) on the Gaudin model and opers. We define a commutative subalgebra in the tensor power of the universal enveloping algebra of a simple Lie algebra $\mathfrak{g}$. This algebra includes the Hamiltonians of the Gaudin model, hence we call it the Gaudin algebra. It is constructed as a quotient of the center of the completed enveloping algebra of the affine Kac-Moody algebra $\widehat{\mathfrak{g}}$ at the critical level. We identify the spectrum of the Gaudin algebra with the space of opers associated to the Langlands dual Lie algebra ${}^L\mathfrak{g}$ on the projective line with regular singularities at the marked points. Next, we recall the construction of the eigenvectors of the Gaudin algebra using the Wakimoto modules over $\widehat{\mathfrak{g}}$ of critical level. The Wakimoto modules are naturally parameterized by Miura opers (or, equivalently, Cartan connections), and the action of the center on them is given by the Miura transformation. This allows us to relate solutions of the Bethe Ansatz equations to Miura opers and ultimately to the flag varieties associated to the Langlands dual Lie algebra ${}^L\mathfrak{g}$.

Mathematics Subject Classification (2000). 17B67 and 82B23.

Keywords. Gaudin model, Bethe Ansatz, oper, Miura transformation, Wakimoto module.

Introduction

Let $\mathfrak{g}$ be a finite-dimensional simple Lie algebra over $\mathbb{C}$ and $U(\mathfrak{g})$ its universal enveloping algebra. Choose a basis $\{J_a\}, a = 1, \ldots, d$, of $\mathfrak{g}$, and let $\{J^a\}$ the dual basis with respect to a non-degenerate invariant bilinear form on $\mathfrak{g}$. Let $z_1, \ldots, z_N$ be a collection of distinct complex numbers.

The *Gaudin Hamiltonians* are the following elements of the algebra $U(\mathfrak{g})^{\otimes N}$:

$$\Xi_i = \sum_{j \neq i} \sum_{a=1}^{d} \frac{J_a^{(i)} J^{a(j)}}{z_i - z_j}, \qquad i = 1, \ldots, N, \tag{0.1}$$

Partially supported by grants from the NSF and DARPA.

where for $A \in \mathfrak{g}$ we denote by $A^{(i)}$ the element of $U(\mathfrak{g})^{\otimes N}$ which is the tensor product of A in the ith factor and 1 in all other factors. One checks easily that these elements commute with each other and are invariant with respect to the diagonal action of $\mathfrak{g}$ on $U(\mathfrak{g})^{\otimes N}$.

For any collection $M_1, \ldots, M_N$ of $\mathfrak{g}$-modules, the Gaudin Hamiltonians give rise to commuting linear operators on $\bigotimes_{i=1}^N M_i$. We are interested in the diagonalization of these operators. More specifically, we will consider the following two cases: when all of the M_i's are Verma modules and when they are finite-dimensional irreducible modules.

It is natural to ask: are there any other elements in $U(\mathfrak{g})^{\otimes N}$ which commute with the Gaudin Hamiltonians? Clearly, the N-fold tensor product $Z(\mathfrak{g})^{\otimes N}$ of the center $Z(\mathfrak{g})$ of $U(\mathfrak{g})$ is the center of $U(\mathfrak{g})^{\otimes N}$, and its elements obviously commute with the Ξ_i's. As shown in [FFR], if $\mathfrak{g}$ has rank grater than one, then in addition to the Gaudin Hamiltonians and the central elements there are other elements in $U(\mathfrak{g})^{\otimes N}$ of orders higher than two which commute with the Gaudin operators and with each other (but explicit formulas for them are much more complicated and unknown in general). Adjoining these "higher Gaudin Hamiltonians" to the Ξ_i's together with the center $Z(\mathfrak{g})^{\otimes N}$, we obtain a large commutative subalgebra of $U(\mathfrak{g})^{\otimes N}$. We will call it the *Gaudin algebra* and denote it by $\mathcal{Z}_{(z_i)}(\mathfrak{g})$.

The construction of $\mathcal{Z}_{(z_i)}(\mathfrak{g})$ will be recalled in Section 2. The key point is the realization of $U(\mathfrak{g})^{\otimes N}$ as the space of *coinvariants* of induced modules over the affine Kac-Moody algebra $\widehat{\mathfrak{g}}$ on the projective line. Using this realization, we obtain a surjective map from the center of the completed universal enveloping algebra of $\widehat{\mathfrak{g}}$ at the *critical level* onto $\mathcal{Z}_{(z_i)}(\mathfrak{g})$.

The next natural question is what is the spectrum of $\mathcal{Z}_{(z_i)}(\mathfrak{g})$, i.e., the set of all maximal ideals of $\mathcal{Z}_{(z_i)}(\mathfrak{g})$, or equivalently, algebra homomorphisms $\mathcal{Z}_{(z_i)}(\mathfrak{g}) \to \mathbb{C}$. Knowing the answer is important, because then we will know how to think about the common eigenvalues of the higher Gaudin operators on the tensor products $\bigotimes_{i=1}^N M_i$ of $\mathfrak{g}$-modules. These common eigenvalues correspond to points of the spectrum of $\mathcal{Z}_{(z_i)}(\mathfrak{g})$.

The answer comes from the description of the center of the completed universal enveloping algebra of $\widehat{\mathfrak{g}}$ at the critical level. In [FF2, F2] it is shown that the spectrum of this center (more precisely, the center of the corresponding vertex algebra) is canonically identified with the space of the so-called ${}^L G$*-opers*, where ${}^L G$ is the *Langlands dual Lie group* to $\mathfrak{g}$ (of adjoint type), on the formal disc. This result leads us to the following description of the spectrum of the algebra $\mathcal{Z}_{(z_i)}(\mathfrak{g})$ of higher Gaudin Hamiltonians: it is the space of ${}^L G$-opers on $\mathbb{P}^1$ with regular singularities at the points $z_1, \ldots, z_N$ and ∞.

We obtain this description from some basic facts about the spaces of coinvariants from [FB]. Recall that the space of coinvariants $H_V(X; (x_i); (M_i))$ is defined in [FB] for any (quasi-conformal) vertex algebra V, a smooth projective curve X, a collection $x_1, \ldots, x_n$ of distinct points of X and a collection of V-modules $M_1, \ldots, M_n$ attached to those points. Suppose that V is a commutative vertex algebra, and so in particular it is a commutative algebra. Suppose that the spectrum of V is the space $S(D)$ of certain geometric objects, such as ${}^LG$-opers, on the disc $D = \operatorname{Spec} \mathbb{C}[[t]]$. Then a V-module is the same as a smooth module over the complete topological algebra $U(V)$ of functions on $S(D^\times)$, which is the space of our objects (such as ${}^LG$-opers) on the punctured disc $D^\times = \operatorname{Spec} \mathbb{C}((t))$. Suppose in addition that each V-module M_i is the space of functions on a subspace S_i of $S(D^\times)$ (with its natural $\operatorname{Fun} S(D^\times)$-module structure). Then the space of coinvariants $H_V(X;(x_i);(M_i))$ is naturally a commutative algebra, and its spectrum is the space of our objects (such as ${}^LG$-opers) on $X \backslash \{x_1, \ldots, x_n\}$ whose restriction to the punctured disc $D^\times_{x_i}$ around x_i belongs to $S_i \subset S(D^\times_{x_i}), i = 1, \ldots, n$.

For example, if $\mathfrak{g} = \mathfrak{sl}_2$, then ${}^LG = PGL_2$, and PGL_2-opers are the same as second order differential operators $\partial_t^2 - q(t)$ acting from sections of the line bundle $\Omega^{-1/2}$ to sections of $\Omega^{3/2}$. A PGL_2-oper on $\mathbb{P}^1$ with regular singularities at $z_1, \ldots, z_N$ and ∞ may be written as the Fuchsian differential operator of second order with regular singularities at $z_1, \ldots, z_N$,

$$\partial_t^2 - \sum_{i=1}^N \frac{c_i}{(t-z_i)^2} - \sum_{i=1}^N \frac{\mu_i}{t-z_i},$$

satisfying the condition $\sum_{i=1}^N \mu_i = 0$ that insures that it also has regular singularity at ∞. Defining such an operator is the same as giving a collection of numbers $c_i, \mu_i, i = 1, \ldots, N$, such that $\sum_{i=1}^N \mu_i = 0$. The set

$$\left\{ c_i, \mu_i, i = 1, \ldots, N \,\middle|\, \sum_{i=1}^N \mu_i = 0 \right\},$$

is then the spectrum of the Gaudin algebra $\mathcal{Z}_{(z_i)}(\mathfrak{g})$, which in this case is the polynomial algebra generated by the Casimir operators $C_i = \frac{1}{2} \sum_a J_a^{(i)} J^{a(i)}, i = 1, \ldots, N$, and the Gaudin Hamiltonians Ξ_i, subject to the relation $\sum_{i=1}^N \Xi_i = 0$. In other words, the numbers c_i record the eigenvalues of the C_i's, while the numbers μ_i record the eigenvalues of the Ξ_i's.

For a general simple Lie algebra $\mathfrak{g}$, the Gaudin algebra $\mathcal{Z}_N(\mathfrak{g})$ has many more generators, and its spectrum does not have such a nice system of coordinates as the c_i's and the μ_i's in the above example. Therefore the description of the spectrum as a space of ${}^LG$-opers is very useful. In particular, we obtain that common eigenvalues of the higher Gaudin Hamiltonians are encoded by ${}^LG$-opers on $\mathbb{P}^1$, with regular singularities at prescribed points. These ${}^LG$-opers appear as generalizations of the above second order Fuchsian operators.

Next, we ask which points in the spectrum of $\mathcal{Z}_{(z_i)}(\mathfrak{g})$ might occur as the common eigenvalues on particular tensor products $\bigotimes_{i=1}^N M_i$. We answer this question first in the case when each M_i admits a central character: namely, it turns out that the central character of M_i fixes the residue of the ${}^LG$-oper at the point z_i. We then show that if all $\mathfrak{g}$-modules M_i are finite-dimensional and irreducible, then the ${}^LG$-opers encoding possible eigenvalues of the higher Gaudin Hamiltonians in $\bigotimes_{i=1}^N M_i$ necessarily have *trivial monodromy representation.*

We conjecture that there is a bijection between the eigenvalues of the Gaudin Hamiltonians on $\bigotimes_{i=1}^N M_i$, where the M_i's are irreducible finite-dimensional $\mathfrak{g}$-modules, and ${}^L G$-opers on $\mathbb{P}^1$ with prescribed singularities at $z_1, \dots, z_N, \infty$ and trivial monodromy.

Thus, we obtain a correspondence between two seemingly unrelated objects: the eigenvalues of the generalized Gaudin Hamiltonians and the ${}^L G$-opers on $\mathbb{P}^1$. The connection between the eigenvalues of the Gaudin operators and differential operators of some sort has been observed previously, but it was not until [FFR, F1] that this phenomenon was explained conceptually.

We present a more geometric description the ${}^L G$-opers without monodromy (which occur as the eigenvalues of the Gaudin Hamiltonians) as isomorphism classes of holomorphic maps from $\mathbb{P}^1$ to ${}^L G/{}^L B$, the flag manifold of ${}^L G$, satisfying a certain transversality condition. For example, if ${}^L G = PGL_2$, they may be described as holomorphic maps $\mathbb{P}^1 \to \mathbb{P}^1$ whose derivative vanishes to prescribed orders at the marked points $z_1, \dots, z_N$ and ∞, and does not vanish anywhere else (these orders correspond to the highest weights of the finite-dimensional representations inserted at those points).

If the ${}^L G$-oper is non-degenerate (in the sense explained in Section 5.2), then we can associate to it an eigenvector of the Gaudin Hamiltonians called a *Bethe vector*. The procedure to construct eigenvectors of the Gaudin Hamiltonians that produces these vectors is known as the *Bethe Ansatz*. In [FFR] we explained that this procedure can also be understood in the framework of coinvariants of $\widehat{\mathfrak{g}}$-modules of critical level. We need to use a particular class of $\widehat{\mathfrak{g}}$-modules, called the *Wakimoto modules*.

Let us recall that the Wakimoto modules of critical level are naturally parameterized by objects closely related to opers, which we call *Miura opers*. They may also be described more explicitly as certain connections on a particular ${}^L H$-bundle $\Omega^{-\rho}$ on the punctured disc, where ${}^L H$ is the Cartan subgroup of ${}^L G$. The center acts on the Wakimoto module corresponding to a Cartan connection by the Miura transformation of this connection (see [F2]). The idea of [FFR] was to use the spaces of coinvariants of the tensor product of the Wakimoto modules to construct eigenvectors of the generalized Gaudin Hamiltonians. We found in [FFR] that the eigenvalues of the Gaudin Hamiltonians on these vectors are encoded by the ${}^L G$-opers which are obtained by applying the Miura transformation to certain Cartan connections on $\mathbb{P}^1$.

More precisely, the Bethe vector depends on an m-tuple of complex numbers w_j, where $j = 1, \dots, m$, with an extra datum attached to each of them, $i_j \in I$, where I is the set of nodes of the Dynkin diagram of $\mathfrak{g}$ (or equivalently, the set of simple roots of $\mathfrak{g}$). These numbers have to be distinct from the z_i's and satisfy the following system of *Bethe Ansatz equations*:

$$\sum_{i=1}^N \frac{\langle \lambda_i, \check{\alpha}_{i_j} \rangle}{w_j - z_i} - \sum_{s \neq j} \frac{\langle \alpha_{i_s}, \check{\alpha}_{i_j} \rangle}{w_j - w_s} = 0, \quad j = 1, \dots, m, \tag{0.2}$$

where λ_i denotes the highest weight of the finite-dimensional $\mathfrak{g}$-module $M_i = V_{\lambda_i}, i = 1, \ldots, N$.

We can compute explicitly the ${}^L G$-oper encoding the eigenvalues of the generalized Gaudin Hamiltonians on this vector. As shown in [FFR], this ${}^L G$-oper is obtained by applying the Miura transformation of the connection

$$\partial_t + \sum_{i=1}^{N} \frac{\lambda_i}{t - z_i} - \sum_{j=1}^{m} \frac{\alpha_{i_j}}{t - w_j} \tag{0.3}$$

on the ${}^L H$-bundle $\Omega^{-\rho}$ on $\mathbb{P}^1$. This ${}^L G$-oper automatically has trivial monodromy.

The Bethe vector corresponding to a solution of the system (0.2) is a highest weight vector in $\bigotimes_{i=1}^{N} V_{\lambda_i}$ of weight

$$\mu = \sum_{i=1}^{N} \lambda_i - \sum_{j=1}^{m} \alpha_{i_j},$$

so it can only be non-zero if μ is a dominant integral weight of $\mathfrak{g}$. But it is still interesting to describe the set of *all* solutions of the Bethe Ansatz equations (0.2), even for non-dominant weights μ.

While the eigenvalues of the Gaudin Hamiltonians are parameterized by ${}^L G$-opers, it turns out that the solutions of the Bethe Ansatz equations are parameterized by the (non-degenerate) Miura ${}^L G$-opers. As mentioned above, those may in turn be related to very simple objects, namely, connections on an ${}^L H$-bundle $\Omega^{-\rho}$ of the kind given above in formula (0.3).

A ${}^L G$-oper on a curve X is by definition a triple $(\mathcal{F}, \nabla, \mathcal{F}_{{}^L B})$, where $\mathcal{F}$ is a ${}^L G$-bundle on X, ∇ is a connection on $\mathcal{F}$ and $\mathcal{F}_{{}^L B}$ is a reduction of $\mathcal{F}$ to a Borel subgroup ${}^L B$ of ${}^L G$, which satisfies a certain transversality condition with ∇. A Miura ${}^L G$-oper is by definition a quadruple $(\mathcal{F}, \nabla, \mathcal{F}_{{}^L B}, \mathcal{F}'_{{}^L B})$ where $\mathcal{F}'_{{}^L B}$ is another ${}^L B$-reduction of $\mathcal{F}$, which is preserved by ∇. The space of Miura opers on a curve X whose underlying oper has regular singularities and trivial monodromy representation (so that $\mathcal{F}$ is isomorphic to the trivial bundle) is isomorphic to the *flag manifold* ${}^L G/{}^L B$ of ${}^L G$. Indeed, in order to define the ${}^L B$-reduction $\mathcal{F}'_{{}^L B}$ of such $\mathcal{F}$ everywhere, it is sufficient to define it at one point $x \in X$ and then use the connection to "spread" it around. But choosing a ${}^L B$-reduction at one point means choosing an element of the twist of ${}^L G/{}^L B$ by $\mathcal{F}_x$, and so we see that the space of all reductions is isomorphic to the flag manifold of ${}^L G$.

The relative position of the two reductions $\mathcal{F}_{{}^L B}$ and $\mathcal{F}'_{{}^L B}$ at each point of X is measured by an element w of the Weyl group W of G. The two reductions are in generic relative position (corresponding to $w = 1$) almost everywhere on X. The mildest possible non-generic relative positions correspond to the simple reflections s_i from W. We call a Miura oper on $\mathbb{P}^1$ with marked points $z_1, \ldots, z_N$ *non-degenerate* if the two reductions $\mathcal{F}_B$ and $\mathcal{F}'_B$ are in generic position at $z_1, \ldots, z_N$, and elsewhere on $\mathbb{P}^1$ their relative position is either generic or corresponds to a simple reflection. We denote the points where the relative position is not generic by $w_j, j = 1, \ldots, m$; each point w_j comes together with a simple reflection s_{i_j}, or equivalently a simple root α_{i_j} attached to it.

It is then easy to see that this collection satisfies the equations (0.2), and conversely any solution of (0.2) corresponds to a non-degenerate Miura oper (or to an ${}^L H$-connection (0.3)). Thus, we obtain that there is a bijection between the set of solutions of (0.2) (for all possible collections $\{i_1, \ldots, i_m\}$) and the set of non-degenerate Miura ${}^L G$-opers such that the underlying ${}^L G$-opers have prescribed residues at the points $z_1, \ldots, z_N, \infty$ and trivial monodromy.

Now let us fix $\lambda_1, \ldots, \lambda_N$ and μ. Then every ${}^L G$-oper τ on $\mathbb{P}^1$ with regular singularities at $z_1, \ldots, z_N$ and ∞ and with prescribed residues corresponding to $\lambda_1, \ldots, \lambda_N$ and μ and trivial monodromy admits a horizontal ${}^L B$-reduction $\mathcal{F}'_{{}^L B}$ satisfying the conditions of a non-degenerate Miura oper. Since these are open conditions, we find that for such $z_1, \ldots, z_N$ the non-degenerate Miura oper structures on a particular ${}^L G$-oper τ on $\mathbb{P}^1$ form an open dense subset in the set of all Miura oper structures on τ. But the set of all Miura structures on a given ${}^L G$-oper τ is isomorphic to the flag manifold ${}^L G/{}^L B$. Therefore we find that the set of non-degenerate Miura oper structures on τ is an open dense subset of ${}^L G/{}^L B$!

Recall that the set of all solutions of the Bethe Ansatz equations (0.2) is the union of the sets of non-degenerate Miura oper structures on all ${}^L G$-opers with trivial monodromy. Hence it is naturally a disjoint union of subsets, parameterized by these ${}^L G$-opers. We have now identified each of these sets with an open and dense subset of the flag manifold ${}^L G/{}^L B$.

Let us summarize our results:

- the eigenvalues of the Hamiltonians of the Gaudin model associated to a simple Lie algebra $\mathfrak{g}$ on the tensor product of finite-dimensional representations are encoded by ${}^L G$-opers on $\mathbb{P}^1$, where ${}^L G$ is the Langlands dual group of G, which have regular singularities at the marked points $z_1, \ldots, z_N, \infty$ and trivial monodromy;
- if such an oper τ is non-degenerate, then we can associte to it a solution of the Bethe Ansatz equations (0.2), which gives rise to the Bethe eigenvector of dominant integral weight whose eigenvalues are encoded by τ;
- there is a one-to-one correspondence between the set of all solutions of the Bethe Ansatz equations (0.2) and the set of non-degenerate Miura opers corresponding to a fixed ${}^L G$-oper;
- the set of non-degenerate Miura opers corresponding to the same underlying ${}^L G$-oper is an open dense subset of the flag manifold ${}^L G/{}^L B$ of the Langlands dual group, and therefore the set of all solutions of the Bethe Ansatz equations (0.2) is the union of certain open dense subsets of the flag manifold of the Langlands dual group, one for each ${}^L G$-oper.

Finally, to a degenerate ${}^L G$-oper we can also attach, at least in some cases, an eigenvector of the Gaudin hamiltonians by generalizing the Bethe Ansatz procedure, as explained in Section 5.5.

The paper is organized as follows. In Section 1 we introduce opers and discuss their basic properties. We define opers with regular singularities and their residues.

In Section 2, following [FFR], we define the Gaudin algebra using the coinvariants of the affine Kac-Moody algebra $\widehat{\mathfrak{g}}$ of critical level. We recall the results of [FF2, F2] on the isomorphism of the center of the completed universal enveloping algebra of $\widehat{\mathfrak{g}}$ at the critical level and the algebra of functions on the space of ${}^LG$-opers on the punctured disc. Using these results and general facts about the spaces of coinvariants from [FB], we describe the spectrum of the Gaudin algebra. In Section 3 we introduce Miura opers, Cartan connections and the Miura transformation and describe their properties, following [F2, F3]. We use these results in the next section, Section 4, to describe the Bethe Ansatz, a construction of eigenvectors of the Gaudin algebra. We introduce the Wakimoto modules of critical level, following [FF1, F2]. The Wakimoto modules are naturally parameterized by the Cartan connections on the punctured disc introduced in Section 3. The action of the center on the Wakimoto modules is given by the Miura transformation. We construct the Bethe vectors, following [FFR], using the coinvariants of the Wakimoto modules. We show that the Bethe Ansatz equations which ensure that this vector is an eigenvector of the Gaudin Hamiltonians coincide with the requirement that the Miura transformation of the Cartan connection on $\mathbb{P}^1$ encoding the Wakimoto modules has no singularities at the points $w_1, \dots, w_m$. Finally, in Section 5 we consider the Gaudin model in the case when all modules M_i finite-dimensional modules. We describe the precise connection between the spectrum of the Gaudin algebra on the tensor product of finite-dimensional modules and the set of ${}^LG$-opers with prescribed singularities at $z_1, \dots, z_N, \infty$ and trivial monodromy.

1. Opers

1.1. Definition of opers

Let G be a simple algebraic group of adjoint type, B a Borel subgroup and $N = [B,B]$ its unipotent radical, with the corresponding Lie algebras $\mathfrak{n} \subset \mathfrak{b} \subset \mathfrak{g}$. The quotient $H = B/N$ is a torus. Choose a splitting $H \to B$ of the homomorphism $B \to H$ and the corresponding splitting $\mathfrak{h} \to \mathfrak{b}$ at the level of Lie algebras. Then we will have a Cartan decomposition $\mathfrak{g} = \mathfrak{n}_- \oplus \mathfrak{h} \oplus \mathfrak{n}$. We will choose generators $\{e_i\}, i = 1, \dots, \ell$, of $\mathfrak{n}$ and generators $\{f_i\}, i = 1, \dots, \ell$ of $\mathfrak{n}_-$ corresponding to simple roots, and denote by $\check{\rho} \in \mathfrak{h}$ the sum of the fundamental coweights of $\mathfrak{g}$. Then we will have the following relations: $[\check{\rho}, e_i] = 1, [\check{\rho}, f_i] = -1$.

A G-oper on a smooth curve X (or a disc $D \simeq \operatorname{Spec} \mathbb{C}[[t]]$ or a punctured disc $D^\times = \operatorname{Spec} \mathbb{C}((t))$) is by definition a triple $(\mathcal{F}, \nabla, \mathcal{F}_B)$, where $\mathcal{F}$ is a principal G-bundle $\mathcal{F}$ on X, ∇ is a connection on $\mathcal{F}$ and $\mathcal{F}_B$ is a B-reduction of $\mathcal{F}$ such that locally on X (with respect to a local coordinate t and a local trivialization of $\mathcal{F}_B$) the connection has the form

$$\nabla = \partial_t + \sum_{i=1}^{\ell} \psi_i(t) f_i + \mathbf{v}(t), \tag{1.1}$$

where each $\psi_i(t)$ is a nowhere vanishing function, and $\mathbf{v}(t)$ is a $\mathfrak{b}$-valued function. The space of G-opers on X is denoted by $\mathrm{Op}_G(X)$.

This definition is due to A. Beilinson and V. Drinfeld [BD1] (in the case when X is the punctured disc opers were first introduced in [DS]).

In particular, if $U = \operatorname{Spec} R$ is an affine curve with the ring of functions R and t is a global coordinate on U (for example, if $U = \operatorname{Spec} \mathbb{C}[[t]]$), then $\mathrm{Op}_G(U)$ is isomorphic to the quotient of the space of operators of the form

$$\nabla = \partial_t + \sum_{i=1}^{\ell} f_i + \mathbf{v}(t), \qquad \mathbf{v}(t) \in \mathfrak{b}(R), \tag{1.2}$$

by the action of the group $N(R)$ (we use the action of $H(R)$ to make all functions $\psi_i(t)$ equal to 1). Recall that the gauge transformation of an operator $\partial_t + A(t)$, where $A(t) \in \mathfrak{g}(R)$ by $g(t) \in G(R)$ is given by the formula

$$g \cdot (\partial_t + A(t)) = \partial_t + gA(t)g^{-1} - \partial_t g \cdot g^{-1}.$$

The operator $\operatorname{ad} \check{\rho}$ defines the principal gradation on $\mathfrak{b}$, with respect to which we have a direct sum decomposition $\mathfrak{b} = \bigoplus_{i \geq 0} \mathfrak{b}_i$. Set

$$p_{-1} = \sum_{i=1}^{\ell} f_i.$$

Let p_1 be the unique element of degree 1 in $\mathfrak{n}$, such that $\{p_{-1}, 2\check{\rho}, p_1\}$ is an $\mathfrak{sl}_2$-triple. Let $V_{\mathrm{can}} = \oplus_{i \in E} V_{\mathrm{can},i}$ be the space of $\operatorname{ad} p_1$-invariants in $\mathfrak{n}$. Then p_1 spans $V_{\mathrm{can},1}$. Choose a linear generator p_j of V_{can,d_j} (if the multiplicity of d_j is greater than one, which happens only in the case $\mathfrak{g} = D_{2n}^{(1)}, d_j = 2n$, then we choose linearly independent vectors in V_{can,d_j}). The following result is due to Drinfeld and Sokolov [DS] (the proof is reproduced in Lemma 2.1 of [F3]).

Lemma 1.1. *The gauge action of $N(R)$ on $\widetilde{\mathrm{Op}}_G(\operatorname{Spec} R)$ is free, and each gauge equivalence class contains a unique operator of the form $\nabla = \partial_t + p_{-1} + \mathbf{v}(t)$, where $\mathbf{v}(t) \in V_{\mathrm{can}}(R)$, so that we can write*

$$\mathbf{v}(t) = \sum_{j=1}^{\ell} v_j(t) \cdot p_j. \tag{1.3}$$

1.2. PGL_2-opers

For $\mathfrak{g} = \mathfrak{sl}_2, G = PGL_2$ we obtain an identification of the space of PGL_2-opers with the space of operators of the form

$$\partial_t + \begin{pmatrix} 0 & v(t) \\ 1 & 0 \end{pmatrix}.$$

If we make a change of variables $t = \varphi(s)$, then the corresponding connection operator will become

$$\partial_s + \begin{pmatrix} 0 & \varphi'(s)v(\varphi(s)) \\ \varphi'(s) & 0 \end{pmatrix}.$$

Applying the B-valued gauge transformation with

$$\begin{pmatrix} 1 & \frac{1}{2}\frac{\varphi''(s)}{\varphi'(s)} \\ 0 & 1 \end{pmatrix} \begin{pmatrix} (\varphi'(s))^{1/2} & 0 \\ 0 & (\varphi'(s))^{1/2} \end{pmatrix},$$

we obtain the operator

$$\partial_s + \begin{pmatrix} 0 & v(\varphi(s))\varphi'(s)^2 - \frac{1}{2}\{\varphi, s\} \\ 1 & 0 \end{pmatrix},$$

where

$$\{\varphi, s\} = \frac{\varphi'''}{\varphi'} - \frac{3}{2}\left(\frac{\varphi''}{\varphi'}\right)^2$$

is the Schwarzian derivative. Thus, under the change of variables $t = \varphi(s)$ we have

$$v(t) \mapsto v(\varphi(s))\varphi'(s)^2 - \frac{1}{2}\{\varphi, s\}.$$

this coincides with the transformation properties of the second order differential operators $\partial_t^2 - v(t)$ acting from sections of $\Omega^{-1/2}$ to sections of $\Omega^{3/2}$, where Ω is the canonical line bundle on X. Such operators are known as *projective connections* on X (see, e.g., [FB], Sect. 9.2), and so PGL_2-opers are the same as projective connections.

For a general $\mathfrak{g}$, the first coefficient function $v_1(t)$ in (1.3) transforms as a projective connection, and the coefficient $v_i(t)$ with $i > 1$ transforms as a $(d_i + 1)$-differential on X. Thus, we obtain an isomorphism

$$\mathrm{Op}_G(X) \simeq \mathcal{P}roj(X) \times \bigoplus_{i=2}^{\ell} \Gamma(X, \Omega^{(d_i+1)}). \tag{1.4}$$

1.3. Opers with regular singularities

Let x be a point of a smooth curve X and $D_x = \operatorname{Spec} \mathcal{O}_x, D_x^\times = \operatorname{Spec} \mathcal{K}_x$, where $\mathcal{O}_x$ is the completion of the local ring of x and $\mathcal{K}_x$ is the field of fractions of $\mathcal{O}_x$. Choose a formal coordinate t at x, so that $\mathcal{O}_x \simeq \mathbb{C}[[t]]$ and $\mathcal{K}_x = \mathbb{C}((t))$. Recall that the space $\mathrm{Op}_G(D_x)$ (resp., $\mathrm{Op}_G(D_x^\times)$) of G-opers on D_x (resp., $D_x^\times$) is the quotient of the space of operators of the form (1.1) where $\psi_i(t)$ and $\mathbf{v}(t)$ take values in $\mathcal{O}_x$ (resp., in $\mathcal{K}_x$) by the action of $B(\mathcal{O}_x)$ (resp., $B(\mathcal{K}_x)$).

A G-oper on D_x with regular singularity at x is by definition (see [BD1], Sect. 3.8.8) a $B(\mathcal{O}_x)$-conjugacy class of operators of the form

$$\nabla = \partial_t + t^{-1}\left(\sum_{i=1}^{\ell} \psi_i(t) f_i + \mathbf{v}(t)\right), \tag{1.5}$$

where $\psi_i(t) \in \mathcal{O}_x, \psi_i(0) \neq 0$, and $\mathbf{v}(t) \in \mathfrak{b}(\mathcal{O}_x)$. Equivalently, it is an $N(\mathcal{O}_x)$-equivalence class of operators

$$\nabla = \partial_t + \frac{1}{t}\left(p_{-1} + \mathbf{v}(t)\right), \qquad \mathbf{v}(t) \in \mathfrak{b}(\mathcal{O}_x). \tag{1.6}$$

Denote by $\mathrm{Op}_G^{\mathrm{RS}}(D_x)$ the space of opers on D_x with regular singularity. It is easy to see that the natural map $\mathrm{Op}_G^{\mathrm{RS}}(D_x) \to \mathrm{Op}_G(D_x^\times)$ is injective. Therefore an oper with regular singularity may be viewed as an oper on the punctured disc. But to an oper with regular singularity one can unambiguously attach a point in

$$\mathfrak{g}/G := \operatorname{Spec} \mathbb{C}[\mathfrak{g}]^G \simeq \mathbb{C}[\mathfrak{h}]^W =: \mathfrak{h}/W,$$

its residue, which in our case is equal to $p_{-1} + \mathbf{v}(0)$.

Given $\check{\lambda} \in \mathfrak{h}$, we denote by $\mathrm{Op}_G^{\mathrm{RS}}(D_x)_{\check{\lambda}}$ the subvariety of $\mathrm{Op}_G^{\mathrm{RS}}(D_x)$ which consists of those opers that have residue $\varpi(-\check{\lambda} - \check{\rho}) \in \mathfrak{h}/W$, where ϖ is the projection $\mathfrak{h} \to \mathfrak{h}/W$.

In particular, the residue of a regular oper $\partial_t + p_{-1} + \mathbf{v}(t)$, where $\mathbf{v}(t) \in \mathfrak{b}(\mathcal{O}_x)$, is equal to $\varpi(-\check{\rho})$ (see [BD1]). Indeed, a regular oper may be brought to the form (1.6) by using the gauge transformation with $\check{\rho}(t) \in B(\mathcal{K}_x)$, after which it takes the form

$$\partial_t + \frac{1}{t}\left(p_{-1} - \check{\rho} + t \cdot \check{\rho}(t)(\mathbf{v}(t))\check{\rho}(t)^{-1}\right).$$

If $\mathbf{v}(t)$ is regular, then so is $\check{\rho}(t)(\mathbf{v}(t))\check{\rho}(t)^{-1}$. Therefore the residue of this oper in $\mathfrak{h}/W$ is equal to $\varpi(-\check{\rho})$, and so $\mathrm{Op}_G(D_x) = \mathrm{Op}_G^{\mathrm{RS}}(D_x)_0$.

For $G = PGL_2$ we identify $\mathfrak{h}$ with $\mathbb{C}$ and so $\check{\lambda}$ with a complex number. Then one finds that $\mathrm{Op}_{PGL_2}^{\mathrm{RS}}(D_x)_{\check{\lambda}}$ is the space of second order operators of the form

$$\partial_t^2 - \frac{\check{\lambda}(\check{\lambda}+2)/4}{t^2} - \sum_{n \geq -1} v_n t^n. \tag{1.7}$$

Now suppose that $\check{\lambda}$ is a dominant integral coweight of $\mathfrak{g}$. Following Drinfeld, introduce the variety $\mathrm{Op}_G(D_x)_{\check{\lambda}}$ as the quotient of the space of operators of the form

$$\nabla = \partial_t + \sum_{i=1}^{\ell} \psi_i(t) f_i + \mathbf{v}(t), \tag{1.8}$$

where

$$\psi_i(t) = t^{\langle \alpha_i, \check{\lambda} \rangle}(\kappa_i + t(\ldots)) \in \mathcal{O}_x, \qquad \kappa_i \neq 0$$

and $\mathbf{v}(t) \in \mathfrak{b}(\mathcal{O}_x)$, by the gauge action of $B(\mathcal{O}_x)$. Equivalently, $\mathrm{Op}_G(D_x)_{\check{\lambda}}$ is the quotient of the space of operators of the form

$$\nabla = \partial_t + \sum_{i=1}^{\ell} t^{\langle \alpha_i, \check{\lambda} \rangle} f_i + \mathbf{v}(t), \tag{1.9}$$

where $\mathbf{v}(t) \in \mathfrak{b}(\mathcal{O}_x)$, by the gauge action of $N(\mathcal{O}_x)$. Considering the $N(\mathcal{K}_x)$-class of such an operator, we obtain an oper on $D_x^\times$. Thus, we have a map $\mathrm{Op}_G(D_x)_{\check{\lambda}} \to \mathrm{Op}_G(D_x^\times)$.

Lemma 1.2 ([F3], Lemma 2.4). *The map $\mathrm{Op}_G(D_x)_{\check{\lambda}} \to \mathrm{Op}_G(D_x^\times)$ is injective and its image is contained in the subvariety $\mathrm{Op}_G^{\mathrm{RS}}(D_x)_{\check{\lambda}}$. Moreover, the points of $\mathrm{Op}_G(D_x)_{\check{\lambda}}$ are precisely those G-opers with regular singularity and residue $\check{\lambda}$ which have no monodromy around x.*

The space $\mathrm{Op}_{PGL_2}(D_x)_{\check\lambda}$ is the subspace of codimension one in $\mathrm{Op}_{PGL_2}(D_x)^{\mathrm{RS}}_{\check\lambda}$. In terms of the coefficients $v_n, n \geq -1$, appearing in formula (1.7) the corresponding equation has the form $P_\lambda(v_n) = 0$, where P_λ is a polynomial of degree $\lambda+1$, where we set $\deg v_n = n+2$. For example, $P_0 = v_{-1}, P_2 = 2v_{-1}^2 - v_0$, etc. In general, the subspace $\mathrm{Op}_G(D_x)_{\check\lambda} \subset \mathrm{Op}_G^{\mathrm{RS}}(D_x)_{\check\lambda}$ is defined by $\dim N$ polynomial equations, where N is the unipotent subgroup of G.

2. The Gaudin model

Let $\mathfrak{g}$ be a simple Lie algebra. The *Langlands dual Lie algebra* ${}^L\mathfrak{g}$ is by definition the Lie algebra whose Cartan matrix is the transpose of that of $\mathfrak{g}$. We will identify the set of roots of $\mathfrak{g}$ with the set of coroots of ${}^L\mathfrak{g}$ and the set of weights of $\mathfrak{g}$ with the set of coweights of ${}^L\mathfrak{g}$. The results on opers from the previous sections will be applied here to the Lie algebra ${}^L\mathfrak{g}$. Thus, in particular, ${}^LG$ will denote the adjoint group of ${}^L\mathfrak{g}$.

2.1. The definition of the Gaudin model

Here we recall the definition of the Gaudin model and the realization of the Gaudin Hamiltonians in terms of the spaces of conformal blocks for affine Kac-Moody algebras of critical level. We follow closely the paper [FFR].

Choose a non-degenerate invariant inner product κ_0 on $\mathfrak{g}$. Let $\{J_a\}, a = 1,\ldots,d$, be a basis of $\mathfrak{g}$ and $\{J^a\}$ the dual basis with respect to κ_0. Denote by Δ the quadratic Casimir operator from the center of $U(\mathfrak{g})$:

$$\Delta = \frac{1}{2}\sum_{a=1}^{d} J_a J^a.$$

The *Gaudin Hamiltonians* are the elements

$$\Xi_i = \sum_{j\neq i}\sum_{a=1}^{d} \frac{J_a^{(i)} J^{a(j)}}{z_i - z_j}, \qquad i = 1,\ldots,N, \tag{2.1}$$

of the algebra $U(\mathfrak{g})^{\otimes N}$. Note that they commute with the diagonal action of $\mathfrak{g}$ on $U(\mathfrak{g})^{\otimes N}$ and that

$$\sum_{i=1}^{N} \Xi_i = 0.$$

2.2. Gaudin model and coinvariants

Let $\widehat{\mathfrak{g}}_{\kappa_c}$ be the affine Kac-Moody algebra corresponding to $\mathfrak{g}$. It is the extension of the Lie algebra $\mathfrak{g}\otimes\mathbb{C}((t))$ by the one-dimensional center $\mathbb{C}K$. The commutation relations in $\widehat{\mathfrak{g}}_{\kappa_c}$ read

$$[A\otimes f(t), B\otimes g(t)] = [A,B]\otimes fg - \kappa_c(A,B)\,\mathrm{Res}_{t=0}\, f dg\cdot K, \tag{2.2}$$

where κ_c is the *critical* invariant inner product on $\mathfrak{g}$ defined by the formula

$$\kappa_c(A,B) = -\frac{1}{2}\operatorname{Tr}_{\mathfrak{g}} \operatorname{ad} A \operatorname{ad} B.$$

Note that $\kappa_c = -h^\vee \kappa_0$, where κ_0 is the inner product normalized as in [K] and $h^\vee$ is the dual Coxeter number.

Denote by $\widehat{\mathfrak{g}}_+$ the Lie subalgebra $\mathfrak{g} \otimes \mathbb{C}[[t]] \oplus \mathbb{C}K$ of $\widehat{\mathfrak{g}}_{\kappa_c}$. Let M be a $\mathfrak{g}$-module. We extend the action of $\mathfrak{g}$ on M to $\mathfrak{g} \otimes \mathbb{C}[[t]]$ by using the evaluation at zero homomorphism $\mathfrak{g} \otimes \mathbb{C}[[t]] \to \mathfrak{g}$ and to $\widehat{\mathfrak{g}}_+$ by making K act as the identity. Denote by $\mathbb{M}$ the corresponding induced $\widehat{\mathfrak{g}}_{\kappa_c}$-module

$$\mathbb{M} = U(\widehat{\mathfrak{g}}_{\kappa_c}) \underset{U(\widehat{\mathfrak{g}}_+)}{\otimes} M.$$

By construction, K acts as the identity on this module. We call such modules the $\widehat{\mathfrak{g}}_{\kappa_c}$-modules of *critical level*. For example, for $\lambda \in \mathfrak{h}^*$ let $\mathbb{C}_\lambda$ be the one-dimensional $\mathfrak{b}$-module on which $\mathfrak{h}$ acts by the character $\lambda : \mathfrak{h} \to \mathbb{C}$ and $\mathfrak{n}$ acts by 0. Let M_λ be the Verma module over $\mathfrak{g}$ of highest weight λ,

$$M_\lambda = U(\mathfrak{g}) \underset{U(\mathfrak{b})}{\otimes} \mathbb{C}_\lambda.$$

The corresponding induced module $\mathbb{M}_\lambda$ is the Verma module over $\widehat{\mathfrak{g}}_{\kappa_c}$ with highest weight λ.

For a dominant integral weight $\lambda \in \mathfrak{h}^*$ denote by V_λ the irreducible finite-dimensional $\mathfrak{g}$-module of highest weight λ. The corresponding induced module $\mathbb{V}_\lambda$ is called the Weyl module over $\widehat{\mathfrak{g}}_{\kappa_c}$ with highest weight λ.

Consider the projective line $\mathbb{P}^1$ with a global coordinate t and N distinct finite points $z_1, \ldots, z_N \in \mathbb{P}^1$. In the neighborhood of each point z_i we have the local coordinate $t - z_i$ and in the neighborhood of the point ∞ we have the local coordinate t^{-1}. Set $\widetilde{\mathfrak{g}}(z_i) = \mathfrak{g} \otimes \mathbb{C}((t - z_i))$ and $\widetilde{\mathfrak{g}}(\infty) = \mathfrak{g} \otimes \mathbb{C}((t^{-1}))$. Let $\widehat{\mathfrak{g}}_N$ be the extension of the Lie algebra $\bigoplus_{i=1}^N \widetilde{\mathfrak{g}}(z_i) \oplus \widetilde{\mathfrak{g}}(\infty)$ by a one-dimensional center $\mathbb{C}K$ whose restriction to each summand $\widetilde{\mathfrak{g}}(z_i)$ or $\widetilde{\mathfrak{g}}(\infty)$ coincides with the above central extension.

Suppose we are given a collection $M_1, \ldots, M_N$ and M_∞ of $\mathfrak{g}$-modules. Then the Lie algebra $\widehat{\mathfrak{g}}_N$ naturally acts on the tensor product $\bigotimes_{i=1}^N \mathbb{M}_i \otimes \mathbb{M}_\infty$ (in particular, K acts as the identity).

Let $\mathfrak{g}_{(z_i)} = \mathfrak{g}_{z_1,\ldots,z_N}$ be the Lie algebra of $\mathfrak{g}$-valued regular functions on $\mathbb{P}^1 \backslash \{z_1, \ldots, z_N, \infty\}$ (i.e., rational functions on $\mathbb{P}^1$, which may have poles only at the points $z_1, \ldots, z_N$ and ∞). Clearly, such a function can be expanded into a Laurent power series in the corresponding local coordinates at each point z_i and at ∞. Thus, we obtain an embedding

$$\mathfrak{g}_{(z_i)} \hookrightarrow \bigoplus_{i=1}^N \widetilde{\mathfrak{g}}(z_i) \oplus \widetilde{\mathfrak{g}}(\infty).$$

It follows from the residue theorem and formula (2.2) that the restriction of the central extension to the image of this embedding is trivial. Hence this embedding lifts to the embedding $\mathfrak{g}_{(z_i)} \to \widehat{\mathfrak{g}}_N$.

Denote by $H(M_1, \ldots, M_N, M_\infty)$ the space of coinvariants of $\bigotimes_{i=1}^N \mathbb{M}_i \otimes \mathbb{M}_\infty$ with respect to the action of the Lie algebra $\mathfrak{g}_{(z_i)}$. By construction, we have a canonical embedding of a $\mathfrak{g}$-module M into the induced $\widehat{\mathfrak{g}}_{\kappa_c}$-module $\mathbb{M}$:

$$x \in M \to 1 \otimes x \in \mathbb{M},$$

which commutes with the action of $\mathfrak{g}$ on both spaces (where $\mathfrak{g}$ is embedded into $\widehat{\mathfrak{g}}_{\kappa_c}$ as the constant subalgebra). Thus we have an embedding

$$\bigotimes_{i=1}^N M_i \otimes M_\infty \hookrightarrow \bigotimes_{i=1}^N \mathbb{M}_i \otimes \mathbb{M}_\infty.$$

The following result is proved in the same way as Lemma 1 in [FFR].

Lemma 2.1. *The composition of this embedding and the projection*

$$\bigotimes_{i=1}^N \mathbb{M}_i \otimes \mathbb{M}_\infty \twoheadrightarrow H(M_1, \ldots, M_N, M_\infty)$$

gives rise to an isomorphism

$$H(M_1, \ldots, M_N, M_\infty) \simeq (\bigotimes_{i=1}^N M_i \otimes M_\infty)/\mathfrak{g}_{\mathrm{diag}}.$$

Let $\mathbb{V}_0$ be the induced $\widehat{\mathfrak{g}}_{\kappa_c}$-module of critical level, which corresponds to the one-dimensional trivial $\mathfrak{g}$-module V_0; it is called the *vacuum module.* Denote by v_0 the generating vector of $\mathbb{V}_0$. We assign the vacuum module to a point $u \in \mathbb{P}^1$ which is different from $z_1, \ldots, z_N, \infty$. Denote by $H(M_1, \ldots, M_N, M_\infty, \mathbb{C})$ the space of $\mathfrak{g}_{(z_i),u}$-invariant functionals on $\bigotimes_{i=1}^N \mathbb{M}_i \otimes \mathbb{M}_\infty \otimes \mathbb{V}_0$ with respect to the Lie algebra $\mathfrak{g}_{(z_i),u}$. Lemma 2.1 tells us that we have a canonical isomorphism

$$H(M_1, \ldots, M_N, M_\infty, \mathbb{C}) \simeq H(M_1, \ldots, M_N, M_\infty).$$

Now observe that by functoriality any endomorphism $X \in \mathrm{End}_{\widehat{\mathfrak{g}}_{\kappa_c}} \mathbb{V}_0$ gives rise to an endomorphism of the space of coinvariants $H(M_1, \ldots, M_N, M_\infty, \mathbb{C})$, and hence of $H(M_1, \ldots, M_N, M_\infty)$. Thus, we obtain a homomorphism of algebras

$$\mathrm{End}_{\widehat{\mathfrak{g}}_{\kappa_c}} \mathbb{V}_0 \to \mathrm{End}_{\mathbb{C}} H(M_1, \ldots, M_N, M_\infty).$$

Let us compute this homomorphism explicitly.

First of all, we identify the algebra $\mathrm{End}_{\widehat{\mathfrak{g}}_{\kappa_c}} \mathbb{V}_0$ with the space

$$\mathfrak{z}(\widehat{\mathfrak{g}}) = \mathbb{V}_0^{\mathfrak{g}[[t]]}$$

of $\mathfrak{g}[[t]]$-invariant vectors in $\mathbb{V}_0$. Indeed, a $\mathfrak{g}[[t]]$-invariant vector v gives rise to an endomorphism of $\mathbb{V}_0$ commuting with the action of $\widehat{\mathfrak{g}}_{\kappa_c}$ which sends the generating vector v_0 to v. Conversely, any $\widehat{\mathfrak{g}}_{\kappa_c}$-endomorphism of V_0 is uniquely determined by the image of v_0 which necessarily belongs to $\mathfrak{z}(\widehat{\mathfrak{g}})$. Thus, we obtain an isomorphism $\mathfrak{z}(\widehat{\mathfrak{g}}) \simeq \mathrm{End}_{\widehat{\mathfrak{g}}_{\kappa_c}}(\mathbb{V}_0)$ which gives $\mathfrak{z}(\widehat{\mathfrak{g}})$ an algebra structure. The opposite algebra structure on $\mathfrak{z}(\widehat{\mathfrak{g}})$ coincides with the algebra structure induced by the identification of $\mathbb{V}_0$ with the algebra $U(\mathfrak{g} \otimes t^{-1}\mathbb{C}[t^{-1}])$. But we will see in the next section that the algebra $\mathfrak{z}(\widehat{\mathfrak{g}})$ is commutative and so the two algebra structures on it coincide.

Now let $v \in \mathfrak{z}(\widehat{\mathfrak{g}}) \subset \mathbb{V}_0$. For any

$$x \in (\bigotimes_{i=1}^{N} M_i \otimes M_\infty)/\mathfrak{g}_{\mathrm{diag}} \simeq H(M_1, \dots, M_N, M_\infty)$$

take a lifting $\widetilde{x}$ to $\bigotimes_{i=1}^{N} M_i \otimes M_\infty$. By Lemma 2.1, the projection of the vector

$$\widetilde{x} \otimes v \in \bigotimes_{i=1}^{N} \mathbb{M}_i \otimes \mathbb{M}_\infty \otimes \mathbb{V}_0$$

onto

$$H(M_1, \dots, M_N, M_\infty, \mathbb{C}) \simeq (\bigotimes_{i=1}^{N} M_i \otimes M_\infty)/\mathfrak{g}_{\mathrm{diag}}$$

is equal to the projection of a vector of the form $(\Psi_u(v) \cdot \widetilde{x}) \otimes v_0$, where

$$\Psi_u(v) \cdot \widetilde{x} \in (\bigotimes_{i=1}^{N} M_i \otimes M_\infty)/\mathfrak{g}_{\mathrm{diag}}.$$

For $A \in \mathfrak{g}$ and $n \in \mathbb{Z}$, denote by A_n the element $A \otimes t^n \in \widehat{\mathfrak{g}}_{\kappa_c}$. Then $\mathbb{V}_0 \simeq U(\mathfrak{g} \otimes t^{-1}\mathbb{C}[t^{-1}])v_0$ has a basis of lexicographically ordered monomials of the form $J^{a_1}_{n_1} \dots J^{a_m}_{n_m} v_0$ with $n_i < 0$. Let us set

$$\mathbb{J}^a_n(u) = -\sum_{i=1}^{N} \frac{J^{a(i)}}{(z_i - u)^n}.$$

Define an anti-homomorphism

$$\Phi_u : U(\mathfrak{g} \otimes t^{-1}\mathbb{C}[t^{-1}]) \to U(\mathfrak{g})^{\otimes N} \otimes \mathbb{C}[(u - z_i)^{-1}]_{i=1,\dots,N}$$

by the formula

$$\Phi_u(J^{a_1}_{n_1} \dots J^{a_m}_{n_m} v_0) = \mathbb{J}^{a_m}_{n_m}(u)\mathbb{J}^{a_{m-1}}_{n_{m-1}}(u) \dots \mathbb{J}^{a_1}_{n_1}(u). \tag{2.3}$$

According to the computation presented in the proof of Proposition 1 of [FFR], we have

$$\Psi_u(v) \cdot \widetilde{x} = \Phi_u(v) \cdot \widetilde{x}.$$

In general, $\Phi_v(u)$ does not commute with the diagonal action of $\mathfrak{g}$, and so $\Psi_v(u) \cdot \widetilde{x}$ depends on the choice of the lifting $\widetilde{x}$. But if $v \in \mathfrak{z}(\widehat{\mathfrak{g}}) \subset \mathbb{V}_0$, then $\Phi_v(u)$ commutes with the diagonal action of $\mathfrak{g}$ and hence gives rise to a well-defined endomorphism $(\bigotimes_{i=1}^N M_i \otimes M_\infty)/\mathfrak{g}_{\mathrm{diag}}$.

Thus, we restrict Φ_u to $\mathfrak{z}(\widehat{\mathfrak{g}})$. This gives us a homomorphism of algebras

$$\mathfrak{z}(\widehat{\mathfrak{g}}) \to \left(U(\mathfrak{g})^{\otimes N}\right)^G \otimes \mathbb{C}[(u-z_i)^{-1}]_{i=1,\dots,N},$$

which we also denote by Φ_u.

For example, consider the Segal-Sugawara vector in $\mathbb{V}_0$:

$$S = \frac{1}{2} \sum_{a=1}^{d} J_{a,-1} J^a_{-1} v_0. \tag{2.4}$$

One shows (see, e.g., [FB]) that this vector belongs to $\mathfrak{z}(\widehat{\mathfrak{g}})$. Consider the corresponding element $\Phi_u(S)$.

Denote by Δ the Casimir operator $\frac{1}{2}\sum_a J_a J^a$ from $U(\mathfrak{g})$.

Proposition 2.2 ([FFR], **Proposition 1**). *We have*

$$\Phi_u(S) = \sum_{i=1}^N \frac{\Xi_i}{u-z_i} + \sum_{i=1}^N \frac{\Delta^{(i)}}{(u-z_i)^2},$$

where the Ξ_i*'s are the Gaudin operators* (2.1).

We wish to study the algebra generated by the image of the map Φ_u.

Proposition 2.3 ([FFR], **Proposition 2**). *For any* $Z_1, Z_2 \in \mathfrak{z}(\widehat{\mathfrak{g}})$ *and any points* $u_1, u_2 \in \mathbb{P}^1 \backslash \{z_1, \dots, z_N, \infty\}$ *the linear operators* $\Psi_{Z_1}(u_1)$ *and* $\Psi_{Z_2}(u_2)$ *commute.*

Let $\mathcal{Z}_{(z_i)}(\mathfrak{g})$ be the span in $U(\mathfrak{g})^{\otimes N}$ of the coefficients in front of the monomials $\prod_{i=1}^N (u-z_i)^{n_i}$ of the series $\Phi_u(v), v \in \mathfrak{z}(\widehat{\mathfrak{g}})$. Since Φ_u is an algebra homomorphism, we find that $\mathcal{Z}_{(z_i)}(\mathfrak{g})$ is a subalgebra of $\left(U(\mathfrak{g})^{\otimes N}\right)^G$, which is commutative by Proposition 2.3. We call it the *Gaudin algebra* associated to $\mathfrak{g}$ and the collection $z_1, \dots, z_N$, and its elements the *generalized Gaudin Hamiltonians.*

2.3. The center of $\mathbb{V}_0$ and ${}^L G$-opers

In order to describe the Gaudin algebra $\mathcal{Z}_{(z_i)}(\mathfrak{g})$ and its spectrum we need to recall the description of $\mathfrak{z}(\widehat{\mathfrak{g}})$. According to [FF2, F2], $\mathfrak{z}(\widehat{\mathfrak{g}})$ is identified with the algebra $\operatorname{Fun}\operatorname{Op}_{{}^L G}(D)$ of (regular) functions on the space $\operatorname{Op}_{{}^L G}(D)$ of ${}^L G$-opers on the disc $D = \operatorname{Spec}\mathbb{C}[[t]]$, where ${}^L G$ is the *Langlands dual group* to G. Since we have assumed that G is simply-connected, ${}^L G$ may be defined as the adjoint group of the Lie algebra ${}^L\mathfrak{g}$ whose Cartan matrix is the transpose of that of $\mathfrak{g}$.

This isomorphism satisfies various properties, one of which we will now recall. Let $\operatorname{Der}\mathcal{O} = \mathbb{C}[[t]]\partial_t$ be the Lie algebra of continuous derivations of the topological algebra $\mathcal{O} = \mathbb{C}[[t]]$. The action of its Lie subalgebra $\operatorname{Der}_0\mathcal{O} = t\mathbb{C}[[t]]\partial_t$ on $\mathcal{O}$ exponentiates to an action of the group $\operatorname{Aut}\mathcal{O}$ of formal changes of variables.

Both Der $\mathcal{O}$ and Aut $\mathcal{O}$ naturally act on V_0 in a compatible way, and these actions preserve $\mathfrak{z}(\widehat{\mathfrak{g}})$. They also act on the space $\mathrm{Op}_{^L G}(D)$.

Denote by $\mathrm{Fun}\,\mathrm{Op}_{^L G}(D)$ the algebra of regular functions on $\mathrm{Op}_{^L G}(D)$. In view of Lemma 1.1, it is isomorphic to the algebra of functions on the space of ℓ-tuples $(v_1(t), \ldots, v_\ell(t))$ of formal Taylor series, i.e., the space $\mathbb{C}[[t]]^\ell$. If we write $v_i(t) = \sum_{n \geq 0} v_{i,n} t^n$, then we obtain

$$\mathrm{Fun}\,\mathrm{Op}_{^L G}(D) \simeq \mathbb{C}[v_{i,n}]_{i \in I, n \geq 0}. \tag{2.5}$$

Note that the vector field $-t\partial_t$ acts naturally on $\mathrm{Op}_{^L G}(D)$ and defines a $\mathbb{Z}$-grading on $\mathrm{Fun}\,\mathrm{Op}_{^L G}(D)$ such that $\deg v_{i,n} = d_i + n + 1$. The vector field $-\partial_t$ acts as a derivation such that $-\partial_t \cdot v_{i,n} = -(d_i + n + 1) v_{i,n+1}$.

Theorem 2.4 ([FF2, F2]). *There is a canonical isomorphism*

$$\mathfrak{z}(\widehat{\mathfrak{g}}) \simeq \mathrm{Fun}\,\mathrm{Op}_{^L G}(D)$$

of algebras which is compatible with the action of Der $\mathcal{O}$ *and* Aut $\mathcal{O}$.

We use this result to describe the twist of $\mathfrak{z}(\widehat{\mathfrak{g}})$ by the Aut $\mathcal{O}$-torsor $\mathcal{A}ut_x$ of formal coordinates at a smooth point x of an algebraic curve X,

$$\mathfrak{z}(\widehat{\mathfrak{g}})_x = \mathcal{A}ut_x \underset{\mathrm{Aut}\,\mathcal{O}}{\times} \mathfrak{z}(\widehat{\mathfrak{g}})$$

(see Ch. 6 of [FB] for more details). It follows from the definition that the corresponding twist of $\mathrm{Fun}\,\mathrm{Op}_{^L G}(D)$ by $\mathcal{A}ut_x$ is nothing but $\mathrm{Fun}\,\mathrm{Op}_{^L G}(D_x)$, where D_x is the disc around x. Therefore we obtain from Theorem 2.4 an isomorphism

$$\mathfrak{z}(\widehat{\mathfrak{g}})_x \simeq \mathrm{Fun}\,\mathrm{Op}_{^L G}(D_x). \tag{2.6}$$

The module $\mathbb{V}_0$ has a natural $\mathbb{Z}$-grading defined by the formulas $\deg v_0 = 0, \deg J^a_n = -n$, and it carries a translation operator T defined by the formulas $Tv_0 = 0, [T, J^a_n] = -nJ^a_{n-1}$. Theorem 2.4 and the isomorphism (2.5) imply that there exist non-zero vectors $S_i \in \mathbb{V}_0^{\mathfrak{g}[[t]]}$ of degrees $d_i + 1, i \in I$, such that

$$\mathfrak{z}(\widehat{\mathfrak{g}}) = \mathbb{C}[T^n S_i]_{i \in I, n \geq 0} v_0.$$

Then under the isomorphism of Theorem 2.4 we have $S_i \mapsto v_{i,0}$, the $\mathbb{Z}$-gradings on both algebras get identified and the action of T on $\mathfrak{z}(\widehat{\mathfrak{g}})$ becomes the action of $-\partial_t$ on $\mathrm{Fun}\,\mathrm{Op}_{^L G}(D)$. Note that the vector S_1 is nothing but the vector (2.4), up to a non-zero scalar.

Recall from [FB] that $\mathbb{V}_0$ is a vertex algebra, and $\mathfrak{z}(\widehat{\mathfrak{g}})$ is its commutative vertex subalgebra; in fact, it is the center of $\mathbb{V}_0$. We will also need the center of the completed universal enveloping algebra of $\widehat{\mathfrak{g}}$ of critical level. This algebra is defined as follows.

Let $U_{\kappa_c}(\widehat{\mathfrak{g}})$ be the quotient of the universal enveloping algebra $U(\widehat{\mathfrak{g}}_{\kappa_c})$ of $\widehat{\mathfrak{g}}_{\kappa_c}$ by the ideal generated by $(K - 1)$. Define its completion $\widetilde{U}_{\kappa_c}(\widehat{\mathfrak{g}})$ as follows:

$$\widetilde{U}_{\kappa_c}(\widehat{\mathfrak{g}}) = \varprojlim U_{\kappa_c}(\widehat{\mathfrak{g}}) / U_{\kappa_c}(\widehat{\mathfrak{g}}) \cdot (\mathfrak{g} \otimes t^N \mathbb{C}[[t]]).$$

It is clear that $\widetilde{U}_{\kappa_c}(\widehat{\mathfrak{g}})$ is a topological algebra which acts on all smooth $\widehat{\mathfrak{g}}_{\kappa_c}$-module. By definition, a smooth $\widehat{\mathfrak{g}}_{\kappa_c}$-module is a $\widehat{\mathfrak{g}}_{\kappa_c}$-module such that any vector is annihilated by $\mathfrak{g} \otimes t^N \mathbb{C}[[t]]$ for sufficiently large N, and K acts as the identity.

Let $Z(\widehat{\mathfrak{g}})$ be the center of $\widetilde{U}_{\kappa_c}(\widehat{\mathfrak{g}})$.

Denote by $\operatorname{Fun} \operatorname{Op}_{^LG}(D^\times)$ the algebra of regular functions on the space $\operatorname{Op}_{^LG}(D^\times)$ of ^LG-opers on the punctured disc $D^\times = \operatorname{Spec} \mathbb{C}((t))$. In view of Lemma 1.1, it is isomorphic to the algebra of functions on the space of ℓ-tuples $(v_1(t), \dots, v_\ell(t))$ of formal Laurent series, i.e., the ind-affine space $\mathbb{C}((t))^\ell$. If we write $v_i(t) = \sum_{n \in \mathbb{Z}} v_{i,n} t^n$, then we obtain that $\operatorname{Fun} \operatorname{Op}_{^LG}(D)$ is isomorphic to the completion of the polynomial algebra $\mathbb{C}[v_{i,n}]_{i \in I, n \in \mathbb{Z}}$ with respect to the topology in which the basis of open neighborhoods of zero is formed by the ideals generated by $v_{i,n}, i \in I, n \leq N$, for $N \leq 0$.

Theorem 2.5 ([F2]). *There is a canonical isomorphism*

$$Z(\widehat{\mathfrak{g}}) \simeq \operatorname{Fun} \operatorname{Op}_{^LG}(D^\times)$$

of complete topological algebras which is compatible with the action of $\operatorname{Der} \mathcal{O}$ *and* $\operatorname{Aut} \mathcal{O}$.

If M is a smooth $\widehat{\mathfrak{g}}_{\kappa_c}$-module, then the action of $Z(\widehat{\mathfrak{g}})$ on M gives rise to a homomorphism

$$Z(\widehat{\mathfrak{g}}) \to \operatorname{End}_{\widehat{\mathfrak{g}}_{\kappa_c}} M.$$

For example, if $M = \mathbb{V}_0$, then using Theorems 2.4 and 2.5 we identify this homomorphism with the surjection

$$\operatorname{Fun} \operatorname{Op}_{^LG}(D^\times) \twoheadrightarrow \operatorname{Fun} \operatorname{Op}_{^LG}(D)$$

induced by the natural embedding

$$\operatorname{Op}_{^LG}(D) \hookrightarrow \operatorname{Op}_{^LG}(D^\times).$$

Recall that the Harish-Chandra homomorphism identifies the center $Z(\mathfrak{g})$ of $U(\mathfrak{g})$ with the algebra $(\operatorname{Fun} \mathfrak{h}^*)^W$ of polynomials on $\mathfrak{h}^*$ which are invariant with respect to the action of the Weyl group W. Therefore a character $Z(\mathfrak{g}) \to \mathbb{C}$ is the same as a point in $\operatorname{Spec}(\operatorname{Fun} \mathfrak{h}^*)^W$ which is the quotient $\mathfrak{h}^*/W$. For $\lambda \in \mathfrak{h}^*$ we denote by $\varpi(\lambda)$ its projection onto $\mathfrak{h}^*/W$. In particular, $Z(\mathfrak{g})$ acts on M_λ and V_λ via its character $\varphi(\lambda + \rho)$. We also denote by I_λ the maximal ideal of $Z(\mathfrak{g})$ equal to the kernel of the homomorphism $Z(\mathfrak{g}) \to \mathbb{C}$ corresponding to the character $\varphi(\lambda + \rho)$.

In what follows we will use the canonical identification between $\mathfrak{h}^*$ and the Cartan subalgebra $^L\mathfrak{h}$ of the Langlands dual Lie algebra $^L\mathfrak{g}$. Recall that in Section 1.3 we defined the space $\operatorname{Op}^{\mathrm{RS}}_{^LG}(D)$ of ^LG-opers on $D^\times$ with regular singularity and its subspace $\operatorname{Op}^{\mathrm{RS}}_{^LG}(D)_\lambda$ of opers with residue $\varpi(-\lambda - \rho)$. We also defined the subspace

$$\operatorname{Op}_{^LG}(D)_{\check{\lambda}} \subset \operatorname{Op}^{\mathrm{RS}}_{^LG}(D)_{\check{\lambda}} \subset \operatorname{Op}_{^LG}(D^\times)$$

of those ^LG-opers which have trivial monodromy. Here we identify the coweights of the group ^LG with the weights of G.

The following result is obtained by combining Theorem 12.4, Lemma 9.4 and Proposition 12.8 of [F2].

Theorem 2.6.

(1) *Let* $\mathbb{U}$ *be the* $\widehat{\mathfrak{g}}_{\kappa_c}$*-module induced from the* $\mathfrak{g}[[t]] \oplus \mathbb{C}K$*-module* $U(\mathfrak{g})$*. Then the homomorphism* $Z(\widehat{\mathfrak{g}}) \to \operatorname{End}_{\widehat{\mathfrak{g}}_{\kappa_c}} \mathbb{U}$ *factors as*
$$Z(\widehat{\mathfrak{g}}) \simeq \operatorname{Fun} \operatorname{Op}_{{}^L G}(D^\times) \twoheadrightarrow \operatorname{Fun} \operatorname{Op}^{\mathrm{RS}}_{{}^L G}(D) \to \operatorname{End}_{\widehat{\mathfrak{g}}_{\kappa_c}} \mathbb{U}.$$

(2) *Let* M *be a* $\mathfrak{g}$*-module on which the center* $Z(\mathfrak{g})$ *acts via its character* $\varpi(\lambda + \rho)$*, and let* $\mathbb{M}$ *be the induced* $\widehat{\mathfrak{g}}_{\kappa_c}$*-module. Then the homomorphism* $Z(\widehat{\mathfrak{g}}) \to \operatorname{End}_{\widehat{\mathfrak{g}}_{\kappa_c}} \mathbb{M}$ *factors as follows*
$$Z(\widehat{\mathfrak{g}}) \simeq \operatorname{Fun} \operatorname{Op}_{{}^L G}(D^\times) \twoheadrightarrow \operatorname{Fun} \operatorname{Op}^{\mathrm{RS}}_{{}^L G}(D)_\lambda \to \operatorname{End}_{\widehat{\mathfrak{g}}_{\kappa_c}} \mathbb{M}.$$
Moreover, if $M = M_\lambda$*, then the last map is an isomorphism*
$$\operatorname{End}_{\widehat{\mathfrak{g}}_{\kappa_c}} \mathbb{M} \simeq \operatorname{Fun} \operatorname{Op}^{\mathrm{RS}}_{{}^L G}(D)_\lambda.$$

(3) *For an integral dominant weight* $\lambda \in \mathfrak{h}^*$ *the homomorphism*
$$\operatorname{Fun} \operatorname{Op}_{{}^L G}(D^\times) \to \operatorname{End}_{\widehat{\mathfrak{g}}_{\kappa_c}} \mathbb{V}_\lambda$$
factors as
$$\operatorname{Fun} \operatorname{Op}_{{}^L G}(D^\times) \to \operatorname{Fun} \operatorname{Op}_{{}^L G}(D)_\lambda \to \operatorname{End}_{\widehat{\mathfrak{g}}_{\kappa_c}} \mathbb{V}_\lambda,$$
and the last map is an isomorphism
$$\operatorname{End}_{\widehat{\mathfrak{g}}_{\kappa_c}} \mathbb{V}_\lambda \simeq \operatorname{Fun} \operatorname{Op}_{{}^L G}(D)_\lambda.$$

2.4. Example

Let us consider the case $\mathfrak{g} = \mathfrak{sl}_2$ in more detail. Introduce the Segal-Sugawara operators $S_n, n \in \mathbb{Z}$, by the formula
$$S(z) = \sum_{n \in \mathbb{Z}} S_n z^{-n-2} = \frac{1}{2} \sum_a {:}J^a(z) J_a(z){:}\,,$$
where the normal ordering is defined as in [FB]. Then the center $Z(\mathfrak{sl}_2)$ is the completion $\mathbb{C}[S_n]^\sim_{n \in \mathbb{Z}}$ of the polynomial algebra $\mathbb{C}[S_n]_{n \in \mathbb{Z}}$ with respect to the topology in which the basis of open neighborhoods of zero is formed by the ideals of $S_n, n > N$, for $N \geq 0$.

We have the following diagram of (vertical) isomorphisms and (horizontal) surjections
$$\begin{array}{ccccccc}
Z(\widehat{\mathfrak{sl}}_2) & \longrightarrow & \operatorname{End}_{\widehat{\mathfrak{g}}_{\kappa_c}} \mathbb{U} & \longrightarrow & \operatorname{End}_{\widehat{\mathfrak{g}}_{\kappa_c}} \mathbb{M}_\lambda & \longrightarrow & \operatorname{End}_{\widehat{\mathfrak{g}}_{\kappa_c}} \mathbb{V}_\lambda \\
\downarrow & & \downarrow & & \downarrow & & \downarrow \\
\mathbb{C}[S_n]^\sim_{n \in \mathbb{Z}} & \longrightarrow & \mathbb{C}[S_n]_{n \leq 0} & \longrightarrow & \mathbb{C}[S_n]_{n \leq 0}/J_\lambda & \longrightarrow & \mathbb{C}[S_n]_{n \leq 0}/J'_\lambda
\end{array}$$
where J_λ is the ideal generated by $(S_0 - \frac{1}{4}\lambda(\lambda+2))$ and J'_λ is the ideal generated by I_λ and the polynomial P_λ introduced at the end of Section 1.3.

The space of PGL_2-opers on $D^\times$ is identified with the space of projective connections of the form $\partial_t^2 - \sum_{n\in\mathbb{Z}} v_n t^n$. The isomorphism of Theorem 2.5 sends S_n to v_{-n-2}. The relevant spaces of PGL_2-opers with regular singularities were described at the end of Section 1.3, and these descriptions agree with the above diagram and Theorem 2.6.

2.5. The Gaudin algebra

Now we are ready to identify the Gaudin algebra $\mathcal{Z}_{(z_i)}(\mathfrak{g})$ with the algebra of functions on a certain space of opers on $\mathbb{P}^1$.

Let $\mathrm{Op}^{\mathrm{RS}}_{^LG}(\mathbb{P}^1)_{(z_i),\infty}$ be the space of ${}^LG$-opers on $\mathbb{P}^1$ with regular singularities at $z_1,\ldots,z_N$ and ∞. For an arbitrary collection of weights $\lambda_1,\ldots,\lambda_N$ and λ_∞, let

$$\mathrm{Op}^{\mathrm{RS}}_{^LG}(\mathbb{P}^1)_{(z_i),\infty;(\lambda_i),\lambda_\infty}$$

be its subspace of those opers whose residue at the point z_i (resp., ∞) is equal to $\varpi(-\lambda_i-\rho), i=1,\ldots,N$ (resp., $\varpi(-\lambda_\infty-\rho)$). Finally, if all of the weights $\lambda_1,\ldots,\lambda_N,\lambda_\infty$ are dominant integral, we introduce a subset

$$\mathrm{Op}_{^LG}(\mathbb{P}^1)_{(z_i),\infty;(\lambda_i),\lambda_\infty} \subset \mathrm{Op}^{\mathrm{RS}}_{^LG}(\mathbb{P}^1)_{(z_i),\infty;(\lambda_i),\lambda_\infty}$$

which consists of those ${}^LG$-opers which have trivial monodromy representation.

On the other hand, for each collection of points $z_1,\ldots,z_N$ on $\mathbb{P}^1\backslash\infty$ we have the Gaudin algebra

$$\mathcal{Z}_{(z_i)}(\mathfrak{g}) \subset \left(U(\mathfrak{g})^{\otimes N}\right)^G \simeq \left(U(\mathfrak{g})^{\otimes(N+1)}/\mathfrak{g}_{\mathrm{diag}}\right)^G,$$

where the second isomorphism is obtained by identifying $U(\mathfrak{g})^{\otimes(N+1)}/\mathfrak{g}_{\mathrm{diag}}$ with $U(\mathfrak{g})^{\otimes N}\otimes 1$.

We have a homomorphism $c_i : Z(\mathfrak{g}) \to U(\mathfrak{g}) \to U(\mathfrak{g})^{\otimes(N+1)}$ corresponding to the ith factor, for all $i=1,\ldots,N$, and a homomorphism $c_\infty : Z(\mathfrak{g}) \to U(\mathfrak{g})^{\otimes(N+1)}$ corresponding to the $(N+1)$st factor. It is easy to see that the images of $c_i, i = 1,\ldots,N$, and c_∞ belong to $\mathcal{Z}_{(z_i)}(\mathfrak{g})$. For a collection of weights $\lambda_1,\ldots,\lambda_N$ and λ_∞, let $I_{(\lambda_i),\lambda_\infty}$ be the ideal of $\mathcal{Z}_{(z_i)}(\mathfrak{g})$ generated by $c_i(I_{\lambda_i}), i=1,\ldots,N$, and $c_\infty(I_{\lambda_\infty})$. Let $\mathcal{Z}_{(z_i),\infty;(\lambda_i),\lambda_\infty}$ be the quotient of $\mathcal{Z}_{(z_i)}(\mathfrak{g})$ by $I_{(\lambda_i),\lambda_\infty}$.

The algebra $\mathcal{Z}_{(z_i),\infty;(\lambda_i),\lambda_\infty}(\mathfrak{g})$ acts on the space of $\mathfrak{g}$-coinvariants in

$$\bigotimes_{i=1}^N M_i \otimes M_\infty,$$

where M_i is a $\mathfrak{g}$-module with central character $\varpi(\lambda_i+\rho), i=1,\ldots,N$, and M_∞ is a $\mathfrak{g}$-module with central character $\varpi(\lambda_\infty+\rho)$. In particular, if all the weights $\lambda_1,\ldots,\lambda_N,\lambda_\infty$ are dominant integral, then we can take as the M_i's the finite-dimensional irreducible modules V_{λ_i} for $i=1,\ldots,N$, and as M_∞ the module V_{λ_∞}. The corresponding space of $\mathfrak{g}$-coinvariants is isomorphic to the space

$$\left(\bigotimes_{i=1}^N V_{\lambda_i}\otimes V_{\lambda_\infty}\right)^G$$

of G-invariants in $\bigotimes_{i=1}^N V_{\lambda_i} \otimes V_{\lambda_\infty}$. Let $\overline{\mathcal{Z}}_{(z_i),\infty;(\lambda_i),\lambda_\infty}(\mathfrak{g})$ be the image of the algebra $\mathcal{Z}_{(z_i),\infty;(\lambda_i),\lambda_\infty}$ in $\operatorname{End}\left(\bigotimes_{i=1}^N V_{\lambda_i} \otimes V_{\lambda_\infty}\right)^G$.

We have the following result.

Theorem 2.7.

(1) *The algebra* $\mathcal{Z}_{(z_i)}(\mathfrak{g})$ *is isomorphic to the algebra of functions on the space* $\operatorname{Op}^{\mathrm{RS}}_{{}^L G}(\mathbb{P}^1)_{(z_i),\infty}$.

(2) *The algebra* $\mathcal{Z}_{(z_i),\infty;(\lambda_i),\lambda_\infty}(\mathfrak{g})$ *is isomorphic to the algebra of functions on the space* $\operatorname{Op}^{\mathrm{RS}}_{{}^L G}(\mathbb{P}^1)_{(z_i),\infty;(\lambda_i),\lambda_\infty}$.

(3) *For a collection of dominant integral weights* $\lambda_1, \ldots, \lambda_N, \lambda_\infty$, *there is a surjective homomorphism from the algebra of functions* $\operatorname{Op}_{{}^L G}(\mathbb{P}^1)_{(z_i),\infty;(\lambda_i),\lambda_\infty}$ *to the algebra* $\overline{\mathcal{Z}}_{(z_i),\infty;(\lambda_i),\lambda_\infty}(\mathfrak{g})$.

Proof. In [FB] we defined, for any quasi-conformal vertex algebra V, a smooth projective curve X, a set of points $x_1, \ldots, x_N \in X$ and a collection of V-modules $M_1, \ldots, M_N$, the space of coinvariants $H_V(X,(x_i),(M_i))$, which is the quotient of $\bigotimes_{i=1}^N M_i$ by the action of a certain Lie algebra. This construction (which is recalled in the proof of Theorem 4.7 in [F3]) is functorial in the following sense. Suppose that we are given a homomorphism $W \to V$ of vertex algebras (so that each M_i becomes a V-module), a collection $R_1, \ldots, R_N$ of W-modules and a collection of homomorphisms of W-modules $M_i \to R_i$ for all $i = 1, \ldots, N$. Then the corresponding map $\bigotimes_{i=1}^N R_i \to \bigotimes_{i=1}^N M_i$ gives rise to a map of the corresponding spaces of coinvariants

$$H_W(X,(x_i),(R_i)) \to H_V(X,(x_i),(M_i)).$$

Suppose now that W is the center of V (see [FB]). Then the action of W on any V-module M factors through a homomorphism $\widetilde{U}(W) \to \operatorname{End}_{\mathbb{C}} M$, where $\widetilde{U}(W)$ is the enveloping algebra of W (see [FB], Sect. 4.3). Let $W(M)$ be the image of this homomorphism. Then $H_W(X,(x_i),(W(M_i)))$ is an algebra, and we obtain a natural homomorphism of algebras

$$H_W(X,(x_i),(W(M_i))) \to \operatorname{End}_{\mathbb{C}} H_V(X,(x_i),(M_i)). \tag{2.7}$$

If $V = \mathbb{V}_0$, then the center of V is precisely the subspace $\mathfrak{z}(\widehat{\mathfrak{g}})$ of $\mathfrak{g}[[t]]$-invariant vectors in $\mathbb{V}_0$ (see [FB]). In particular, $\mathfrak{z}(\widehat{\mathfrak{g}})$ is a commutative vertex subalgebra of $\mathbb{V}_0$. A module over the vertex algebra $\mathfrak{z}(\widehat{\mathfrak{g}})$ is the same as a module over the topological algebra $\widetilde{U}(\mathfrak{z}(\widehat{\mathfrak{g}}))$ which is nothing but the center $Z(\widehat{\mathfrak{g}})$ of $\widetilde{U}_{\kappa_c}(\widehat{\mathfrak{g}})$ (see [F2], Sect. 11). The action of $Z(\widehat{\mathfrak{g}})$ on any $\widehat{\mathfrak{g}}_{\kappa_c}$-module M factors through the homomorphism $Z(\widehat{\mathfrak{g}}) \to \operatorname{End}_{\widehat{\mathfrak{g}}_{\kappa_c}} M$. Let $Z(M)$ denote the image of this homomorphism. Recall that we have identified $Z(\widehat{\mathfrak{g}})$ with $\operatorname{Fun}\operatorname{Op}_{{}^L G}(D^\times)$ in Theorem 2.5. For each $\widehat{\mathfrak{g}}_{\kappa_c}$-module M, the algebra $Z(M)$ is a quotient of $\operatorname{Fun}\operatorname{Op}_{{}^L G}(D^\times)$, and hence $\operatorname{Spec} Z(M)$ is a subscheme in $\operatorname{Op}_{{}^L G}(D^\times)$ which we denote by $\operatorname{Op}^M_{{}^L G}(D^\times)$.

The space of coinvariants $H_{\mathfrak{z}(\widehat{\mathfrak{g}})}(X,(x_i),Z(M_i))$ is computed in the same way as in Theorem 4.7 of [F3]:

$$H_{\mathfrak{z}(\widehat{\mathfrak{g}})}(X,(x_i),Z(M_i)) \simeq \operatorname{Fun}\operatorname{Op}_{^LG}(X,(x_i),(M_i)) \tag{2.8}$$

where $\operatorname{Op}_{^LG}(X,(x_i),(M_i))$ is the space of ^LG-opers on X which are regular on $X\backslash\{x_1,\dots,x_N\}$ and such that their restriction to $D_x^\times$ belongs to $\operatorname{Op}^{M_i}_{^LG}(D^\times_{x_i})$ for all $i=1,\dots,N$.

Let u be an additional point of X, different from $x_1,\dots,x_N$, and let us insert $\mathfrak{z}(\widehat{\mathfrak{g}}) \simeq Z(\mathbb{V}_0)$ at this point. Then by Theorem 10.3.1 of [FB] we have an isomorphism

$$H_{\mathfrak{z}(\widehat{\mathfrak{g}})}(X,(x_i),(Z(M_i))) \simeq H_{\mathfrak{z}(\widehat{\mathfrak{g}})}(X;(z_i),u;(Z(M_i)),Z(\mathbb{V}_0)).$$

Hence we obtain a homomorphism

$$\mathfrak{z}(\widehat{\mathfrak{g}})_u \simeq Z(\mathbb{V}_0)_u \to H_{\mathfrak{z}(\widehat{\mathfrak{g}})}(X,(x_i),(Z(M_i))). \tag{2.9}$$

The corresponding homomorphism

$$\operatorname{Fun}\operatorname{Op}_{^LG}(D_u) \to \operatorname{Fun}\operatorname{Op}_{^LG}(X,(x_i),(M_i))$$

(see formula (2.6)) is induced by the embedding

$$\operatorname{Op}_{^LG}(X,(x_i),(M_i)) \hookrightarrow \operatorname{Op}_{^LG}(D_u)$$

obtained by restricting an oper to D_u.

We apply this construction in the case when the curve X is $\mathbb{P}^1$, the points are $z_1,\dots,z_N$ and ∞, and the modules are $\widehat{\mathfrak{g}}_{\kappa_c}$-modules $M_1,\dots,M_N$ and M_∞. It is proved in [FB] (see Theorem 9.3.3 and Remark 9.3.10) that the corresponding space of coinvariants is the space of $\mathfrak{g}_{(z_i)}$-coinvariants of $\bigotimes_{i=1}^N \mathbb{M}_i \otimes \mathbb{M}_\infty$, which is the space $H(M_1,\dots,M_N,M_\infty)$ that we have computed in Lemma 2.1.

The homomorphism (2.7) specializes to a homomorphism

$$H_{\mathfrak{z}(\widehat{\mathfrak{g}})}(\mathbb{P}^1;(z_i),\infty;(Z(M_i)),Z(M_\infty)) \to \operatorname{End}_{\mathbb{C}} H(M_1,\dots,M_N,M_\infty). \tag{2.10}$$

Observe that by its very definition the homomorphism

$$\Phi_u : \mathfrak{z}(\widehat{\mathfrak{g}})_u \to \operatorname{End}_{\mathbb{C}} H(M_1,\dots,M_N,M_\infty)$$

constructed in Section 2.2 factors through the homomorphisms (2.10) and (2.9). Hence the image of Φ_u is a quotient of the algebra

$$H_{\mathfrak{z}(\widehat{\mathfrak{g}})}(\mathbb{P}^1;(z_i),\infty;(Z(M_i)),Z(M_\infty)) \simeq \operatorname{Fun}\operatorname{Op}_{^LG}(\mathbb{P}^1;(z_i),\infty;(Z(M_i)),Z(M_\infty)),$$

according to the isomorphism (2.8).

Let us specialize this result to our setting. First we suppose that all M_i's and M_∞ are equal to $U(\mathfrak{g})$. By Theorem 2.6,(1), we have

$$\operatorname{Op}^{U(\mathfrak{g})}_{^LG}(D^\times) = \operatorname{Op}^{\mathrm{RS}}_{^LG}(D).$$

Therefore we find that

$$H_{\mathfrak{z}(\widehat{\mathfrak{g}})}(\mathbb{P}^1;(z_i),\infty;(U(\mathfrak{g})),U(\mathfrak{g})) \simeq \operatorname{Fun}\operatorname{Op}^{\mathrm{RS}}_{^LG}(\mathbb{P}^1)_{(z_i),\infty}. \tag{2.11}$$

Next, by Theorem 2.6,(2), we have

$$\mathrm{Op}_{^L G}^{U(\mathfrak{g})/I_\lambda}(D^\times) = \mathrm{Op}_{^L G}^{\mathrm{RS}}(D)_\lambda.$$

Therefore

$$H_{\mathfrak{z}(\widehat{\mathfrak{g}})}(\mathbb{P}^1; (z_i), \infty; (U(\mathfrak{g})/I_{\lambda_i}), U(\mathfrak{g})/I_{\lambda_\infty}) \simeq \mathrm{Fun}\,\mathrm{Op}_{^L G}^{\mathrm{RS}}(\mathbb{P}^1)_{(z_i),\infty;(\lambda_i),\lambda_\infty}. \tag{2.12}$$

Finally, by Theorem 2.6,(3), we have

$$\mathrm{Op}_{^L G}^{\mathbb{V}_\lambda}(D^\times) = \mathrm{Op}_{^L G}(D)_\lambda.$$

Therefore

$$H_{\mathfrak{z}(\widehat{\mathfrak{g}})}(\mathbb{P}^1; (z_i), \infty; (\mathbb{V}_{\lambda_i}), \mathbb{V}_{\lambda_\infty}) \simeq \mathrm{Fun}\,\mathrm{Op}_{^L G}(\mathbb{P}^1)_{(z_i),\infty;(\lambda_i),\lambda_\infty}. \tag{2.13}$$

Moreover, in all three cases the homomorphism Φ_u is just the natural homomorphism from $\mathrm{Fun}\,\mathrm{Op}_{^L G}(D_u)$ to the above algebras of functions that is induced by the restrictions of the corresponding opers to the disc D_u.

Now consider the homomorphism (2.10) in the case of the space of coinvariants given by the left-hand side of formula (2.11),

$$\mathrm{Fun}\,\mathrm{Op}_{^L G}^{\mathrm{RS}}(\mathbb{P}^1)_{(z_i),\infty} \to \mathrm{End}_{\mathbb{C}}\, U(\mathfrak{g})^{\otimes(N+1)}/\mathfrak{g}_{\mathrm{diag}} \simeq \mathrm{End}_{\mathbb{C}}\, U(\mathfrak{g})^{\otimes N},$$

where the second isomorphism we use the identification of $U(\mathfrak{g})^{\otimes(N+1)}/\mathfrak{g}_{\mathrm{diag}}$ with $U(\mathfrak{g})^{\otimes N}$ corresponding to the first N factors. It follows from the explicit computation of this map given above that its image belongs to

$$\left(U(\mathfrak{g})^{\otimes N}\right)^G \subset \mathrm{End}_{\mathbb{C}}\, U(\mathfrak{g})^{\otimes N},$$

and so we have a homomorphism

$$\mathrm{Fun}\,\mathrm{Op}_{^L G}^{\mathrm{RS}}(\mathbb{P}^1)_{(z_i),\infty} \to \left(U(\mathfrak{g})^{\otimes N}\right)^G. \tag{2.14}$$

By definition, the Gaudin algebra $\mathcal{Z}_{(z_i)}(\mathfrak{g})$ is the image of this homomorphism.

Likewise, we obtain from formula (2.12) that the homomorphism (2.10) gives rise to a homomorphism

$$\mathrm{Fun}\,\mathrm{Op}_{^L G}^{\mathrm{RS}}(\mathbb{P}^1)_{(z_i),\infty;(\lambda_i),\lambda_\infty} \to \left(U(\mathfrak{g})^{\otimes(N+1)}/\mathfrak{g}_{\mathrm{diag}}\right)^G / I_{(\lambda_i),\lambda_\infty}, \tag{2.15}$$

whose image is the algebra $\mathcal{Z}_{(z_i),\infty;(\lambda_i),\lambda_\infty}(\mathfrak{g})$.

Finally, formula (2.13) gives us a homomorphism

$$\mathrm{Fun}\,\mathrm{Op}_{^L G}(\mathbb{P}^1)_{(z_i),\infty;(\lambda_i),\lambda_\infty} \to \mathrm{End}_{\mathbb{C}}\left(\bigotimes_{i=1}^N V_{\lambda_i} \otimes V_{\lambda_\infty}\right),$$

whose image is the algebra $\overline{\mathcal{Z}}_{(z_i),\infty;(\lambda_i),\lambda_\infty}(\mathfrak{g})$. This proves part (3) of the theorem.

To prove parts (1) and (2), it remains to show that the homomorphisms (2.14) and (2.15) are injective. It is sufficient to prove that the latter is injective. To see that, we pass to the associate graded spaces on both sides with respect to natural filtrations which we now describe.

According to the identification given in formula (1.4), the algebra of functions on $\mathrm{Op}^{\mathrm{RS}}_{^LG}(\mathbb{P}^1)_{(z_i),\infty;(\lambda_i),\lambda_\infty}$ is filtered, and the corresponding associated graded algebra is the algebra of functions on the vector space

$$C^{\mathrm{RS}}_{(z_i),\infty} = \bigoplus_{i=1}^{\ell} \Gamma(\mathbb{P}^1, \Omega^{\otimes(d_i+1)}(-d_i z_1 - \ldots - d_i z_N - d_i \infty)),$$

where Ω is the canonical line bundle on $\mathbb{P}^1$. The algebra

$$\left(U(\mathfrak{g})^{\otimes(N+1)}/\mathfrak{g}_{\mathrm{diag}}\right)^G / I_{(\lambda_i),\lambda_\infty}$$

carries a PBW filtration, and the associated graded is the algebra of functions on the space $\mu^{-1}((T^*G/B)^{N+1})/G$, where $\mu : (T^*G/B)^{N+1} \to \mathfrak{g}^*$ is the moment map corresponding to the diagonal action of G on $(T^*G/B)^{N+1}$. The two filtrations are compatible according to [F2]. The corresponding homomorphism of the associate graded algebras

$$\mathrm{Fun}\, C^{\mathrm{RS}}_{(z_i),\infty} \to \mathrm{Fun}\, \mu^{-1}((T^*G/B)^{N+1})/G \tag{2.16}$$

is induced by a map

$$h_{(z_i),\infty} : \mu^{-1}((T^*G/B)^{N+1})/G \to C^{\mathrm{RS}}_{(z_i),\infty},$$

that we now describe.

Let us identify the tangent space to a point $gB \subset G/B$ with $(\mathfrak{g}/g\mathfrak{b}g^{-1})^* \simeq g\mathfrak{n}g^{-1}$. Then a point in $\mu^{-1}((T^*G/B)^{N+1})/G$ consists of an $(N+1)$-tuple of points g_iB of G/B and an $(N+1)$-tuple of vectors (η_i), where $\eta_i \in g_i\mathfrak{n}g_i^{-1} \subset \mathfrak{g}$ such that $\sum_{i=1}^{N+1} \eta_i = 0$, considered up to simultaneous conjugation by G.

We attach to it the $\mathfrak{g}$-valued one-form

$$\eta = \sum_{i=1}^{N} \frac{\eta_i}{t - z_i} dt$$

on $\mathbb{P}^1$ with poles at $z_1, \ldots, z_N, \infty$. Let $P_1, \ldots, P_\ell$ be generators of the algebra of G-invariant polynomials on $\mathfrak{g}$ of degrees $d_i + 1$. Then

$$h_{(z_i),\infty}((g_i),(\eta_i)) = (P_i(\eta))_{i=1}^{\ell} \in C^{\mathrm{RS}}_{(z_i),\infty}.$$

The space $\mu^{-1}((T^*G/B)^{N+1})/G$ is identified with the moduli space of Higgs fields on the trivial G-bundle with parabolic structures at $z_1, \ldots, z_N, \infty$, and the map $h_{(z_i),\infty}$ is nothing but the Hitchin map (see [ER]). The Hitchin map is known to be proper, so in particular it is surjective (see, e.g., [M]). Therefore the corresponding homomorphism (2.16) of algebras of functions is injective. This implies that the homomorphism (2.15) is also injective and completes the proof of the theorem. □

This theorem has an important application to the question of simultaneous diagonalization of generalized Gaudin Hamiltonians, or equivalently, of the commutative algebra $\mathcal{Z}_{(z_i)}(\mathfrak{g})$, on the tensor product $\bigotimes_{i=1}^N M_i$ of $\mathfrak{g}$-modules. Indeed, the joint eigenvalues of the generalized Gaudin Hamiltonians on any eigenvector in $\bigotimes_{i=1}^N M_i$ correspond to a point in the spectrum of the algebra $\mathcal{Z}_{(z_i)}(\mathfrak{g})$, which, according to Theorem 2.7,(1), is a point of the space $\mathrm{Op}^{\mathrm{RS}}_{^L G}(\mathbb{P}^1)_{(z_i),\infty}$.

If we assume in addition that each of the modules M_i admits a central character $\varpi(\lambda_i+\rho)$ (for instance, if M_i is the Verma module M_{λ_i}) and we are looking for eigenvectors in the component of $\bigotimes_{i=1}^N M_i$ corresponding to the central character $\varpi(-\lambda_\infty-\rho)$ with respect to the diagonal action of $\mathfrak{g}$, then the joint eigenvalues define a point in the spectrum of the algebra $\mathcal{Z}_{(z_i),\infty;(\lambda_i),\lambda_\infty}(\mathfrak{g})$, i.e., a point of $\mathrm{Op}^{\mathrm{RS}}_{^L G}(\mathbb{P}^1)_{(z_i),\infty;(\lambda_i),\lambda_\infty}$.

Finally, for a collection of dominant integral weights $\lambda_1,\dots,\lambda_N,\lambda_\infty$, the joint eigenvalues of the generalized Gaudin Hamiltonians on $(\bigotimes_{i=1}^N V_{\lambda_i}\otimes V_{\lambda_\infty})^G$ is a point in the spectrum of the algebra $\overline{\mathcal{Z}}_{(z_i),\infty;(\lambda_i),\lambda_\infty}(\mathfrak{g})$, which is a point of $\mathrm{Op}_{^L G}(\mathbb{P}^1)_{(z_i),\infty;(\lambda_i),\lambda_\infty}$.

A natural question is whether, conversely, one can attach to a $^L G$-oper on $\mathbb{P}^1$ with regular singularities at $z_1,\dots,z_N,\infty$ (and satisfying additional conditions as above) an eigenvector in $\bigotimes_{i=1}^N M_i$ with such eigenvalues. It turns out that for general modules this is not true, but if these modules are finite-dimensional, then we conjecture that it is true. In order to construct the eigenvectors we use the procedure called Bethe Ansatz. As shown in [FFR], this procedure may be cast in the framework of coinvariants that we have discussed in this section, using the Wakimoto modules over $\widehat{\mathfrak{g}}_{\kappa_c}$. We will explain that in Section 4 and Section 5.5. But first we need to introduce Miura opers and Cartan connections.

3. Miura opers and Cartan connections

By definition (see [F2], Sect. 10.3), a *Miura G-oper* on X (which is a smooth curve or a disc) is a quadruple $(\mathcal{F},\nabla,\mathcal{F}_B,\mathcal{F}'_B)$, where $(\mathcal{F},\nabla,\mathcal{F}_B)$ is a G-oper on X and $\mathcal{F}'_B$ is another B-reduction of $\mathcal{F}$ which is preserved by ∇.

We denote the space of Miura G-opers on X by $\mathrm{MOp}_G(X)$.

3.1. Miura opers and flag manifolds

A B-reduction of $\mathcal{F}$ which is preserved by the connection ∇ is uniquely determined by a B-reduction of the fiber $\mathcal{F}_x$ of $\mathcal{F}$ at any point $x\in X$ (in the case when $U=D$, x has to be the origin $0\in D$). The set of such reductions is the $\mathcal{F}_x$-twist

$$(G/B)_{\mathcal{F}_x}=\mathcal{F}_x\underset{G}{\times}G/B=\mathcal{F}_{B,x}\underset{B}{\times}G/B=(G/B)_{\mathcal{F}_{B,x}} \tag{3.1}$$

of the flag manifold G/B. If X is a curve or a disc and the oper connection has a regular singularity and trivial monodromy representation, then this connection gives us a global (algebraic) trivialization of the bundle $\mathcal{F}$. Then any B-reduction of the fiber $\mathcal{F}_x$ gives rise to a global (algebraic) B-reduction of $\mathcal{F}$. Thus, we obtain:

Lemma 3.1. *Suppose that we are given an oper τ on a curve X (or on the disc) such that the oper connection has a regular singularity and trivial monodromy.*

Then for each $x \in X$ there is a canonical isomorphism between the space of Miura opers with the underlying oper τ and the twist $(G/B)_{\mathcal{F}_{B,x}}$.

Recall that the B-orbits in G/B, known as the *Schubert cells*, are parameterized by the Weyl group W of G. Let w_0 be the longest element of the Weyl group of G. Denote the orbit $Bw^{-1}w_0B \subset G/B$ by S_w (so that S_1 is the open orbit). We obtain from the second description of $(G/B)_{\mathcal{F}_x}$ given in formula (3.1) that $(G/B)_{\mathcal{F}_x}$ decomposes into a union of locally closed subvarieties $S_{w,\mathcal{F}_{B,x}}$, which are the $\mathcal{F}_{B,x}$-twists of the Schubert cells S_w. The B-reduction $\mathcal{F}_{B,x}$ defines a point in $(G/B)_{\mathcal{F}_{B,x}}$. We will say that the B-reductions $\mathcal{F}_{B,x}$ and $\mathcal{F}'_{B,x}$ are in *relative position* w with if $\mathcal{F}'_{B,x}$ belongs to $S_{w,\mathcal{F}_{B,x}}$. In particular, if it belongs to the open orbit $S_{1,\mathcal{F}_{B,x}}$, we will say that $\mathcal{F}_{B,x}$ and $\mathcal{F}'_{B,x}$ are in generic position.

A Miura G-oper is called *generic* at the point $x \in X$ if the B-reductions $\mathcal{F}_{B,x}$ and $\mathcal{F}'_{B,x}$ of $\mathcal{F}_x$ are in generic position. In other words, $\mathcal{F}'_{B,x}$ belongs to the stratum $\mathrm{Op}_G(X) \times S_{1,\mathcal{F}_{B,x}} \subset \mathrm{MOp}_G(X)$. Being generic is an open condition. Therefore if a Miura oper is generic at $x \in X$, then there exists an open neighborhood U of x such that it is also generic at all other points of U. We denote the space of generic Miura opers on U by $\mathrm{MOp}_G(U)_{\mathrm{gen}}$.

Lemma 3.2. *Suppose we are given a Miura oper on the disc D_x around a point $x \in X$. Then its restriction to the punctured disc $D_x^\times$ is generic.*

Proof. Since being generic is an open condition, we obtain that if a Miura oper is generic at x, it is also generic on the entire D_x. Hence we only need to consider the situation where the Miura oper is not generic at x, i.e., the two reductions $\mathcal{F}_{B,x}$ and $\mathcal{F}'_{B,x}$ are in relative position $w \neq 1$. Let us trivialize the B-bundle $\mathcal{F}_B$, and hence the G-bundle $\mathcal{F}_G$ over D_x. Then ∇ gives us a connection on the trivial G-bundle which we can bring to the canonical form

$$\nabla = \partial_t + p_{-1} + \sum_{j=1}^{\ell} v_j(t) \cdot p_j$$

(see Lemma 1.1). It induces a connection on the trivial G/B-bundle. We are given a point gB in the fiber of the latter bundle which lies in the orbit $S_w = Bw_0wB$, where $w \neq 1$. Consider the horizontal section whose value at x is gB, viewed as a map $D_x \to G/B$. We need to show that the image of this map lies in the open B-orbit $S_1 = Bw_0B$ over $D_x^\times$, i.e., it does not lie in the orbit S_y for any $y \neq 1$.

Suppose that this is not so, and the image of the horizontal section actually lies in the orbit S_y for some $y \neq 1$. Since all B-orbits are H-invariant, we obtain that the same would be true for the horizontal section with respect to the connection $\nabla' = h\nabla h^{-1}$ for any constant element of H. Choosing $h = \check{\rho}(a)$ for $a \in \mathbb{C}^\times$, we can bring the connection to the form

$$\partial_t + a^{-1}p_{-1} + \sum_{j=1}^{\ell} a^{d_j} v_j(t) \cdot p_j.$$

Changing the variable t to $s = a^{-1}t$, we obtain the connection

$$\partial_s + p_{-1} + \sum_{j=1}^{\ell} a^{d_j+1} v_j(t),$$

so choosing small a we can make the functions $v_j(t)$ arbitrarily small. Therefore without loss of generality we can consider the case when our connection operator is $\nabla = \partial_t + p_{-1}$.

In this case our assumption that the horizontal section lies in $S_y, y \neq 1$, means that the vector field $\xi_{p_{-1}}$ corresponding to the infinitesimal action of p_{-1} on G/B is tangent to an orbit $S_y, y \neq 1$, in the neighborhood of some point gB of $S_w \subset G/B, w \neq 1$. But then, again because of the H-invariance of the B-orbits, the vector field $\xi_{hp_{-1}h^{-1}}$ is also tangent to this orbit for any $h \in H$. For any $i = 1 \ldots, \ell$, there exists a one-parameter subgroup $h_\epsilon^{(i)}, \epsilon \in \mathbb{C}^\times$ in H, such that $\lim_{\epsilon \to 0} \epsilon p_{-1} \epsilon^{-1} = f_i$. Hence we obtain that each of the vector fields $\xi_{f_i}, i = 1 \ldots, \ell$, is tangent to the orbit $S_y, y \neq 1$, in the neighborhood of $gB \in S_w, w \neq 1$. But then all commutators of these vectors fields are also tangent to this orbit. Hence we obtain that all vector fields of the form $\xi_p, p \in \mathfrak{n}_-$, are tangent to S_y in the neighborhood of $gB \in S_w$.

Consider any point of G/B that does not belong to the open dense orbit S_1. Then the quotient of the tangent space to this point by the tangent space to the B-orbit passing through this point is non-zero and the vector fields from the Lie algebra $\mathfrak{n}_-$ map surjectively onto this quotient. Therefore they cannot be tangent to the orbit $S_y, y \neq 1$, in a neighborhood of gB. Therefore our Miura oper is generic on $D_x^\times$. □

This lemma shows that any Miura oper on any smooth curve X is generic over an open dense subset.

3.2. Cartan connections

Introduce the H-bundle $\Omega^{\check{\rho}}$ on X which is uniquely determined by the following property: for any character $\lambda : H \to \mathbb{C}^\times$, the line bundle $\Omega^{\check{\rho}} \underset{H}{\times} \lambda$ associated to the corresponding one-dimensional representation of H is $\Omega^{\langle \lambda, \check{\rho} \rangle}$.

Explicitly, connections on $\Omega^{\check{\rho}}$ may be described as follows. If we choose a local coordinate t on X, then we trivialize $\Omega^{\check{\rho}}$ and represent the connection as an operator $\partial_t + \mathbf{u}(t)$, where $\mathbf{u}(t)$ is an $\mathfrak{h}$-valued function on X. If s is another coordinate such that $t = \varphi(s)$, then this connection will be represented by the operator

$$\partial_s + \varphi'(s)\mathbf{u}(\varphi(s)) - \check{\rho} \cdot \frac{\varphi''(s)}{\varphi'(s)}. \tag{3.2}$$

Let $\mathrm{Conn}(\Omega^{\check{\rho}})_X$ be the space of connections on the H-bundle $\Omega^{\check{\rho}}$ on X. When no confusion can arise, we will simply write Conn_X. We define a map

$$\mathbf{b}_X : \mathrm{Conn}_X \to \mathrm{MOp}_G(X)_{\mathrm{gen}}.$$

Suppose we are given a connection $\overline{\nabla}$ on the H-bundle $\Omega^{\check{\rho}}$ on D. We associate to it a generic Miura oper as follows. Let us choose a splitting $H \to B$ of the homomorphism $B \to H$ and set $\mathcal{F} = \Omega^{\check{\rho}} \underset{H}{\times} G, \mathcal{F}_B = \Omega^{\check{\rho}} \underset{H}{\times} B$, where we consider the adjoint action of H on G and on B obtained through the above splitting. The choice of the splitting also gives us the opposite Borel subgroup B_-, which is the unique Borel subgroup in generic position with B containing H. Let again w_0 be the longest element of the Weyl group of $\mathfrak{g}$. Then $w_0 B$ is a B-torsor equipped with a left action of H, so we define the B-subbundle $\mathcal{F}'_B$ of $\mathcal{F}$ as $\Omega^{\check{\rho}} \underset{H}{\times} w_0 B$.

Observe that the space of connections on $\mathcal{F}$ is isomorphic to the direct product

$$\mathrm{Conn}_X \times \bigoplus_{\alpha \in \Delta} \Gamma(X, \Omega^{\alpha(\check{\rho})+1}).$$

Its subspace corresponding to negative simple roots is isomorphic to the tensor product of $\left(\bigoplus_{i=1}^{\ell} \mathfrak{g}_{-\alpha_i}\right)$ and $\mathrm{Fun}\, X$. Having chosen a basis element f_i of $\mathfrak{g}_{-\alpha_i}$ for each $i = 1, \ldots, \ell$, we now construct an element $p_{-1} = \sum_{i=1}^{\ell} f_i \otimes 1$ of this space. Now we set $\nabla = \overline{\nabla} + p_{-1}$. By construction, ∇ has the correct relative position with the B-reduction $\mathcal{F}_B$ and preserves the B-reduction $\mathcal{F}'_B$. Therefore the quadruple $(\mathcal{F}, \nabla, \mathcal{F}_B, \mathcal{F}'_B)$ is a generic Miura oper on X. We define the morphism $\mathbf{b}_X$ by setting $\mathbf{b}_X(\overline{\nabla}) = (\mathcal{F}, \nabla, \mathcal{F}_B, \mathcal{F}'_B)$.

This map is independent of the choice of a splitting $H \to B$ and of the generators $f_i, i = 1, \ldots, \ell$.

Proposition 3.3 ([F2],**Proposition 10.4**). *The map $\mathbf{b}_X$ is an isomorphism of algebraic varieties*

$$\mathrm{Conn}_X \to \mathrm{MOp}_G(X)_{\mathrm{gen}}.$$

Thus, generic Miura opers are the same as Cartan connections, which are much simpler objects than opers.

The composition $\overline{\mathbf{b}}_X$ of $\mathbf{b}_X$ and the forgetful map $\mathrm{MOp}_G(X)_{\mathrm{gen}} \to \mathrm{Op}_G(X)$ is called the *Miura transformation*.

For example, in the case of $\mathfrak{g} = \mathfrak{sl}_2$, we have a connection $\overline{\nabla} = \partial_t - u(t)$ on the line bundle $\Omega^{1/2}$ (equivalently, a connection $\overline{\nabla}^t = \partial_t + u(t)$ on $\Omega^{-1/2}$), and the Miura transformation assigns to this connection the PGL_2-oper

$$\partial_t + \begin{pmatrix} -u(t) & 0 \\ 1 & u(t) \end{pmatrix}.$$

The oper B reduction $\mathcal{F}_B$ corresponds to the upper triangular matrices, and the Miura B-reduction $\mathcal{F}'_B$ corresponds to the lower triangular matrices. The corresponding projective connection is

$$\partial_t^2 - v(t) = (\partial_t - u(t))(\partial_t + u(t)),$$

i.e.,

$$u(t) \mapsto v(t) = u(t)^2 - u'(t).$$

3.3. Singularities of Cartan connections

Consider the Miura transformation $\overline{\mathbf{b}}_{D_x^\times}$ in the case of the punctured disc $D_x^\times$,

$$\overline{\mathbf{b}}_{D_x^\times} : \mathrm{Conn}(\Omega^{\check{\rho}})_{D_x^\times} \to \mathrm{Op}_G(D_x^\times).$$

Let $\mathrm{Conn}(\Omega^{\check{\rho}})^{\mathrm{RS}}_{D_x} \subset \mathrm{Conn}(\Omega^{\check{\rho}})_{D_x^\times}$ be the space of all connections on the H-bundle $\Omega^{\check{\rho}}$ on D_x with regular singularity, i.e., those for which the connection operator has the form

$$\overline{\nabla} = \partial_t + \frac{\check{\lambda}}{t} + \sum_{n\geq 0} u_n t^n.$$

We define a map

$$\mathrm{res}_{\mathfrak{h}} : \mathrm{Conn}(\Omega^{\check{\rho}})^{\mathrm{RS}}_{D} \to \mathfrak{h}$$

assigning to such a connection its residue $\check{\lambda}$.

It follows from the definition of the Miura transformation $\overline{\mathbf{b}}_{D_x^\times}$ that its restriction to $\mathrm{Conn}(\Omega^{\check{\rho}})^{\mathrm{RS}}_{D_x} \subset \mathrm{Conn}(\Omega^{\check{\rho}})_{D_x^\times}$ takes values in $\mathrm{Op}_G^{\mathrm{RS}}(D_x)$. Hence we obtain a morphism

$$\overline{\mathbf{b}}_x^{\mathrm{RS}} : \mathrm{Conn}(\Omega^{\check{\rho}})^{\mathrm{RS}}_{D_x} \to \mathrm{Op}_G^{\mathrm{RS}}(D_x).$$

Explicitly, after choosing a coordinate t on D_x, we can write $\overline{\nabla}$ as $\partial_t + t^{-1}\mathbf{u}(t)$, where $\mathbf{u}(t) \in \mathfrak{h}[[t]]$. Its residue is $\mathbf{u}(0)$. Then the corresponding oper with regular singularity is by definition the $N((t))$-equivalence class of the operator

$$\nabla = \partial_t + p_{-1} + t^{-1}\mathbf{u}(t),$$

which is the same as the $N[[t]]$-equivalence class of the operator

$$\check{\rho}(t)\nabla\check{\rho}(t)^{-1} = \partial_t + t^{-1}(p_{-1} - \check{\rho} + \mathbf{u}(t)),$$

so it is indeed an oper with regular singularity.

Therefore it follows from the definition that we have a commutative diagram

$$\begin{array}{ccc} \mathrm{Conn}(\Omega^{\check{\rho}})^{\mathrm{RS}}_{D_x} & \xrightarrow{\overline{\mathbf{b}}_x^{\mathrm{RS}}} & \mathrm{Op}_G^{\mathrm{RS}}(D_x) \\ {\scriptstyle \mathrm{res}_{\mathfrak{h}}}\downarrow & & \downarrow{\scriptstyle \mathrm{res}} \\ \mathfrak{h} & \longrightarrow & \mathfrak{h}/W \end{array} \tag{3.3}$$

where the lower horizontal map is the composition of the map $\check{\lambda} \mapsto \check{\lambda} - \check{\rho}$ and the projection $\varpi : \mathfrak{h} \to \mathfrak{h}/W$.

Now let $\mathrm{Conn}^{\mathrm{reg}}_{D_x,\check{\lambda}}$ be the preimage under $\overline{\mathbf{b}}_x^{\mathrm{RS}}$ of the subspace $\mathrm{Op}_G(D_x)_{\check{\lambda}} \subset \mathrm{Op}_G^{\mathrm{RS}}(D_x)$ of G-opers on $D_x^\times$ with regular singularity, residue $\varpi(-\check{\lambda} - \check{\rho})$ and trivial monodromy. By the commutativity of the above diagram, a connection in $\mathrm{Conn}^{\mathrm{reg}}_{D_x,\check{\lambda}}$ necessarily has residue of the form $-w(\check{\lambda} + \check{\rho}) + \check{\rho}$ for some element w of the Weyl group of G, so that $\mathrm{Conn}^{\mathrm{reg}}_{D_x,\check{\lambda}}$ is the disjoint union of its subsets $\mathrm{Conn}^{\mathrm{reg}}_{D_x,\check{\lambda},w}$ consisting of connections with residue $-w(\check{\lambda} + \check{\rho}) + \check{\rho}$.

The restriction of $\overline{\mathbf{b}}_x^{\mathrm{RS}}$ to $\mathrm{Conn}^{\mathrm{reg}}_{D_x,\check{\lambda},w}$ is a map

$$\overline{\mathbf{b}}_{\check{\lambda},w} : \mathrm{Conn}^{\mathrm{reg}}_{D_x,\check{\lambda},w} \to \mathrm{Op}_G(D_x)_{\check{\lambda}}.$$

Let us recall that by construction of the Miura transformation $\overline{\mathbf{b}}$, each oper on $D_x^\times$ which lies in the image of the map $\overline{\mathbf{b}}$ (hence in particular, in the image of $\overline{\mathbf{b}}_{\check{\lambda},w}$) carries a canonical horizontal B-reduction

$$\mathcal{F}'_B = \Omega^{\check{\rho}} \underset{H}{\times} w_0 B$$

(i.e., it carries a canonical structure of Miura oper on $D_x^\times$). But if this oper is in the image of $\overline{\mathbf{b}}_{\check{\lambda},w}$, i.e., belongs to $\mathrm{Op}_G(D_x)_{\check{\lambda}}$, then the oper B-reduction $\mathcal{F}_B$ (and hence the oper bundle $\mathcal{F}$) has a canonical extension to a B-bundle on the entire disc D_x, namely, one for which the oper connection has the form (1.8). Therefore the B-reduction $\mathcal{F}'_B = \Omega^{\check{\rho}} \underset{H}{\times} w_0 B$ may also be extended to D_x.

Therefore we can lift $\overline{\mathbf{b}}_{\check{\lambda},w}$ to a map

$$\mathbf{b}_{\check{\lambda},w} : \mathrm{Conn}^{\mathrm{reg}}_{D_x,\check{\lambda},w} \to \mathrm{MOp}_G(D_x)_{\check{\lambda}}.$$

Let $\mathrm{MOp}_G(D_x)_{\check{\lambda},w} \subset \mathrm{MOp}_G(D_x)_{\check{\lambda}}$ be the subvariety of those Miura opers of coweight $\check{\lambda}$ which have relative position w at x. Then

$$\mathrm{MOp}_G(D_x)_{\check{\lambda},w} \simeq \mathrm{Op}_G(D_x)_{\check{\lambda}} \times S_{w,\mathcal{F}'_{B,x}}.$$

The following result is due to D. Gaitsgory and myself [FG] (see [F3], Proposition 2.9).

Proposition 3.4. *For each $w \in W$ the morphism $\mathbf{b}_{\check{\lambda},w}$ is an isomorphism between the varieties $\mathrm{Conn}^{\mathrm{reg}}_{D_x,\check{\lambda},w}$ and $\mathrm{MOp}_G(D_x)_{\check{\lambda},w}$.*

Proof. First we observe that at the level of points the map defined by $\mathbf{b}_{\check{\lambda},w}, w \in W$, from the union of $\mathrm{Conn}^{\mathrm{reg}}_{D_x,\check{\lambda},w}, w \in W$, to $\mathrm{MOp}_G(D_x)_{\check{\lambda}}$, is a bijection. Indeed, by Proposition 3.3 we have a map taking a Miura oper from $\mathrm{MOp}_G(D_x)_{\check{\lambda}}$, considered as a Miura oper on the punctured disc $D_x^\times$, to a connection $\overline{\nabla}$ on the H-bundle $\Omega^{\check{\rho}}$ over $D_x^\times$. We have shown above that $\overline{\nabla}$ has regular singularity at x and that its residue is of the form $-w(\check{\lambda}+\check{\rho})+\check{\rho}, w \in W$. Thus, we obtain a map from the set of points of $\mathrm{MOp}_G(D_x)_{\check{\lambda}}$ to the union of $\mathrm{Conn}^{\mathrm{reg}}_{D_x,\check{\lambda},w}, w \in W$, and by Proposition 3.3 it is a bijection.

It remains to show that if the Miura oper belongs to $\mathrm{MOp}_G(D_x)_{\check{\lambda},w}$, then the corresponding connection has residue precisely $-w(\check{\lambda}+\check{\rho})+\check{\rho}$.

Thus, we are given a G-oper $(\mathcal{F},\nabla,\mathcal{F}_B,\mathcal{F}'_B)$ of coweight $\check{\lambda}$. Let us choose a trivialization of the B-bundle $\mathcal{F}_B$. Then the connection operator reads

$$\nabla = \partial_t + \sum_{i=1}^{\ell} t^{\langle \alpha_i,\check{\lambda}\rangle} f_i + \mathbf{v}(t), \qquad \mathbf{v}(t) \in \mathfrak{b}[[t]]. \tag{3.4}$$

Suppose that the horizontal B-reduction $\mathcal{F}'_B$ of our Miura oper has relative position w with $\mathcal{F}_B$ at x (see Section 3 for the definition of relative position). We need to show that the corresponding connection on $\mathcal{F}'_H \simeq \Omega^{\check{\rho}}$ has residue $-w(\check{\lambda}+\check{\rho})+\check{\rho}$.

This is equivalent to the following statement. Let $\Phi(t)$ be the G-valued solution of the equation

$$\left(\partial_t + \sum_{i=1}^{\ell} t^{\langle \alpha_i, \check{\lambda}\rangle} f_i + \mathbf{v}(t)\right)\Phi(t) = 0, \tag{3.5}$$

such that $\Phi(0) = 1$. Since the connection operator is regular at $t=0$, this solution exists and is unique. Then $\Phi(t)w^{-1}w_0$ is the unique solution of the equation (3.5) whose value at $t=0$ is equal to $w^{-1}w_0$.

By Lemma 3.2, we have

$$\Phi(t)w^{-1}w_0 = X_w(t)Y_w(t)Z_w(t)w_0,$$

where

$$X_w(t) \in N((t)), \qquad Y_w(t) \in H((t)), \qquad Z_w(t) \in N_-((t)).$$

We can write $Y_w(t) = \check{\mu}_w(t)\widetilde{Y}_w(t)$, where $\check{\mu}_w$ is a coweight and $\widetilde{Y}_w(t) \in H[[t]]$.

Since the connection ∇ preserves

$$\Phi(t)w_0\mathfrak{b}_+w_0\Phi(t)^{-1} = \Phi(t)\mathfrak{b}_-\Phi(t)^{-1},$$

the connection $X(t)_w^{-1}\nabla X_w(t)$ preserves

$$Y_w(t)Z_w(t)\mathfrak{b}_-Z_w(t)^{-1}Y_w(t)^{-1} = \mathfrak{b}_-,$$

and therefore has the form

$$\partial_t + \sum_{i=1}^{\ell} t^{\langle \alpha_i, \check{\lambda}\rangle} f_i - \frac{\check{\mu}_w}{t} + \mathbf{u}(t), \qquad \mathbf{u}(t) \in \mathfrak{h}[[t]].$$

By conjugating it with $\check{\lambda}(t)$ we obtain a connection

$$\partial_t + p_{-1} - \frac{\check{\lambda}+\check{\mu}_w}{t} + \mathbf{u}(t), \qquad \mathbf{u}(t) \in \mathfrak{h}[[t]].$$

Therefore we need to show that

$$\check{\mu}_w = w(\check{\lambda}+\check{\rho}) - (\check{\lambda}+\check{\rho}). \tag{3.6}$$

To see that, let us apply the identity $\Phi(t)w^{-1} = X_w(t)Y_w(t)Z_w(t)$ to a non-zero vector $v_{w_0(\nu)}$ of weight $w_0(\nu)$ in a finite-dimensional irreducible $\mathfrak{g}$-module V_ν of highest weight ν (so that $v_{w_0(\nu)}$ is a lowest weight vector and hence is unique up to scalar). The right-hand side will then be equal to a $P(t)v_{w_0(\nu)}$ plus the sum of terms of weights greater than $w_0(\nu)$, where $P(t) = ct^{\langle w_0(\nu), \check{\mu}_w\rangle}, c \neq 0$, plus the sum of terms of higher degree in t. Applying the left-hand side to $v_{w_0(\nu)}$, we obtain $\Phi(t)v_{w^{-1}w_0(\nu)}$, where $v_{w^{-1}w_0(\nu)} \in V_\nu$ is a non-zero vector of weight $w^{-1}w_0(\nu)$ which is also unique up to a scalar.

Thus, we need to show that the coefficient with which $v_{w_0(\nu)}$ enters the expression $\Phi(t)v_{w^{-1}w_0(\nu)}$ is a polynomial in t whose lowest degree is equal to

$$\langle w_0(\nu), w(\check{\lambda}+\check{\rho}) - (\check{\lambda}+\check{\rho})\rangle,$$

because if this is so for all dominant integral weights ν, then we obtain the desired equality (3.6). But this formula is easy to establish. Indeed, from the form (3.4) of the oper connection ∇ it follows that we can obtain a vector proportional to v_{w_0} by applying the operators $\frac{1}{\langle\alpha_i,\check{\lambda}\rangle+1}t^{\langle\alpha_i,\check{\lambda}\rangle+1}f_i, i = 1,\ldots,\ell$, to $v_{w^{-1}w_0(\nu)}$ in some order. The linear combination of these monomials appearing in the solution is the term of the lowest degree in t with which $v_{w_0(\nu)}$ enters $\Phi(t)v_{w^{-1}w_0(\nu)}$. It follows from Lemma 3.2 that it is non-zero. The corresponding power of t is nothing but the difference between the $(\check{\lambda}+\check{\rho})$-degrees of the vectors $v_{w^{-1}w_0}$ and v_{w_0}, i.e.,

$$\langle w^{-1}w_0(\nu), \check{\lambda}+\check{\rho}\rangle - \langle w_0(\nu), \check{\lambda}+\check{\rho}\rangle = \langle w_0(\nu), w(\check{\lambda}+\check{\rho}) - (\check{\lambda}+\check{\rho})\rangle,$$

as desired. This completes the proof. □

Thus, we have identified the space $\operatorname{Conn}^{\mathrm{reg}}_{D_x,\check{\lambda},w}$ of connections on the H-bundle $\Omega^{\check{\rho}}$ on $D_x^\times$ of the form with the space of Miura opers on $D_x^\times$ such that the underlying oper belongs to $\operatorname{Op}_G(D_x)_{\check{\lambda}}$ and the corresponding B-reductions $\mathcal{F}_B$ and $\mathcal{F}'_B$ have relative position w at x.

The condition that the image under $\mathbf{b}_{\check{\lambda},w}$ of a connection of the form

$$\partial_t - \frac{w(\check{\lambda}+\check{\rho})-\check{\rho}}{t} + \mathbf{u}(t), \qquad \mathbf{u}(t) \in \mathfrak{h}[[t]],$$

is an oper without monodromy imposes polynomial equations on the coefficients of the series $\mathbf{u}(t)$. Consider the simplest case when $\check{\lambda} = 0$ and $w = s_i$, the ith simple reflection. Then $-w(\check{\lambda}+\check{\rho})+\check{\rho} = \check{\alpha}_i$, so we write this connection as

$$\overline{\nabla} = \partial_t + \frac{\check{\alpha}_i}{t} + \mathbf{u}(t), \qquad \mathbf{u}(t) \in \mathfrak{h}[[t]]. \tag{3.7}$$

Lemma 3.5 ([F3],Lemma 2.10). *A connection of the form* (3.7) *belongs to* $\operatorname{Conn}^{\mathrm{reg}}_{D_x,s_i}$ *(i.e., the corresponding G-oper is regular at x) if and only if* $\langle\alpha_i, \mathbf{u}(0)\rangle = 0$.

4. Wakimoto modules and Bethe Ansatz

In this section we explain how to construct Bethe eigenvectors of the generalized Gaudin Hamiltonians. For that we utilize the Wakimoto modules over $\widehat{\mathfrak{g}}_{\kappa_c}$ which are parameterized by Cartan connections on the punctured disc. As the result, the eigenvectors will be parameterized by the Cartan connections on $\mathbb{P}^1$ with regular singularities at $z_1,\ldots,z_N,\infty$ and some additional points $w_1,\ldots,w_m$ with residues $\lambda_1,\ldots,\lambda_N,\lambda_\infty$ and $-\alpha_{i_1},\ldots,-\alpha_{i_m}$ and whose Miura transformation is an oper that has no singularities at $w_1,\ldots,w_m$.

4.1. Definition of Wakimoto modules

We recall some of the results of [FF1, F2] on the construction of the Wakimoto realization.

Let $\mathcal{A}^{\mathfrak{g}}$ be the Weyl algebra with generators $a_{\alpha,n}, a^*_{\alpha,n}$, $\alpha \in \Delta_+$, $n \in \mathbb{Z}$, and relations

$$[a_{\alpha,n}, a^*_{\beta,m}] = \delta_{\alpha,\beta}\delta_{n,-m}, \qquad [a_{\alpha,n}, a_{\beta,m}] = [a^*_{\alpha,n}, a^*_{\beta,m}] = 0. \tag{4.1}$$

Introduce the generating functions

$$a_\alpha(z) = \sum_{n\in\mathbb{Z}} a_{\alpha,n} z^{-n-1}, \tag{4.2}$$

$$a^*_\alpha(z) = \sum_{n\in\mathbb{Z}} a^*_{\alpha,n} z^{-n}. \tag{4.3}$$

Let $M_{\mathfrak{g}}$ be the Fock representation of $\mathcal{A}^{\mathfrak{g}}$ generated by a vector $|0\rangle$ such that

$$a_{\alpha,n}|0\rangle = 0, \quad n \geq 0; \qquad a^*_{\alpha,n}|0\rangle = 0, \quad n > 0.$$

It carries a vertex algebra structure (see [F2]).

Let π_0 be the commutative algebra $\mathbb{C}[b_{i,n}]_{i=1,\ldots,\ell;n<0}$ with the derivation T given by the formula

$$T \cdot b_{i_1,n_1} \ldots b_{i_m,n_m} = -\sum_{j=1}^{m} n_j b_{i_1,n_1} \ldots b_{i_j,n_j-1} \cdots b_{i_m,n_m}.$$

Then π_0 is naturally a commutative vertex algebra (see [FB], § 2.3.9). In particular, we have

$$Y(b_{i,-1}, z) = b_i(z) = \sum_{n<0} b_{i,n} z^{-n-1}.$$

Recall that $\mathbb{V}_0$ carries the structure of a vertex algebra. Its vacuum vector will be denoted by v_{κ_c}.

In the next theorem describing the Wakimoto realization we will need the formulas defining the action of the Lie algebra $\mathfrak{g}$ on the space of functions on the big cell U of the flag manifold G/B_+. Here U is the open orbit of the unipotent subgroup $N_+ = [B_+, B_+]$ which is isomorphic to N_+. Since the exponential map $\mathfrak{n}_+ \to N_+$ is an isomorphism, we obtain a system of coordinates $\{y_\alpha\}_{\alpha\in\Delta_+}$ on U corresponding to a fixed basis of root vectors $\{e_\alpha\}_{\alpha\in\Delta_+}$ in $\mathfrak{n}_+$.

For each $\chi \in \mathfrak{h}^*$ we have a homomorphism $\rho_\chi : \mathfrak{g} \hookrightarrow \mathcal{D}_{\leq 1}(N_+)$ under which the $\mathcal{D}_{\leq 1}(N_+)$-module $\operatorname{Fun} U$ becomes isomorphic to the $\mathfrak{g}$-module M^*_χ that is contragredient to the Verma module of highest weight χ.

Explicitly, ρ_χ looks as follows. Let $e_i, h_i, f_i, i = 1\dots,\ell$, be the generators of $\mathfrak{g}$. Then

$$\rho_\chi(e_i) = \frac{\partial}{\partial y_{\alpha_i}} + \sum_{\beta\in\Delta_+} P^i_\beta(y_\alpha)\frac{\partial}{\partial y_\beta}, \tag{4.4}$$

$$\rho_\chi(h_i) = -\sum_{\beta\in\Delta_+} \beta(h_i) y_\beta \frac{\partial}{\partial y_\beta} + \chi(h_i), \tag{4.5}$$

$$\rho_\chi(f_i) = \sum_{\beta\in\Delta_+} Q^i_\beta(y_\alpha)\frac{\partial}{\partial y_\beta} + \chi(h_i) y_{\alpha_i}, \tag{4.6}$$

for some polynomials P^i_β, Q^i_β in $y_\alpha, \alpha \in \Delta_+$.

In addition, we have a Lie algebra anti-homomorphism $\rho^R : \mathfrak{n}_+ \to \mathcal{D}_{\leq 1}(N_+)$ which corresponds to the *right* action of $\mathfrak{n}_+$ on N_+. The differential operators $\rho^R(x), x \in \mathfrak{n}_+$, commute with the differential operators $\rho_\chi(x'), x' \in \mathfrak{n}_+$ (though their commutation relations with $\rho_\chi(x'), x' \notin \mathfrak{n}_+$, are complicated in general). We have

$$\rho^R(e_i) = \frac{\partial}{\partial y_{\alpha_i}} + \sum_{\beta\in\Delta_+} P^{R,i}_\beta(y_\alpha)\frac{\partial}{\partial y_\beta}$$

for some polynomials $P^{R,i}_\beta, Q^i_\beta$ in $y_\alpha, \alpha \in \Delta_+$. We let

$$e^R_i(z) = \sum_{n\in\mathbb{Z}} e^R_{i,n} z^{-n-1} = a_{\alpha_i}(z) + \sum_{\beta\in\Delta_+} P^{R,i}_\beta(a^*_\alpha(z)) a_\beta(z). \tag{4.7}$$

Now we can state the main result, due to [FF1, F2], concerning the Wakimoto realization at the critical level.

Theorem 4.1. *There exists a homomorphism of vertex algebras*

$$w_{\kappa_c} : \mathbb{V}_0 \to M_{\mathfrak{g}} \otimes \pi_0$$

such that

$$e_i(z) \mapsto a_{\alpha_i}(z) + \sum_{\beta\in\Delta_+} {:}P^i_\beta(a^*_\alpha(z)) a_\beta(z){:},$$

$$h_i(z) \mapsto -\sum_{\beta\in\Delta_+} \beta(h_i){:}a^*_\beta(z) a_\beta(z){:} + b_i(z),$$

$$f_i(z) \mapsto \sum_{\beta\in\Delta_+} {:}Q^i_\beta(a^*_\alpha(z)) a_\beta(z){:} + c_i \partial_z a^*_{\alpha_i}(z) + b_i(z) a^*_{\alpha_i}(z),$$

where the polynomials P^i_β, Q^i_β are introduced in formulas (4.4)–(4.6).

In order to make the homomorphism w_{κ_c} coordinate-independent, we need to define the actions of the group $\operatorname{Aut}\mathcal{O}$ on $\mathbb{V}_0, M_{\mathfrak{g}}$ and π_0 which are intertwined by w_{κ_c}. We already have a natural action of $\operatorname{Aut}\mathcal{O}$ on $\mathbb{V}_0$ which is induced by its action on $\widehat{\mathfrak{g}}_{\kappa_c}$ preserving the subalgebra $\mathfrak{g}[[t]]$. Next, we define the $\operatorname{Aut}\mathcal{O}$-action on

$M_{\mathfrak{g}}$ by stipulating that for each $\alpha \in \Delta_+$ the generating functions $a_\alpha(z)$ and $a^*_\alpha(z)$ transform as a one-form and a functions on $D^\times$, respectively.

Finally, we identify π_0 with the algebra $\operatorname{Fun}\operatorname{Conn}(\Omega^{-\rho})_D$, where $\operatorname{Conn}(\Omega^{-\rho})_D$ is the space of connections on the ${}^L H$-bundle $\Omega^{-\rho}$ on D. Namely, we write a connection from $\operatorname{Conn}(\Omega^{-\rho})_D$ as the first order operator $\partial_t + \chi(t)$, where $\chi(t)$ takes values in ${}^L\mathfrak{h} \simeq \mathfrak{h}^*$. Set

$$b_i(t) = \langle \check{\alpha}_i, \chi(t) \rangle = \sum_{n<0} b_{i,n} t^{-n-1}.$$

Then we obtain an identification

$$\operatorname{Fun}\operatorname{Conn}(\Omega^\rho)_D \simeq \mathbb{C}[b_{i,n}]_{i=1,\dots,\ell;n<0} = \pi_0.$$

Now the natural action of $\operatorname{Aut}\mathcal{O}$ on $\operatorname{Fun}\operatorname{Conn}(\Omega^\rho)_D$ gives rise to an action of $\operatorname{Aut}\mathcal{O}$ on π_0.

To write down this action explicitly, suppose that we have a connection on the ${}^L H$-bundle $\Omega^{-\rho}$ on a curve X which with respect to a local coordinate t is represented by the first order operator $\partial_t + \chi(t)$. If s is another coordinate such that $t = \varphi(s)$, then this connection will be represented by the operator

$$\partial_s + \varphi'(s)\chi(\varphi(s)) + \rho \cdot \frac{\varphi''(s)}{\varphi'(s)}. \tag{4.8}$$

Proposition 4.2 ([F2], Cor. 5.4). *The homomorphism w_{κ_c} commutes with the action of* $\operatorname{Aut}\mathcal{O}$.

For a connection $\overline{\nabla}$ on the ${}^L H$-bundle $\Omega^{-\rho}$, let $\overline{\nabla}^*$ be the dual connection on Ω^ρ. Explicitly, if $\overline{\nabla}$ is given by the operator $\partial_t + \mathbf{w}(t)$, then $\overline{\nabla}^*$ is given by the operator $\partial_t - \mathbf{w}(t)$. Sending $\overline{\nabla}$ to $\overline{\nabla}^*$, we obtain an isomorphism $\operatorname{Conn}(\Omega^{-\rho})_U \simeq \operatorname{Conn}(\Omega^\rho)_U$.

Recall the Miura transformation

$$\overline{\mathbf{b}}_{D^\times} : \operatorname{Conn}(\Omega^\rho)_{D^\times} \to \operatorname{Op}_{{}^L G}(D^\times)$$

defined in Section 3.3. Using the isomorphism $\operatorname{Conn}(\Omega^{-\rho})_U \simeq \operatorname{Conn}(\Omega^\rho)_U$, we obtain a map

$$\overline{\mathbf{b}}^*_{D^\times} : \operatorname{Conn}(\Omega^{-\rho})_{D^\times} \to \operatorname{Op}_{{}^L G}(D^\times) \tag{4.9}$$

which by abuse of terminology we will also refer to as Miura transformation.

Consider the center $\mathfrak{z}(\widehat{\mathfrak{g}})$ of the vertex algebra $\mathbb{V}_0$. Recall that we have

$$\mathfrak{z}(\widehat{\mathfrak{g}}) = \mathbb{V}_0^{\mathfrak{g}[[t]]} \simeq \operatorname{End}_{\widehat{\mathfrak{g}}_{\kappa_c}} \mathbb{V}_0.$$

Theorem 4.3 ([F2], Theorem 11.3). *Under the homomorphism w_{κ_c} the center $\mathfrak{z}(\widehat{\mathfrak{g}}) \subset \mathbb{V}_0$ gets mapped to the vertex subalgebra π_0 of $M_{\mathfrak{g}} \otimes \pi_0$. Moreover, we have the following commutative diagram*

$$\begin{array}{ccc} \pi_0 & \xrightarrow{\sim} & \operatorname{Fun}\operatorname{Conn}(\Omega^{-\rho})_D \\ \uparrow & & \uparrow \\ \mathfrak{z}(\widehat{\mathfrak{g}}) & \xrightarrow{\sim} & \operatorname{Fun}\operatorname{Op}_{{}^L G}(D) \end{array} .$$

where the right vertical map is induced by the Miura transformation $\overline{\mathbf{b}}_D^*$ *given by* (4.9).

This theorem implies the following result. Let N be a module over the vertex algebra $M_{\mathfrak{g}}$ and R a module over the vertex algebra $\pi_0 \simeq \operatorname{Fun}\operatorname{Conn}(\Omega^\rho)_D$. Then the homomorphism w_{κ_c} gives rise to the structure of a $\mathbb{V}_0$-module, and hence of a smooth $\widehat{\mathfrak{g}}_{\kappa_c}$-module on the tensor product $N \otimes R$.

It follows from the general theory of [FB], Ch. 5, that a module over the vertex algebra $M_{\mathfrak{g}}$ is the same as a smooth module over the Weyl algebra $\mathcal{A}^{\mathfrak{g}}$, i.e., such that every vector is annihilated by $a_{\alpha,n}, a^*_{\alpha,n}$ for large enough n. Likewise, a module over the commutative vertex algebra $\operatorname{Fun}\operatorname{Conn}(\Omega^{-\rho})_D$ is the same as a smooth module over the commutative topological algebra $\operatorname{Fun}\operatorname{Conn}(\Omega^{-\rho})_{D^\times}$. Note that

$$\operatorname{Fun}\operatorname{Conn}(\Omega^{-\rho})_{D^\times} \simeq \varprojlim \mathbb{C}[b_{i,n}]_{i=1,\dots,\ell;n\in\mathbb{Z}}/I_N,$$

where I_N is the ideal generated by $b_{i,n}, i = 1,\dots,\ell; n \geq N$. A module over this algebra is called smooth if every vector is annihilated by the ideal I_N for large enough N. Thus, for each choice of $\mathcal{A}^{\mathfrak{g}}$-module N we obtain a "semi-infinite induction" functor from the category of smooth $\operatorname{Fun}\operatorname{Conn}(\Omega^{-\rho})_{D^\times}$-modules to the category of smooth $\widehat{\mathfrak{g}}_{\kappa_c}$-modules (on which the central element $\mathbf{1}$ acts as the identity), $R \mapsto N \otimes R$.

Now Theorem 4.3 implies

Corollary 4.4. *The action of* $Z(\widehat{\mathfrak{g}})$ *on the* $\widehat{\mathfrak{g}}_{\kappa_c}$*-module* $N \otimes R$ *is independent of the choice of* N *and factors through the homomorphism* $Z(\widehat{\mathfrak{g}}) \simeq \operatorname{Fun}\operatorname{Op}_{{}^LG}(D^\times) \to \operatorname{Fun}\operatorname{Conn}(\Omega^{-\rho})_{D^\times}$ *induced by the Miura transformation* $\overline{\mathbf{b}}^*_{D^\times}$.

For example, for a connection $\overline{\nabla} = \partial_z + \chi(z)$ in $\operatorname{Conn}(\Omega^{-\rho})_{D^\times}$, let $\mathbb{C}_{\overline{\nabla}} = \mathbb{C}_{\chi(z)}$ be the corresponding one-dimensional module of $\operatorname{Fun}\operatorname{Conn}(\Omega^{-\rho})_{D^\times}$. Then $M_{\mathfrak{g}} \otimes \mathbb{C}_{\chi(z)}$ is called the *Wakimoto module* of critical level corresponding to $\partial_z + \chi(z)$, and is denoted by $W_{\chi(z)}$.

We will use below the following result.

Lemma 4.5 ([FFR], Lemma 2). *Let*

$$\chi(z) = -\frac{\alpha_i}{z} + \sum_{n=0}^{\infty} \chi_n z^n, \qquad \chi_n \in \mathfrak{h}^*.$$

The vector $e^R_{i,-1}|0\rangle \in W_{\chi(z)}$ *is* $\mathfrak{g}[[t]]$*-invariant if and only if* $\langle \check{\alpha}_i, \chi_0 \rangle = 0$.

4.2. Functoriality of coinvariants

We will construct eigenvectors of the generalized Gaudin Hamiltonians using functionals on the space of coinvariants of Wakimoto modules on $\mathbb{P}^1$. First we explain some general facts about the compatibility of various spaces of coinvariants.

Following the general construction of [FB], Ch. 10, we define the spaces of coinvariants $H_{M_{\mathfrak{g}}}(X,(x_i),(M_i))$ (resp., $H_{\pi_0}(X,\{x_i\},(R_i))$) attached to a smooth projective curve X, points $x_1,\dots,x_p$ and $M_{\mathfrak{g}}$-modules $N_1,\dots,N_p$ (resp.,

π_0-modules $R_1, \ldots, R_p$). These spaces are defined in elementary terms in [FFR], Sect. 5, and in [FB], Ch. 13–14. We also define spaces of coinvariants for the tensor product vertex algebra $M_{\mathfrak{g}} \otimes \pi_0$. We have a natural identification

$$H_{M_{\mathfrak{g}}\otimes\pi_0}(X,(x_i),(N_i \otimes R_i)) = H_{M_{\mathfrak{g}}}(X,(x_i),(N_i)) \otimes H_{\pi_0}(X,(x_i),(R_i)).$$

The vertex algebra homomorphism $w_{\kappa_c} : \mathbb{V}_0 \to M_{\mathfrak{g}} \otimes \pi_0$ of Theorem 4.1 gives rise to a surjective map of the spaces of coinvariants

$$H_{\mathbb{V}_0}(X,(x_i),(N_i \otimes R_i)) \twoheadrightarrow H_{M_{\mathfrak{g}}\otimes\pi_0}(X,(x_i),(N_i \otimes R_i))$$

and hence a surjective map

$$H_{\mathbb{V}_0}(X,(x_i),(N_i \otimes R_i)) \twoheadrightarrow H_{M_{\mathfrak{g}}}(X,(x_i),(N_i)) \otimes H_{\pi_0}(X,(x_i),(R_i)). \tag{4.10}$$

Now recall that in the proof of Theorem 2.7 we explained that since $\mathfrak{z}(\widehat{\mathfrak{g}}) \simeq \operatorname{Fun}\operatorname{Op}_{{}^L G}(D)$ is the center of $\mathbb{V}_0$, we have a homomorphism of algebras

$$H_{\mathfrak{z}(\widehat{\mathfrak{g}})}(X,(x_i),(Z(M_i))) \to \operatorname{End}_{\mathbb{C}} H_{\mathbb{V}_0}(X,(x_i),(M_i)). \tag{4.11}$$

In addition, we have an isomorphism

$$H_{\mathfrak{z}(\widehat{\mathfrak{g}})}(X,(x_i),(Z(M_i))) \simeq \operatorname{Fun}\operatorname{Op}_{{}^L G}(X,(x_i),(M_i)), \tag{4.12}$$

where $\operatorname{Op}_{{}^L G}(X,(x_i),(M_i))$ is the space of ${}^L G$-opers on X which are regular on $X \backslash \{x_1, \ldots, x_N\}$ and such that their restriction to $D_x^\times$ belongs to the space $\operatorname{Op}_{{}^L G}^{M_i}(D_{x_i}^\times) = \operatorname{Spec} Z(M_i)$ for all $i = 1, \ldots, N$.

We can describe the space of coinvariants $H_{\pi_0}(X,(x_i),(R_i))$ for the commutative vertex algebra $\pi_0 \simeq \operatorname{Fun}\operatorname{Conn}(\Omega^{-\rho})_D$ in similar terms. If R is a π_0-module, or equivalently, a smooth $\operatorname{Fun}\operatorname{Conn}(\Omega^{-\rho})_{D^\times}$-module, let $\pi_0(R)$ be the image of the corresponding homomorphism

$$\operatorname{Fun}\operatorname{Conn}(\Omega^{-\rho})_{D^\times} \to \operatorname{End}_{\mathbb{C}} R.$$

Let $\operatorname{Conn}(\Omega^{-\rho})(X,(x_i),\ (R_i))$ be the space of connections on $\Omega^{-\rho}$ on $X \backslash \{x_1, \ldots, x_N\}$ whose restriction to $D_{x_i}^\times$ belongs to the space $\operatorname{Conn}(\Omega^{-\rho})_{D_{x_i}^\times}^{R_i} = \operatorname{Spec} \pi_0(R_i)$. Then we have an isomorphism

$$H_{\pi_0}(X,(x_i),(R_i)) \simeq \operatorname{Fun}\operatorname{Conn}(\Omega^{-\rho})(X,(x_i),(R_i)).$$

It follows from Corollary 4.4 that if $M_i = N_i \otimes R_i$, then $Z(N_i \otimes R_i)$ is the image of $Z(\widehat{\mathfrak{g}}) \simeq \operatorname{Fun}_{{}^L G}(D^\times)$ under the homomorphism

$$\operatorname{Fun}_{{}^L G}(D^\times) \xrightarrow{\overline{\mathbf{b}}^*_{D^\times}} \operatorname{Fun}\operatorname{Conn}(\Omega^{-\rho})_{D^\times} \to \operatorname{End}_{\mathbb{C}} R_i,$$

where the first map is induced by the Miura transformation.

Recall that in Section 3.2 we introduced the maps

$$\mathbf{b}_U : \operatorname{Conn}(\Omega^\rho)_U \to \operatorname{MOp}_{{}^L G}(U)_{\text{gen}}, \qquad \overline{\mathbf{b}}_U : \operatorname{Conn}(\Omega^\rho)_U \to \operatorname{Op}_{{}^L G}(U).$$

Using the identification $\operatorname{Conn}(\Omega^\rho)_U \simeq \operatorname{Conn}(\Omega^{-\rho})_U$, we obtain maps

$$\mathbf{b}^*_U : \operatorname{Conn}(\Omega^{-\rho})_U \to \operatorname{MOp}_{{}^L G}(U)_{\text{gen}}, \qquad \overline{\mathbf{b}}^*_U : \operatorname{Conn}(\Omega^{-\rho})_U \to \operatorname{Op}_{{}^L G}(U).$$

The second map is the Miura transformation. In the case when $U=X\backslash\{x_1,\dots,x_N\}$ the Miura transformation $\overline{\mathbf{b}}^*_U$ restricts to a map

$$\mathrm{Conn}(\Omega^{-\rho})(X,(x_i),(R_i)) \to \mathrm{Op}_{^LG}(X,(x_i),(N_i \otimes R_i)),$$

and hence gives rise to a homomorphism

$$\mathrm{Fun}\,\mathrm{Op}_{^LG}(X,(x_i),(N_i \otimes R_i)) \to \mathrm{Fun}\,\mathrm{Conn}(\Omega^{-\rho})(X,(x_i),(R_i)), \tag{4.13}$$

which does not depend on the N_i's.

Now we have an action of the algebra $\mathrm{Op}_{^LG}(X,(x_i),(N_i \otimes R_i))$ on both sides of the map (4.10). The action on the left-hand side comes from the homomorphisms (4.11) and (4.12). The action on the right-hand side comes from the homomorphism (4.13).

The functoriality of the coinvariants implies the following

Lemma 4.6. *The map* (4.10) *commutes with the action on both sides of the algebra* $\mathrm{Op}_{^LG}(X,(x_i),(N_i \otimes R_i))$.

Consider the special case where all modules $R_i, i = 1,\dots,N$, are one-dimensional, i.e., $R_i = \mathbb{C}_{\overline{\nabla}_i}$, where $\overline{\nabla}_i$ is a point in $\mathrm{Conn}(\Omega^{-\rho})_{D^\times_{x_i}}$, and R_i is the representation obtained from the homomorphism $\mathrm{Fun}\,\mathrm{Conn}(\Omega^{-\rho})_{D^\times_{x_i}}$ induced by $\overline{\nabla}_i$. In this case we have the following description of the space $H_{\pi_0}(X,(x_i),(\mathbb{C}_{\overline{\nabla}_i}))$ which follows from the general results on coinvariants of commutative vertex algebras from Sect. 9.4 of [FB] (see also [FFR], Proposition 4, in the case when $X = \mathbb{P}^1$).

Proposition 4.7. *The space* $H_{\pi_0}(X,(x_i),(\mathbb{C}_{\overline{\nabla}_i}))$ *is one-dimensional if and only if there exists a connection* $\overline{\nabla}$ *on the* ^LH*-bundle* $\Omega^{-\rho}$ *on* $X\backslash\{x_1,\dots,x_p,\infty\}$ *such that the restriction of* $\overline{\nabla}$ *to the punctured disc at each* x_i *is equal to* $\overline{\nabla}_i, i = 1,\dots,N$.

Otherwise, $H_{\pi_0}(X,(x_i),(\mathbb{C}_{\overline{\nabla}_i})) = 0$.

Suppose that such a connection $\overline{\nabla}$ exists and the space $H_{\pi_0}(X,(x_i),(\mathbb{C}_{\overline{\nabla}_i}))$ is one-dimensional. Then the map (4.10) reads

$$H_{\mathbb{V}_0}(X,(x_i),(N_i \otimes \mathbb{C}_{\overline{\nabla}_i})) \twoheadrightarrow H_{M_\mathfrak{g}}(X,(x_i),(N_i)).$$

Composing this map with any linear functional on $H_{M_\mathfrak{g}}(X,(x_i),(N_i))$, we obtain a linear functional

$$\phi : H_{\mathbb{V}_0}(X,(x_i),(N_i \otimes \mathbb{C}_{\overline{\nabla}_i})) \to \mathbb{C}.$$

The algebra $\mathfrak{z}(\widehat{\mathfrak{g}})_u = \mathrm{Fun}\,\mathrm{Op}_{^LG}(D_u)$ acts on the space $H_{\mathbb{V}_0}(X,(x_i),(N_i \otimes \mathbb{C}_{\overline{\nabla}_i}))$, and its action factors through the homomorphism (2.9)

$$\mathrm{Fun}\,\mathrm{Op}_{^LG}(D_u) \to \mathrm{Fun}\,\mathrm{Op}_{^LG}(X,(x_i),(N_i \otimes \mathbb{C}_{\overline{\nabla}_i})).$$

According to Lemma 4.6, this action factors through the homomorphism (4.13). In our case $R_i = \mathbb{C}_{\overline{\nabla}_i}$, and so $\mathrm{Conn}(\Omega^{-\rho})(X,(x_i),(R_i))$ is a point corresponding to the connection $\overline{\nabla}$. It follows from the definition that $\mathrm{Op}_{^LG}(X,(x_i),(N_i \otimes \mathbb{C}_{\overline{\nabla}_i}))$ is also a point, corresponding to the oper which is the image of $\overline{\nabla}$ under the Miura transformation $\overline{\mathbf{b}}^*_{X\backslash\{x_1,\dots,x_N\}}$. Thus, we obtain:

Proposition 4.8. *The functional ϕ is an eigenvector of the algebra $\mathfrak{z}(\widehat{\mathfrak{g}})_u \simeq \operatorname{Op}_{^LG}(D_u)$, and the corresponding homomorphism* $\operatorname{Fun}\operatorname{Op}_{^LG}(D_u) \to \mathbb{C}$ *is defined by the point $\overline{\mathbf{b}}^*_{D_u}(\overline{\nabla}|_{D_u}) \in \operatorname{Op}_{^LG}(D_u)$.*

4.3. The case of $\mathbb{P}^1$

We will use Proposition 4.8 in the case when $X = \mathbb{P}^1$, equipped with a global coordinate t. Consider a collection of distinct points $x_1, \ldots, x_p$ and ∞ on $\mathbb{P}^1$. As the $M_{\mathfrak{g}}$-module N_i attached to $x_i, i = 1, \ldots, p$, we will take $M_{\mathfrak{g}}$. As the $M_{\mathfrak{g}}$-module N_∞ attached to the point ∞ we will take another module $M'_{\mathfrak{g}}$ generated by a vector $|0\rangle'$ such that

$$a_{\alpha,n}|0\rangle' = 0, \quad n > 0; \qquad a^*_{\alpha,n}|0\rangle' = 0, \quad n \geq 0.$$

As the π_0-module R_i attached to $x_i, i = 1, \ldots, p$, we will take the one-dimensional module $\mathbb{C}_{\overline{\nabla}_i} = \mathbb{C}_{\chi_i(z)}$, where $\overline{\nabla}_i = \partial_z + \chi_i(z)$, $\chi_i(z) \in \mathfrak{h}^*((z))$ and as module R_∞ attached to the point ∞ we will take $\mathbb{C}_{\chi_\infty(z)}$. Thus, $M_i \otimes R_i$ is the Wakimoto module $W_{\chi_i(z)}$ for all $i = 1, \ldots, p$. We denote the module $M'_{\mathfrak{g}} \otimes \mathbb{C}_{\chi_\infty(z)}$ by $W'_{\chi_\infty(z)}$.

We have the following special case of Proposition 4.7:

Proposition 4.9 ([FFR], Proposition 4). *The space $H_{M_{\mathfrak{g}}}(\mathbb{P}^1; (x_i), \infty; (M_{\mathfrak{g}}), M'_{\mathfrak{g}})$ is one-dimensional and the projection of the vector $|0\rangle^{\otimes N} \otimes |0\rangle'$ on it is non-zero.*

The space $H_{\pi_0}(\mathbb{P}^1, (x_i), \infty; (\mathbb{C}_{\chi_i(z)}), \mathbb{C}_{\chi_\infty}(z))$ is one-dimensional if and only if there exists a connection $\overline{\nabla}$ on the ^LH-bundle $\Omega^{-\rho}$ on $\mathbb{P}^1$ which is regular on $\mathbb{P}^1 \backslash \{x_1, \ldots, x_p, \infty\}$ and such that the restriction of $\overline{\nabla}$ to the punctured disc at each x_i is equal to $\partial_t + \chi_i(t - x_i)$, and its restriction to the punctured disc at ∞ is equal to $\partial_{t^{-1}} + \chi_\infty(t^{-1})$.

Otherwise, $H_{\pi_0}(\mathbb{P}^1, (x_i), \infty; (\mathbb{C}_{\chi_i(z)}), \mathbb{C}_{\chi_\infty}(z)) = 0$.

Formula (4.8) shows that if we have a connection on $\Omega^{-\rho}$ over $\mathbb{P}^1$ whose restriction to $\mathbb{P}^1 \backslash \infty$ is represented by the operator $\partial_t + \chi(t)$, then its restriction to the punctured disc $D^\times_\infty$ at ∞ reads, with respect to the coordinate $u = t^{-1}$, is represented by the operator

$$\partial_u - u^{-2}\chi(u^{-1}) - 2\rho u^{-1}.$$

We will choose $\chi_i(z)$ to be of the form

$$\chi_i(z) = \chi_i + \sum_{n \geq 0} \chi_{i,n} z^n,$$

and $\chi_\infty(z)$ to be of the form

$$\chi_\infty(z) = \chi_\infty + \sum_{n \geq 0} \chi_{\infty,n} z^n.$$

The condition of the proposition is then equivalent to saying that the restriction of $\overline{\nabla}$ to $\mathbb{P}^1 \backslash \infty$ is represented by the operator

$$\partial_t + \sum_{i=1}^{p} \frac{\chi_i}{t - z_i},$$

and that $\chi_i(t-z_i)$ is the expansion of $\chi(t)$ at $z_i, i = 1,\ldots,p$, while $\chi_\infty(t^{-1})$ is the expansion of $-t^2\chi(t) - 2\rho t$ in powers of t^{-1}. Note that then we have $\chi_\infty(u) = \chi_\infty/u + \ldots$, where $\chi_\infty = -2\rho - \sum_{i=1}^p \chi_i$.

By Proposition 4.9, the spaces of coinvariants $H_{M_\mathfrak{g}}(\mathbb{P}^1;(x_i),\infty;(M_\mathfrak{g}),M'_\mathfrak{g})$ and $H_{\pi_0}(\mathbb{P}^1;(x_i),\infty;(\mathbb{C}_{\chi_i(z)}),\mathbb{C}_{\chi_\infty}(z))$ are one-dimensional. Hence the map (4.10) gives rise to a non-zero linear functional

$$\tau_{(x_i)} : H_{\mathbb{V}_0}(\mathbb{P}^1;(x_i),\infty;(W_{\chi_i(z)}),W'_{\chi_\infty(z)}) \to \mathbb{C}. \tag{4.14}$$

We normalize it so that its value on $|0\rangle^{\otimes N} \otimes |0\rangle'$ is equal to 1.

The algebra $\mathfrak{z}(\widehat{\mathfrak{g}})_u = \operatorname{Fun}\operatorname{Op}_{{}^LG}(D_u)$ acts on this space. By Proposition 4.8, $\tau_{(x_i)}$ is an eigenvector of this algebra and the corresponding homomorphism

$$\operatorname{Fun}\operatorname{Op}_{{}^LG}(D_u) \to \mathbb{C}$$

is defined by the point $\overline{\mathbf{b}}^*_{D_u}(\overline{\nabla}|_{D_u}) \in \operatorname{Op}_{{}^LG}(D_u)$.

Let us fix highest weights of $\mathfrak{g}$, $\lambda_1,\ldots,\lambda_N$, and a set of simple roots of $\mathfrak{g}$, $\alpha_{i_1},\ldots,\alpha_{i_m}$. Consider a connection $\overline{\nabla}$ on $\Omega^{-\rho}$ on $\mathbb{P}^1$ whose restriction to $\mathbb{P}^1\backslash\infty$ is equal to $\partial_t + \lambda(t)$, where

$$\lambda(t) = \sum_{i=1}^N \frac{\lambda_i}{t-z_i} - \sum_{j=1}^m \frac{\alpha_{i_j}}{t-w_j}. \tag{4.15}$$

Denote by $\lambda_i(t-z_i)$ the expansions of $\lambda(t)$ at the points $z_i, i = 1,\ldots,N$, and by $\mu_j(t-w_j)$ the expansions of $\lambda(t)$ at the points $w_j, j = 1,\ldots,m$. We have:

$$\lambda_i(z) = \frac{\lambda_i}{z} + \cdots, \qquad \mu_j(z) = -\frac{\alpha_{i_j}}{z} + \mu_{j,0} + \cdots,$$

where

$$\mu_{j,0} = \sum_{i=1}^N \frac{\lambda_i}{w_j - z_i} - \sum_{s\neq j} \frac{\alpha_{i_s}}{w_j - w_s}. \tag{4.16}$$

Finally, let $\lambda_\infty(t^{-1})$ be the expansion of $-t^2\lambda(t) - 2\rho t$ in powers of t^{-1}. Thus we have

$$\lambda_\infty(z) = z^{-1}\left(-\sum_{i=1}^N \lambda_i + \sum_{j=1}^m \alpha_{i_j} - 2\rho\right) + \cdots. \tag{4.17}$$

We have a linear functional $\tau_{(z_i),(w_j)}$ on the corresponding space of coinvariants

$$H_{\mathbb{V}_0}(\mathbb{P}^1;(z_i),(w_j),\infty;(W_{\lambda_i(z)}),(W_{\mu_j(z)}),W'_{\lambda_\infty(z)}).$$

In particular, this space is non-zero. The functional $\tau_{(z_i),(w_j)}$ is an eigenvector of $\mathfrak{z}(\widehat{\mathfrak{g}})_u = \operatorname{Fun}\operatorname{Op}_{{}^LG}(D_u)$, and its eigenvalue is the ${}^LG$-oper on D_u which is the restriction to D_u of the oper in $\operatorname{Op}^{\mathrm{RS}}_{{}^LG}(\mathbb{P}^1)_{(z_i),(w_j),\infty;(\lambda_i),(-\alpha_{i_j}),\lambda_\infty}$ given by the Miura transformation of the connection $\partial_t + \lambda(t)$. We will now use it to construct an eigenvector of the Gaudin algebra.

4.4. Bethe vectors

By Lemma 4.5, the vectors $e^R_{i_j,-1}|0\rangle \in W_{\mu_j(z)}$ are $\mathfrak{g}[[t]]$-invariant if and only if the equations $\langle \check{\alpha}_{i_j}, \mu_{j,0}\rangle = 0$ are satisfied, where $\mu_{j,0}$ is the constant coefficient in the expansion of $\lambda(t)$ at w_j given by formula (4.16), i.e., the following system of equations is satisfied

$$\sum_{i=1}^{N} \frac{\langle \check{\alpha}_{i_j}, \lambda_i\rangle}{w_j - z_i} - \sum_{s\neq j} \frac{\langle \check{\alpha}_{i_j}, \alpha_{i_s}\rangle}{w_j - w_s} = 0, \quad j = 1, \ldots, m. \tag{4.18}$$

They are called the *Bethe Ansatz equations.* This is a system of equations on the complex numbers $w_j, j = 1, \ldots, m$, to each of which we attach a simple root α_{i_j}. We have an obvious action of a product of symmetric groups permuting the points w_j corresponding to simple roots of the same kind. In what follows, by a *solution* of the Bethe Ansatz equations we will understand a solution defined up to these permutations.

Suppose that these equations are satisfied. Then we obtain a homomorphism of $\widehat{\mathfrak{g}}_{\kappa_c}$-modules

$$\bigotimes_{i=1}^{N} W_{\lambda_i(z)} \otimes \mathbb{V}_0^{\otimes m} \otimes W'_{\lambda_\infty(z)} \to \bigotimes_{i=1}^{N} W_{\lambda_i(z)} \otimes \bigotimes_{j=1}^{m} W_{\mu_j(z)} \otimes W'_{\lambda_\infty(z)},$$

which sends the vacuum vector in the jth copy of $\mathbb{V}_0$ to $e^R_{i_j,-1}|0\rangle \in W_{\mu_j(z)}$. Hence we obtain the corresponding map of the spaces of coinvariants. But the insertion of $\mathbb{V}_0$ does not change the space of coinvariants. Hence the first space of coinvariants is isomorphic to

$$H_{\mathbb{V}_0}(\mathbb{P}^1; (z_i), \infty; (W_{\lambda_i(z)}), W'_{\lambda_\infty(z)}) = H((W_{\lambda_i(z)}), W'_{\lambda_\infty(z)}).$$

Therefore, composing this map with $\tau_{(z_i),(w_j)}$, we obtain a linear functional

$$\widetilde{\tau}_{(z_i),(w_j)} : H((W_{\lambda_i(z)}), W'_{\lambda_\infty(z)}) \to \mathbb{C}. \tag{4.19}$$

By construction, $\widetilde{\tau}_{(z_i),(w_j)}$ is an eigenvector of $\mathfrak{z}(\widehat{\mathfrak{g}})_u = \operatorname{Fun}\operatorname{Op}_{{}^LG}(D_u)$, and its eigenvalue is the ${}^LG$-oper on D_u which is the restriction of an oper on $\mathbb{P}^1$ with singularities given by the Miura transformation of the connection $\partial_t + \lambda(t)$.

Let $U = \mathbb{P}^1 \backslash \{z_1, \ldots, z_N, w_1, \ldots, w_m, \infty\}$ and $\operatorname{Conn}(\Omega^{-\rho})^{\text{gen}}_{(z_i),\infty;(\lambda_i),\lambda_\infty}$ be the subset of $\operatorname{Conn}(\Omega^{-\rho})_U$, consisting of all connections on $\Omega^{-\rho}$ on $\mathbb{P}^1$ of the form $\partial_t + \lambda(t)$, where $\lambda(t)$ is given by (4.15), such that their Miura transformation is a ${}^LG$-oper in

$$\operatorname{Op}^{\text{RS}}_{{}^LG}(\mathbb{P}^1)_{(z_i),\infty;(\lambda_i),\lambda_\infty} \subset \operatorname{Op}^{\text{RS}}_{{}^LG}(\mathbb{P}^1)_{(z_i),(w_j),\infty;(\lambda_i),(-\alpha_{i_j}),\lambda_\infty}.$$

In other words, $\operatorname{Conn}(\Omega^{-\rho})^{\text{gen}}_{(z_i),\infty;(\lambda_i),\lambda_\infty}$ consists of those connections $\partial_t + \lambda(t)$, where $\lambda(t)$ is given by (4.15), whose Miura transformation is a ${}^LG$-oper on $\mathbb{P}^1$ that is regular at $w_1, \ldots, w_m$.

Thus, we have a map

$$\overline{\mathbf{b}}^*_{(z_i),\infty;(\lambda_i),\lambda_\infty} : \operatorname{Conn}(\Omega^{-\rho})^{\text{gen}}_{(z_i),\infty;(\lambda_i),\lambda_\infty} \to \operatorname{Op}^{\text{RS}}_{{}^LG}(\mathbb{P}^1)_{(z_i),\infty;(\lambda_i),\lambda_\infty}. \tag{4.20}$$

Proposition 4.10 ([FFR], Proposition 6). *There is a bijection between the set of solutions of the Bethe Ansatz equations* (4.18) *and the set* $\operatorname{Conn}(\Omega^{-\rho})^{\mathrm{gen}}_{(z_i),\infty;(\lambda_i),\lambda_\infty}$.

Proof. Applying the Miura transformation to the connection $\partial_t + \lambda(t)$, where $\lambda(t)$ is given by (4.15), we obtain a ${}^LG$-oper on $\mathbb{P}^1$ which has regular singularities at the points $z_1, \ldots, z_N$, and ∞ with residues $\varpi(\lambda_1 + \rho), \ldots, \varpi(\lambda_N + \rho)$ and $\varpi(\lambda_\infty + \rho)$, respectively, and at the points $w_1, \ldots, w_m$ with residues $\varpi(-\alpha_{i_j} + \rho) = \varpi(\rho)$. But by Lemma 3.5, this oper is regular at the points $w_1, \ldots, w_m$ if and only if the equations $\langle \check{\alpha}_{i_j}, \mu_{j,0} \rangle = 0$ are satisfied, where $\mu_{j,0}$ is the constant coefficient in the expansion of $\lambda(t)$ at w_j given by formula (4.16). These equations are precisely the Bethe Ansatz equations (4.18). □

Now let $w_1, \ldots, w_m$ be a solution of the Bethe Ansatz equations (4.18). Consider the corresponding connection $\partial_t + \lambda(t)$ in $\operatorname{Conn}(\Omega^{-\rho})^{\mathrm{gen}}_{(z_i),\infty;(\lambda_i),\lambda_\infty}$, where $\lambda(t)$ is of the form (4.15). Recall that the linear functional $\widetilde{\tau}_{(z_i),(w_j)}$ from formula (4.19) is an eigenvector of $\mathfrak{z}(\widehat{\mathfrak{g}})_u = \operatorname{Fun}\operatorname{Op}_{{}^LG}(D_u)$, and its eigenvalue is the ${}^LG$-oper on D_u obtained by restriction of the oper in $\operatorname{Op}^{\mathrm{RS}}_{{}^LG}(\mathbb{P}^1)_{(z_i),\infty;(\lambda_i),\lambda_\infty}$ given by the Miura transformation of the connection $\partial_t + \lambda(t)$.

Therefore we obtain that the action of $\mathfrak{z}(\widehat{\mathfrak{g}})_u$ on the space $H((W_{\lambda_i(z)}), W'_{\lambda_\infty(z)})$ factors through the Gaudin algebra $\mathcal{Z}_{(z_i)}(\mathfrak{g})$, which acts on it according to the character corresponding to the ${}^LG$-oper $\overline{\mathbf{b}}^*_{(z_i),\infty;(\lambda_i),\lambda_\infty}(\partial_t + \lambda(t))$.

We wish to use this fact to construct eigenvectors of the algebra $\mathcal{Z}_{(z_i)}(\mathfrak{g})$ on the tensor product of Verma modules $\bigotimes_{i=1}^N M_{\lambda_i}$.

For $\lambda(z) \in \mathfrak{h}^* \otimes z^{-1}\mathbb{C}[[z]]$ denote by λ_{-1} the residue of $\lambda(z)dz$ at $z = 0$. Let $\widetilde{W}_{\lambda(z)}$ the subspace of a Wakimoto module $W_{\lambda(z)}$, which is generated from the vector $|0\rangle$ by the operators $a^*_{\alpha,0}, \alpha \in \Delta_+$. This space is stable under the action of the constant subalgebra $\mathfrak{g}$ of $\widehat{\mathfrak{g}}_{\kappa_c}$. It follows from the formulas defining the action of $\widehat{\mathfrak{g}}_{\kappa_c}$ on $W_{\lambda(z)}$ that as a $\mathfrak{g}$-module, $\widetilde{W}_{\lambda(z)}$ is isomorphic to the module $M^*_{\lambda_{-1}}$ contragredient to the Verma module over $\mathfrak{g}$ with highest weight λ_{-1}. Likewise, the subspace $\widetilde{W}'_{\lambda(z)}$ generated from the vector $|0\rangle'$ by the operators $a_{\alpha,0}, \alpha \in \Delta_+$, is stable under the action of the constant subalgebra $\mathfrak{g}$ of $\widehat{\mathfrak{g}}_{\kappa_c}$ and is isomorphic as a $\mathfrak{g}$-module to the module $M'_{2\rho+\lambda_{-1}}$ the Verma module over $\mathfrak{g}$ with *lowest* weight $2\rho + \lambda_{-1}$.

Let $\mathbb{M}^*_\chi$ be the $\widehat{\mathfrak{g}}_{\kappa_c}$-module induced from M^*_χ, and let $\mathbb{M}'_\chi$ be the $\widehat{\mathfrak{g}}_{\kappa_c}$-module induced from M'_χ (see Section 2.2). The embeddings of $\mathfrak{g}$-modules $M^*_{\lambda_{-1}} \to W_{\lambda(z)}$ and $M'_{2\rho+\lambda_{-1}} \to W'_{\lambda(z)}$ induce homomorphisms of $\widehat{\mathfrak{g}}_{\kappa_c}$-modules $\mathbb{M}^*_{\lambda_{-1}} \to W_{\lambda(z)}$ and $\mathbb{M}'_{2\rho+\lambda_{-1}} \to W'_{\lambda(z)}$. Thus, we obtain a homomorphism

$$\bigotimes_{i=1}^N \mathbb{M}^*_{\lambda_i} \otimes \mathbb{M}'_{2\rho+\lambda_\infty} \to \bigotimes_{i=1}^N W_{\lambda_i(z)} \otimes W'_{\lambda_\infty(z)},$$

where

$$\lambda_\infty = -2\rho - \sum_{i=1}^N \lambda_i + \sum_{j=1}^m \alpha_{i_j}.$$

This homomorphism gives rise to a map of the corresponding spaces of coinvariants

$$H((\mathbb{M}^*_{\lambda_i}), \mathbb{M}'_{2\rho+\lambda_\infty}) \to H((W_{\lambda_i(z)}), W'_{\lambda_\infty(z)}).$$

The composition of this map with the functional $\widetilde{\tau}_{(z_i),(w_j)}$ defined above is a linear functional

$$\overline{\tau}_{(z_i),(w_j)} : H((\mathbb{M}^*_{\lambda_i}), \mathbb{M}'_{2\rho+\lambda_\infty}) \to \mathbb{C}.$$

But we have an isomorphism

$$H((\mathbb{M}^*_{\lambda_i}), \mathbb{M}'_{2\rho+\lambda_\infty}) \simeq (\bigotimes_{i=1}^N M^*_{\lambda_i} \otimes M'_{2\rho+\lambda_\infty})/\mathfrak{g} \simeq (\bigotimes_{i=1}^N M^*_{\lambda_i})_{-2\rho-\lambda_\infty}/\mathfrak{n}_-,$$

where the subscript indicates the subspace of weight $-2\rho - \lambda_\infty = \sum_{i=1}^N \lambda_i - \sum_{j=1}^m \alpha_{i_j}$. Thus, we obtain an $\mathfrak{n}_-$-invariant functional

$$\psi(w_1^{i_1}, \dots, w_m^{i_m}) : (\bigotimes_{i=1}^N M^*_{\lambda_i})_{-2\rho-\lambda_\infty} \to \mathbb{C},$$

or equivalently an $\mathfrak{n}_+$-invariant vector

$$\phi(w_1^{i_1}, \dots, w_m^{i_m}) \in \otimes_{i=1}^N M_{\lambda_i}$$

of weight $\sum_{i=1}^N \lambda_i - \sum_{j=1}^m \alpha_{i_j}$.

Recall that $\widetilde{\tau}_{(z_i),(w_j)}$ is an eigenvector of the algebra $\mathfrak{z}(\widehat{\mathfrak{g}})_u = \operatorname{Fun} \operatorname{Op}_{^LG}(D_u)$, and its eigenvalue is the ^LG-oper on $\mathbb{P}^1$ given by the Miura transformation of the connection $\partial_t + \lambda(t)$, where $\lambda(t)$ is given by (4.15). Hence $\phi(w_1^{i_1}, \dots, w_m^{i_m})$ is also an eigenvector of $\mathfrak{z}(\widehat{\mathfrak{g}})_u$ with the same eigenvalue. But we know from the proof of Theorem 2.7 that the action of $\mathfrak{z}(\widehat{\mathfrak{g}})_u$ on the space of coinvariants $H((\mathbb{M}^*_{\lambda_i}), \mathbb{M}'_{2\rho+\lambda_\infty})$ factors through the Gaudin algebra $\mathcal{Z}_{(z_i)}(\mathfrak{g})$ and further through $\mathcal{Z}_{(z_i),\infty;(\lambda_i),\lambda_\infty}(\mathfrak{g})$. Therefore $\phi(w_1^{i_1}, \dots, w_m^{i_m})$ is an eigenvector of the Gaudin algebra $\mathcal{Z}_{(z_i)}(\mathfrak{g})$, and the eigenvalues of $\mathcal{Z}_{(z_i)}(\mathfrak{g})$ on this vector are encoded by the ^LG-oper obtained by the Miura transformation of the connection $\partial_t + \lambda(t)$, where $\lambda(t)$ is given by (4.15).

Let us summarize our results.

Theorem 4.11 ([FFR], Theorem 3). *If the Bethe Ansatz equations* (4.18) *are satisfied, then the vector*

$$\phi(w_1^{i_1}, \dots, w_m^{i_m}) \in \otimes_{i=1}^N M_{\lambda_i}$$

is an eigenvector of the algebra $\mathcal{Z}_{(z_i),\infty;(\lambda_i),\lambda_\infty}(\mathfrak{g})$. Its eigenvalues define a point in

$$\operatorname{Spec} \mathcal{Z}_{(z_i),\infty;(\lambda_i),\lambda_\infty} = \operatorname{Op}^{\mathrm{RS}}_{^LG}(\mathbb{P}^1)_{(z_i),\infty;(\lambda_i),\lambda_\infty},$$

which is the image of the connection $\partial_t + \lambda(t)$ with $\lambda(t)$ as in formula (4.15), *under the Miura transformation $\overline{\mathbf{b}}^*_{(z_i),\infty;(\lambda_i),\lambda_\infty}$ given by formula* (4.20).

The vector $\phi(w_1^{i_1}, \ldots, w_m^{i_m})$ is called the *Bethe vector* corresponding to a solution of the Bethe Ansatz equations (4.18), or equivalently, an element of the set $\mathrm{Conn}(\Omega^{-\rho})^{\mathrm{gen}}_{(z_i),\infty;(\lambda_i),\lambda_\infty}$. An explicit formula for this vector is given in [FFR], Lemma 3 (see also [ATY]):

$$\phi(w_1^{i_1}, \ldots, w_m^{i_m}) \tag{4.21}$$

$$= (-1)^m \sum_{p=(I^1,\ldots,I^N)} \prod_{j=1}^{N} \frac{F^{(j)}_{i^j_1} F^{(j)}_{i^j_2} \ldots F^{(j)}_{i^j_{a_j}}}{(w_{i^j_1} - w_{i^j_2})(w_{i^j_2} - w_{i^j_3}) \ldots (w_{i^j_{a_j}} - z_j)} v_{\lambda_1} \otimes \ldots v_{\lambda_N}.$$

Here the summation is taken over all *ordered* partitions $I^1 \cup I^2 \cup \ldots \cup I^N$ of the set $\{1, \ldots, m\}$, where $I^j = \{i^j_1, i^j_2, \ldots, i^j_{a_j}\}$.

The fact that this vector is an eigenvector of the Gaudin Hamiltonians Ξ_i has also been established by other methods in [BaFl, RV]. But Theorem 4.11 gives us much more: it tells us that this vector is also an eigenvector of all other generalized Gaudin Hamiltonians, and identifies the ${}^L G$-oper encoding their eigenvalues on this vector as the Miura transformation of the connection $\partial_t + \lambda(t)$. We recall that this last result follows from Corollary 4.4 which states that the central characters on the Wakimoto modules are obtained via the Miura transformation.

5. The case of finite-dimensional $\mathfrak{g}$-modules

In this section we consider the eigenvectors of the Gaudin algebra $\mathcal{Z}_{(z_i)}(\mathfrak{g})$ on the space

$$H((V_{\lambda_i}), V_{\lambda_\infty}) = \left(\bigotimes_{i=1}^{N} V_{\lambda_i} \otimes V_{\lambda_\infty} \right)^G$$

(recall that V_λ denotes the irreducible finite-dimensional $\mathfrak{g}$-module with a dominant integral highest weight λ). Recall that the action of $\mathcal{Z}_{(z_i)}(\mathfrak{g})$ on this space factors through the algebra $\mathcal{Z}_{(z_i),\infty;(\lambda_i),\lambda_\infty}(\mathfrak{g})$, so that we have a homomorphism

$$\mathcal{Z}_{(z_i),\infty;(\lambda_i),\lambda_\infty} \to \mathrm{End}_{\mathbb{C}} \bigotimes_{i=1}^{N} V_{\lambda_i} \otimes V_{\lambda_\infty}.$$

The image of this homomorphism was denoted by $\overline{\mathcal{Z}}_{(z_i),\infty;(\lambda_i),\lambda_\infty}(\mathfrak{g})$. Then the set of all common eigenvalues of the algebra $\mathcal{Z}_{(z_i)}(\mathfrak{g})$ acting on $\left(\bigotimes_{i=1}^{N} V_{\lambda_i} \otimes V_{\lambda_\infty}\right)^G$ (without multiplicities) is precisely the spectrum of the algebra $\overline{\mathcal{Z}}_{(z_i),\infty;(\lambda_i),\lambda_\infty}(\mathfrak{g})$.

By Theorem 2.7,(3), the algebra $\overline{\mathcal{Z}}_{(z_i),\infty;(\lambda_i),\lambda_\infty}(\mathfrak{g})$ is a quotient of the algebra of functions on the space $\mathrm{Op}_{{}^L G}(\mathbb{P}^1)_{(z_i),\infty;(\lambda_i),\lambda_\infty}$, so that its spectrum injects into $\mathrm{Op}_{{}^L G}(\mathbb{P}^1)_{(z_i),\infty;(\lambda_i),\lambda_\infty}$. Thus, we have an injective map

$$\overline{\mathcal{Z}}_{(z_i),\infty;(\lambda_i),\lambda_\infty}(\mathfrak{g}) \hookrightarrow \mathrm{Op}_{{}^L G}(\mathbb{P}^1)_{(z_i),\infty;(\lambda_i),\lambda_\infty}. \tag{5.1}$$

In other words, each common eigenvalue of the Gaudin algebra $\mathcal{Z}_{(z_i)}(\mathfrak{g})$ that occurs in the space $\left(\bigotimes_{i=1}^N V_{\lambda_i} \otimes V_{\lambda_\infty}\right)^G$ is encoded by a point of $\operatorname{Op}_{^L G}(\mathbb{P}^1)_{(z_i),\infty;(\lambda_i),\lambda_\infty}$, i.e., by a ${}^L G$-oper on $\mathbb{P}^1$ with regular singularities at $z_1, \ldots, z_N, \infty$, with residues $\lambda_1, \ldots, \lambda_N, \lambda_\infty$ at those points and *trivial monodromy* representation

$$\pi_1(\mathbb{P}^1 \backslash \{z_1, \ldots, z_N, \infty\}) \to {}^L G.$$

We wish to construct an inverse map from $\operatorname{Op}_{^L G}(\mathbb{P}^1)_{(z_i),\infty;(\lambda_i),\lambda_\infty}$ to the spectrum of $\mathcal{Z}_{(z_i)}(\mathfrak{g})$ on $\left(\bigotimes_{i=1}^N V_{\lambda_i} \otimes V_{\lambda_\infty}\right)^G$.

We start by a more intrinsic geometric description of $\operatorname{Op}_{^L G}(\mathbb{P}^1)_{(z_i),\infty;(\lambda_i),\lambda_\infty}$.

5.1. Geometric description of opers without monodromy

In order to simplify our notation, in this section we will use the symbol G instead of ${}^L G$ to denote our group.

Suppose that we have a G-oper $\tau : (\mathcal{F}, \nabla, \mathcal{F}_B)$ on a curve U. Consider the associated bundle of flag manifolds

$$(G/B)_{\mathcal{F}} = \mathcal{F} \underset{G}{\times} G/B \simeq \mathcal{F}_B \underset{B}{\times} G/B = (G/B)_{\mathcal{F}_B}.$$

The reduction $\mathcal{F}_B$ gives rise to a section of this bundle.

Now suppose that the monodromy representation $\pi_1(U) \to G$ corresponding to this oper is trivial. Picking a point $x \in U$, we may then use the connection ∇ to identify the fibers of the oper bundle $\mathcal{F}$ at all points of U with the fiber $\mathcal{F}_x$ at x. Choosing a trivialization $\imath_x : G \xrightarrow{\sim} \mathcal{F}_x$ of $\mathcal{F}_x$, we then obtain a trivialization of the bundle $\mathcal{F}$, and hence of the bundle $(G/B)_{\mathcal{F}}$. Let us fix $\imath_x$ and the corresponding trivialization of $\mathcal{F}$. Then the B-reduction $\mathcal{F}_B$ gives rise to a map $\phi_\tau : U \to G/B$.

The oper condition may be described as follows. Define an open B-orbit $\mathbf{O} \subset [\mathfrak{n}, \mathfrak{n}]^\perp/\mathfrak{b} \subset \mathfrak{g}/\mathfrak{b}$, consisting of vectors which are stabilized by $N \subset B$, and such that all of their negative simple root components, with respect to the adjoint action of $H = B/N$, are non-zero. This orbit may also be described as the B-orbit of the sum of the projections of simple root generators f_i of any nilpotent subalgebra $\mathfrak{n}_-$, which is in generic position with $\mathfrak{b}$, onto $\mathfrak{g}/\mathfrak{b}$. The torus H acts simply transitively on $\mathbf{O}$, and so $\mathbf{O}$ is an H-torsor.

Define an ℓ-dimensional distribution $T_{\mathrm{Op}} G/B$ in the tangent bundle to G/B as follows. The tangent space to gB is identified with the quotient $\mathfrak{g}/g\mathfrak{b}g^{-1}$. The fiber of $T_{\mathrm{Op}} G/B$ at gB is by definition its subspace $[\mathfrak{n}, \mathfrak{n}]^\perp/\mathfrak{b}$. It contains an open dense subset $g\mathbf{O}g^{-1}$. We denote by $\overset{\circ}{T}_{\mathrm{Op}} G/B$ the open dense subset of $T_{\mathrm{Op}} G/B$ whose fiber at gB is $g\mathbf{O}g^{-1}$.

A map $\phi : U \to G/B$ is said to satisfy the oper condition if at each point of U the tangent vector to ϕ belongs to $\overset{\circ}{T}_{\mathrm{Op}} G/B$.

For example, if $G = PGL_2$, then the flag manifold is $\mathbb{P}^1$, and a map $\phi : U \to \mathbb{P}^1$ satisfies the oper condition if and only if it has a non-vanishing differential everywhere on X.

If τ is an oper, then the map $\phi_\tau : U \to G/B$ satisfies the oper condition (note that $\overset{\circ}{T}_{\mathrm{Op}} G/B$ is G-invariant, and therefore this condition is independent of the trivialization $\imath_x$). Conversely, if we are given a map $\phi : U \to G/B$ satisfying the oper condition, then the triple $(\mathcal{F}, \nabla, \mathcal{F}_B)$, where $\mathcal{F}$ is the trivial G-bundle on U, ∇ is the trivial connection, and $\mathcal{F}_B$ is the B-reduction of $\mathcal{F}$ defined by the map ϕ, is a G-oper. Thus, we obtain a bijection between the set of all G-opers on U with trivial monodromy representation and the set of equivalence classes, with respect to the action of G on G/B, of maps $\phi : U \to G/B$ satisfying the oper condition.

Now let $\check{\lambda}$ be a dominant integral coweight of G. A map $D^\times \to H$ is said to vanish at the origin to order $\check{\lambda}$ if it may be obtained as the composition of an embedding $D^\times \to C^\times$ at the origin and the cocharacter $\check{\lambda} : \mathbb{C}^\times \to H$. We will say that a map $\phi : U \to G/B$ satisfies the $\check{\lambda}$-oper condition at $y \in U$ if the restriction of ϕ to the punctured disc $D_y^\times$ satisfies the oper condition and the corresponding map $\phi_{D_y^\times} : D_y^\times \to \mathbf{O} \simeq H$ vanishes to the order $\check{\lambda}$.

For example, if $G = PGL_2$, then we can identify the dominant integral weights of G with non-negative integers. Then a map $\phi : U \to \mathbb{P}^1$ satisfies the $\check{\lambda}$-oper condition at $y \in U$ if its differential vanishes to the order $\check{\lambda}$ at y.

Recalling the definition of the space $\mathrm{Op}_G(D_x)_{\check{\lambda}}$ given in Section 1.3, we obtain the following result.

Proposition 5.1. *There is a bijection between the set* $\mathrm{Op}_G(\mathbb{P}^1)_{(z_i),\infty;(\check{\lambda}_i),\check{\lambda}_\infty}$ *and the set of equivalence classes, with respect to the action of* G *on* G/B, *of the maps* $\phi : \mathbb{P}^1 \to G/B$ *satisfying the oper condition on* $\mathbb{P}^1 \backslash \{z_1, \ldots, z_N, \infty\})$, *the* $\check{\lambda}_i$*-oper condition at* z_i *and the* $\check{\lambda}_\infty$*-oper condition at* ∞.

In particular, the points of $\mathrm{Op}_{PGL_2}(\mathbb{P}^1)_{(z_i),\infty;(\check{\lambda}_i),\check{\lambda}_\infty}$ are the same as the equivalence classes of maps $\mathbb{P}^1 \to \mathbb{P}^1$ whose differential does not vanish anywhere on $\mathbb{P}^1 \backslash \{z_1, \ldots, z_N, \infty\})$ and vanishes to order $\check{\lambda}_i$ at $z_i, i = 1, \ldots, N$, and to order $\check{\lambda}_\infty$ at ∞.

5.2. From opers without monodromy to Bethe vectors

Let $\tau = (\mathcal{F}, \nabla, \mathcal{F}_{^LB}) \in \mathrm{Op}_{^LG}(\mathbb{P}^1)_{(z_i),\infty;(\lambda_i),\lambda_\infty}$. By Proposition 5.1, we attach to it a map $\phi_\tau : \mathbb{P}^1 \to {}^LG/{}^LB$. A horizontal ${}^LB$-reduction on τ is completely determined by the choice of ${}^LB$-reduction at any given point $x \in \mathbb{P}^1$, which is a point of $({}^LG/{}^LB)_{\mathcal{F}_{^LB,x}} \simeq {}^LG/{}^LB$ (see Lemma 3.1). We will choose the point ∞ as our reference point on $\mathbb{P}^1$. Then we obtain

Lemma 5.2. *For* $\tau \in \mathrm{Op}_{^LG}(\mathbb{P}^1)_{(z_i),\infty;(\lambda_i),\lambda_\infty}$ *the set of Miura opers on* $\mathbb{P}^1$ *with the underlying oper* τ *is in bijection with the set of points of* $({}^LG/{}^LB)_{\mathcal{F}_{^LB,\infty}} \simeq {}^LG/{}^LB$.

Now suppose we are given an oper $\tau \in \mathrm{Op}_{^LG}(\mathbb{P}^1)_{(z_i),\infty;(\lambda_i),\lambda_\infty}$. We fix a trivialization of the fiber $\mathcal{F}_\infty$ of $\mathcal{F}$ at ∞ and consider the corresponding map $\phi_\tau : \mathbb{P}^1 \to {}^LG/{}^LB$. Then the value of ϕ_τ at $\infty \in \mathbb{P}^1$ is the point ${}^LB \in {}^LG/{}^LB$. Consider the Miura oper structure on τ corresponding to ${}^LB \in {}^LG/{}^LB$, considered as a point in the fiber $({}^LG/{}^LB)_{\mathcal{F}_{^LB,\infty}}$. This is the unique Miura oper structure on

τ for which the horizontal Borel reduction $\mathcal{F}'_{^LB,\infty}$ at the point ∞ coincides with the oper reduction $\mathcal{F}_{^LB,\infty}$ (which is the point ${}^LB \in {}^LG/{}^LB$ with respect to our trivialization).

Let us suppose that the oper τ satisfies the following conditions:

(1) $\phi_\tau(z_i)$ is in generic position with ${}^LB$ for all $i = 1, \ldots, N$;

(2) the relative position of $\phi_\tau(x)$ and ${}^LB$ is either generic or corresponds to a simple reflection $s_i \in W$ for all $x \in \mathbb{A}^1 = \mathbb{P}^1 \backslash \infty$.

Then we will call τ a *non-degenerate* oper.

Let τ be a non-degenerate oper in $\operatorname{Op}_{^LG}(\mathbb{P}^1)_{(z_i),\infty;(\lambda_i),\lambda_\infty}$. Consider the unique Miura oper $(\mathcal{F}, \nabla, \mathcal{F}_{^LB}, \mathcal{F}'_{^LB})$ whose underlying oper is τ and such that $\mathcal{F}_{^LB,\infty}$ corresponds to the point ${}^LB \in {}^LG/{}^LB \simeq ({}^LG/{}^LB)_{\mathcal{F}_{^LB,\infty}}$. According to the above conditions, the reductions $\mathcal{F}_{^LB}$ and $\mathcal{F}'_{^LB}$ are in generic position for all but finitely many points of $\mathbb{P}^1$, which are distinct from $z_1, \ldots, z_N, \infty$. Let us denote these points by $w_1, \ldots, w_m$. Then at the point w_j the two reductions have relative position s_{i_j}. According to Proposition 3.3, to this Miura oper corresponds a connection on the ${}^LH$-bundle Ω^ρ, and hence a connection $\overline{\nabla}$ on the dual bundle $\Omega^{-\rho}$ on $\mathbb{P}^1 \backslash \{z_1, \ldots, z_N, w_1, \ldots, w_m, \infty\}$. Moreover, by Proposition 3.4, the connection $\overline{\nabla}$ extends to a connection with regular singularity on $\Omega^{-\rho}$ on the entire $\mathbb{P}^1$, with residues λ_i at $z_i, i = 1, \ldots, N$, $w_0(\lambda_\infty + \rho) - \rho$ at ∞, and $-\alpha_{i_j}$ at $w_j, j = 1, \ldots, m$. (Note that in Proposition 3.4 we described the residues of connections of the bundle Ω^ρ. Since now we consider connections on $\Omega^{-\rho}$ which is dual to Ω^ρ, when applying Proposition 3.4 we need to change the signs of all residues.)

Therefore, the restriction of $\overline{\nabla}$ to $\mathbb{P}^1 \backslash \infty$ reads

$$\overline{\nabla} = \partial_t + \sum_{i=1}^N \frac{\lambda_i}{t - z_i} - \sum_{j=1}^m \frac{\alpha_{i_j}}{t - w_j}. \tag{5.2}$$

Using transformation formula (3.2), we find that the residue of $\overline{\nabla}$ at ∞ is equal to

$$-2\rho - \sum_{i=1}^N \lambda_i + \sum_{j=1}^m \alpha_{i_j}.$$

But by our assumption this residue is equal to $w_0(\lambda_\infty + \rho) - \rho$. Therefore we find that

$$-w_0(\lambda_\infty) = \sum_{i=1}^N \lambda_i - \sum_{j=1}^m \alpha_{i_j}. \tag{5.3}$$

In addition, the condition that the oper τ has trivial monodromy representation and Lemma 3.5 imply that if

$$\partial_t - \alpha_{i_j}/(t - w_j) + \mathbf{u}_j(t - w_j), \qquad \mathbf{u}_j(u) \in {}^L\mathfrak{h}[[u]],$$

is the expansion of the connection (5.2) at the point w_j, then $\langle \check{\alpha}_{i_j}, \mathbf{u}_j(0) \rangle = 0$ for all $j = 1, \ldots, m$. When we write them explicitly, we see immediately that these equations are precisely the Bethe Ansatz equations (4.18)!

Thus, we have attached to a non-degenerate oper $\tau \in \mathrm{Op}_{{}^L G}(\mathbb{P}^1)_{(z_i),\infty;(\lambda_i),\lambda_\infty}$ a connection $\overline{\nabla} = \partial_t + \lambda(t)$ on $\Omega^{-\rho}$ and a solution of the Bethe Ansatz equations. By construction, the Miura transformation of this connection is the oper τ, and so we obtain the following

Proposition 5.3. *Suppose that all opers in* $\mathrm{Op}_{{}^L G}(\mathbb{P}^1)_{(z_i),\infty;(\lambda_i),\lambda_\infty}$ *are non-degenerate. Then there is a bijection between the set of opers* $\mathrm{Op}_{{}^L G}(\mathbb{P}^1)_{(z_i),\infty;(\lambda_i),\lambda_\infty}$ *and the set of solutions of Bethe Ansatz equations* (4.18) *such that the weight* $\sum_{i=1}^N \lambda_i - \sum_{j=1}^m \alpha_{i_j}$ *is dominant.*

Thus, given a non-degenerate oper $\tau \in \mathrm{Op}_{{}^L G}(\mathbb{P}^1)_{(z_i),\infty;(\lambda_i),\lambda_\infty}$, we obtain a solution of Bethe Ansatz equations (4.18). We then follow the procedure of Section 4.4 to assign to the above solution of the Bethe Ansatz equations an eigenvector of the Gaudin algebra $\mathcal{Z}_{(z_i)}(\mathfrak{g})$ in $\bigotimes_{i=1}^N M_{\lambda_i}$, which is an $\mathfrak{n}_+$-invariant vector $\phi(w_1^{i_1},\dots,w_m^{i_m})$ of weight $\sum_{i=1}^N \lambda_i - \sum_{j=1}^m \alpha_{i_j}$. Consider its projection onto $\bigotimes_{i=1}^N V_{\lambda_i}$. By formula (5.3), the weight of this vector is

$$\sum_{i=1}^{N} \lambda_i - \sum_{j=1}^{m} \alpha_{i_j} = -w_0(\lambda_\infty), \tag{5.4}$$

and so it may be viewed as an eigenvector of the Gaudin algebra in the space

$$V^G_{(\lambda_i),\lambda_\infty} = \left(\bigotimes_{i=1}^{N} V_{\lambda_i} \otimes V_{\lambda_\infty} \right)^G .$$

By Theorem 4.11, the eigenvalues of the Gaudin algebra $\mathcal{Z}_{(z_i)}(\mathfrak{g})$ on this vector are encoded by the Miura transformation of the connection $\overline{\nabla}$, which is precisely the oper τ with which we have started. Therefore we obtain the following result.

Lemma 5.4. *The two opers encoding the eigenvalues of the Gaudin algebra on the Bethe vectors corresponding to two different solutions of the Bethe Ansatz equations are necessarily different.*

Suppose that all opers in $\mathrm{Op}_{{}^L G}(\mathbb{P}^1)_{(z_i),\infty;(\lambda_i),\lambda_\infty}$ are *non-degenerate*[1] and that each of the Bethe vectors is *non-zero*. Then we obtain a map

$$\mathrm{Op}_{{}^L G}(\mathbb{P}^1)_{(z_i),\infty;(\lambda_i),\lambda_\infty} \to \operatorname{Spec} \overline{\mathcal{Z}}_{(z_i),\infty;(\lambda_i),\lambda_\infty}(\mathfrak{g})$$

that is inverse to the map (5.1) discussed at the beginning of this section.

Let us summarize our results.

[1] It follows from the results of Mukhin and Varchenko in [MV2] that for some $\lambda_1,\dots,\lambda_N,\lambda_\infty$ it is possible that there exist degenerate opers in $\mathrm{Op}_{{}^L G}(\mathbb{P}^1)_{(z_i),\infty;(\lambda_i),\lambda_\infty}$ even for generic values of $z_1,\dots,z_N$.

Proposition 5.5. *Assume that all ^LG-opers in* $\mathrm{Op}_{^LG}(\mathbb{P}^1)_{(z_i),\infty;(\lambda_i),\lambda_\infty}$ *are non-degenerate and that all Bethe vectors obtained from solutions of the Bethe Ansatz equations* (4.18) *satisfying the condition* (5.4) *are non-zero. Then there is a bijection between the spectrum of the generalized Gaudin Hamiltonians on* $V^G_{(\lambda_i),\lambda_\infty}$ *(not counting multiplicities) and the set* $\mathrm{Op}_{^LG}(\mathbb{P}^1)_{(z_i),\infty;(\lambda_i),\lambda_\infty}$ *of* ^LG*-opers on* $\mathbb{P}^1$ *with regular singularities at* $z_1, \ldots, z_N, \infty$ *and residues* $\varpi(\lambda_1+\rho), \ldots, \varpi(\lambda_N+\rho)$ *and* $\varpi(\lambda_\infty+\rho)$*, respectively, which have trivial monodromy.*

Moreover, if in addition the Gaudin Hamiltonians are diagonalizable and have simple spectrum on $V^G_{(\lambda_i),\lambda_\infty}$*, then the Bethe vectors constitute an eigenbasis of* $V^G_{(\lambda_i),\lambda_\infty}$.

The last statement of Proposition 5.5 that the Bethe vectors constitute an eigenbasis of $V^G_{(\lambda_i),\lambda_\infty}$ is referred to as the completeness of the Bethe Ansatz.

We note that the completeness of the Bethe Ansatz has been previously proved for $\mathfrak{g} = \mathfrak{sl}_2$ (resp., for $\mathfrak{g} = \mathfrak{sl}_n$ when all λ_i's are multiples of the first fundamental weight) and generic values of $z_1, \ldots, z_N$ by Scherbak and Varchenko [SV] (resp., by Scherbak [S2]) by other methods.

We will discuss the degenerate opers and the corresponding eigenvectors of the Gaudin hamiltonians in Section 5.5.

5.3. The case of $\mathfrak{sl}_2$

In the case when $\mathfrak{g} = \mathfrak{sl}_2$, we identify the weights with complex numbers. The Bethe vector has a simple form in this case. We use the standard basis $\{e, h, f\}$ of $\mathfrak{sl}_2$. Then

$$\phi(w_1, \ldots, w_m) = f(w_1) \ldots f(w_m) v_{\lambda_1} \otimes \ldots v_{\lambda_N}, \tag{5.5}$$

where

$$f(w) = \sum_{i=1}^N \frac{f^{(i)}}{w - z_i}.$$

The Bethe Ansatz equations read

$$\sum_{i=1}^N \frac{\lambda_i}{w_j - z_i} = \sum_{s \neq j} \frac{2}{w_j - w_s}.$$

If these equations are satisfied, then this vector is a highest weight vector of weight λ_∞ (with respect to the diagonal action), and it is an eigenvector in $\bigotimes_{i=1}^N M_{\lambda_i}$ of the Gaudin algebra

$$\mathcal{Z}_{(z_i),\infty;(\lambda_i),\lambda_\infty} = \mathbb{C}[\Xi_i]_{i=1,\ldots,N} / \left(\sum_{i=1}^N \Xi_i \right),$$

where Ξ_i is the ith Gaudin Hamiltonian given by formula (0.1).

Its eigenvalue is represented by the PGL_2-oper

$$\partial_t - \sum_{i=1}^N \frac{\lambda_i(\lambda_i+2)/4}{(t-z_i)^2} - \sum_{i=1}^N \frac{c_i}{t - z_i} = (\partial_t - \lambda(t))(\partial_t + \lambda(t)),$$

where

$$\lambda(t) = \sum_{i=1}^{N} \frac{\lambda_i/2}{t - z_i} - \sum_{j=1}^{m} \frac{1}{t - w_j}.$$

In other words, the eigenvalue c_i of Ξ_i on $\phi(w_1, \ldots, w_m)$ is equal to

$$c_i = \lambda_i \left(\sum_{j \neq i} \frac{\lambda_j}{z_i - z_j} - \sum_{j=1}^{m} \frac{1}{z_i - w_j} \right). \tag{5.6}$$

For $\mathfrak{g} = \mathfrak{sl}_2$, all opers all opers in $\mathrm{Op}_{^L G}(\mathbb{P}^1)_{(z_i),\infty;(\lambda_i),\lambda_\infty}$ are non-degenerate for generic values of $z_1, \ldots, z_N$ (for example, this follows from the results of [SV]).

5.4. Solutions of the Bethe Ansatz equations and flag manifolds

In Section 5.2 we considered only those solutions of the Bethe Ansatz equation for which

$$\sum_{i=1}^{N} \lambda_i - \sum_{j=1}^{m} \alpha_{i_j} \tag{5.7}$$

is a dominant integral weight. The reason for that was that we wanted to construct eigenvectors of $\mathcal{Z}_{(z_i)}(\mathfrak{g})$ in $\bigotimes_{i=1}^N V_{\lambda_i}$ of weight (5.7). Since the Bethe vectors are automatically $\mathfrak{n}_+$-invariant, we find that such a vector can be non-zero in $\bigotimes_{i=1}^N V_{\lambda_i}$ only if this weight is dominant integral. In this case we may write it in the form $-w_0(\lambda_\infty)$, where λ_∞ is another dominant integral weight. Then the corresponding Bethe vector gives us an eigenvector of $\mathcal{Z}_{(z_i)}(\mathfrak{g})$ in the space

$$V^G_{(\lambda_i),\lambda_\infty} = \left(\bigotimes_{i=1}^{N} V_{\lambda_i} \otimes V_{\lambda_\infty} \right)^G.$$

We have seen in Section 5.2 that these solutions of the Bethe Ansatz equations are in one-to-one correspondence with the Miura opers on $\mathbb{P}^1$ such that the corresponding oper in $\mathrm{Op}_{^L G}(\mathbb{P}^1)_{(z_i),\infty;(\lambda_i),\lambda_\infty}$ is non-degenerate, and such that the horizontal ${}^L B$-reduction $\mathcal{F}'_{^L B}$ coincides with the oper ${}^L B$-reduction $\mathcal{F}_{^L B}$ at ∞.

It is natural to consider more general solutions of the Bethe Ansatz equations, i.e., those for which the weight (5.7) is not necessarily dominant integral. The first step is to relate these solutions to Miura opers.

Suppose that we are given an arbitrary solution of the Bethe Ansatz equations (4.18). By Proposition 4.10, we attach to it the connection $\overline{\nabla}$ on the ${}^L H$-bundle $\Omega^{-\rho}$ on $\mathbb{P}^1$ whose restriction to $\mathbb{P}^1 \backslash \infty$ reads

$$\overline{\nabla} = \partial_t + \sum_{i=1}^{N} \frac{\lambda_i}{t - z_i} - \sum_{j=1}^{m} \frac{\alpha_{i_j}}{t - w_j}. \tag{5.8}$$

Recall the Miura transformation $\overline{\mathbf{b}}^*_{(z_i),\infty;(\lambda_i),\lambda_\infty}$ introduced in formula (4.20). Applying this map to $\overline{\nabla}$, we obtain a ${}^L G$-oper τ. It has regular singularities at $z_1, \ldots, z_N$ and ∞, and is regular elsewhere on $\mathbb{P}^1$ (recall that the Bethe Ansatz equations ensure that the oper τ is regular at the points $w_1, \ldots, w_m$). Around the point z_i the oper connection reads

$$\nabla = \partial_t + p_{-1} - \frac{\lambda_i}{t - z_i} + \text{reg}\,.$$

and by applying the gauge transformation with $\lambda(t)$ we can bring it to the form (1.9). Therefore, according to Lemma 1.2, τ has no monodromy around $z_i, i = 1, \ldots, N$.

Let us consider the point ∞. Since τ has no monodromy anywhere on $\mathbb{P}^1 \backslash \infty$, it cannot have monodromy at ∞ either. By Lemma 1.2, the restriction of τ to the disc around ∞ belongs to $\text{Op}_G(D_\infty)_{\lambda_\infty}$ for some integral dominant weight λ. For notational convenience we represent λ in the form $-w_0(\lambda_\infty)$. But then the commutative diagram (3.3) implies that the residue of $\overline{\nabla}$ at ∞ is equal to $w'(\lambda + \rho) - \rho$ for some $w' \in W$. Writing $w' = ww_0$, find that the residue is equal to $w(\lambda_\infty + \rho) + \rho$. On the other hand, using transformation formula (3.2), we find that the residue of $\overline{\nabla}$ at ∞ is equal to

$$-2\rho - \sum_{i=1}^{N} \lambda_i + \sum_{j=1}^{m} \alpha_{i_j}.$$

Therefore we must have the equality

$$\sum_{i=1}^{N} \lambda_i - \sum_{j=1}^{m} \alpha_{i_j} = w(\lambda_\infty + \rho) - \rho \tag{5.9}$$

for some dominant integral weight λ_∞ and $w \in W$. Thus, the oper τ belongs to $\text{Op}_{^LG}(\mathbb{P}^1)_{(z_i),\infty;(\lambda_i),\lambda_\infty}$, and so $\overline{\nabla}$ is an element of $\text{Conn}(\Omega^{-\rho})^{\text{gen}}_{(z_i),\infty;(\lambda_i),\lambda_\infty}$.

Recall from Section 4.2 that the Miura transformation $\overline{\mathbf{b}}^*_{(z_i),\infty;(\lambda_i),\lambda_\infty}$ may be lifted to a map

$$\mathbf{b}^*_{(z_i),\infty;(\lambda_i),\lambda_\infty} : \text{Conn}(\Omega^{-\rho})^{\text{gen}}_{(z_i),\infty;(\lambda_i),\lambda_\infty} \to \text{MOp}_{^LG}(\mathbb{P}^1)_{(z_i),\infty;(\lambda_i),\lambda_\infty},$$

where $\text{MOp}_{^LG}(\mathbb{P}^1)_{(z_i),\infty;(\lambda_i),\lambda_\infty}$ is the space of Miura opers on $\mathbb{P}^1$ whose underlying opers belong to $\text{Op}_{^LG}(\mathbb{P}^1)_{(z_i),\infty;(\lambda_i),\lambda_\infty}$.

Recall from Lemma 5.2 that the space of all Miura opers on $\mathbb{P}^1$ corresponding to a fixed ^LG-oper $\tau \in \text{Op}_{^LG}(\mathbb{P}^1)_{(z_i),\infty;(\lambda_i),\lambda_\infty}$ is isomorphic to the flag variety $^LG/^LB$.

Let

$$\text{MOp}_{^LG}(\mathbb{P}^1)^{\text{gen}}_{(z_i),\infty;(\lambda_i),\lambda_\infty} \subset \text{MOp}_{^LG}(\mathbb{P}^1)_{(z_i),\infty;(\lambda_i),\lambda_\infty}$$

be the subvariety of those Miura opers whose horizontal ^LB-reduction satisfies the conditions (1) and (2) of Section 5.2. We call such Miura opers *non-degenerate*. It is clear that this subvariety is open and dense.

It follows from our construction that the image of the map $\mathbf{b}^*_{(z_i),\infty;(\lambda_i),\lambda_\infty}$ is contained in $\text{MOp}^{\text{gen}}_{^LG}(\mathbb{P}^1)_{(z_i),\infty;(\lambda_i),\lambda_\infty}$. Conversely, suppose we are given a Miura oper in $\text{MOp}^{\text{gen}}_{^LG}(\mathbb{P}^1)_{(z_i),\infty;(\lambda_i),\lambda_\infty}$. Then we obtain a connection on $w_0(\Omega^\rho) =$

$\mathcal{F}'_{^LB}/^LN$, and hence a connection on $\Omega^{-\rho}$. It follows from Lemma 3.5 that this connection belongs to $\mathrm{Conn}(\Omega^{-\rho})^{\mathrm{gen}}_{(z_i),\infty;(\lambda_i),\lambda_\infty}$. Moreover, in the same way as in the proof of Proposition 3.3 we find that thus constructed map

$$\mathrm{MOp}^{\mathrm{gen}}_{^LG}(\mathbb{P}^1)_{(z_i),\infty;(\lambda_i),\lambda_\infty} \to \mathrm{Conn}(\Omega^{-\rho})^{\mathrm{gen}}_{(z_i),\infty;(\lambda_i),\lambda_\infty}$$

is inverse to $\mathbf{b}_{(z_i),\infty;(\lambda_i),\lambda_\infty}$. Therefore we obtain

Lemma 5.6. *The map* $\mathbf{b}^*_{(z_i),\infty;(\lambda_i),\lambda_\infty}$ *is a bijection between* $\mathrm{Conn}(\Omega^{-\rho})^{\mathrm{gen}}_{(z_i),\infty;(\lambda_i),\lambda_\infty}$ *and* $\mathrm{MOp}^{\mathrm{gen}}_{^LG}(\mathbb{P}^1)_{(z_i),\infty;(\lambda_i),\lambda_\infty}$.

Combining Lemma 5.6 and Proposition 5.1, we obtain

Theorem 5.7. *The set of solutions of the Bethe Ansatz equations* (4.18) *is in bijection with the set of points of* $\mathrm{MOp}_{^LG}(\mathbb{P}^1)^{\mathrm{gen}}_{(z_i),\infty;(\lambda_i),\lambda_\infty}$, *where* λ_∞ *satisfies* (5.9) *for some* $w \in W$.

Let us fix an oper τ. Then considering the value of the reduction $\mathcal{F}'_B$ at $\infty \in \mathbb{P}^1$ as in Lemma 5.2 above, we obtain an identification of the space

$$\mathrm{MOp}_{^LG}(\mathbb{P}^1)^{\mathrm{gen}}_{(z_i),\infty;(\lambda_i),\lambda_\infty}$$

with an open dense subset of $^LG/^LB$ which we denote by $(^LG/^LB)_\tau$. Furthermore, by Proposition 3.4, those elements of $\mathrm{MOp}_{^LG}(\mathbb{P}^1)^{\mathrm{gen}}_{(z_i),\infty;(\lambda_i),\lambda_\infty}$ which satisfy formula (5.9) correspond to points that lie in the Schubert cell

$$S_w = {}^LBw_0ww_0{}^LB \subset {}^LG/{}^LB.$$

Note that except for the big cell S_{w_0}, the intersection between the Schubert cell S_w and the open dense subset $(^LG/^LB)_\tau \subset {}^LG/^LB$ could be either an open dense subset of S_w or empty.[2] Theorem 5.7 may be made more precise as follows:

Theorem 5.8. *The set of solutions of the Bethe Ansatz equations* (4.18) *decomposes into a union of disjoint subsets labelled by* $\mathrm{Op}_{^LG}(\mathbb{P}^1)_{(z_i),\infty;(\lambda_i),\lambda_\infty}$. *The set of points corresponding to* ^LG*-oper* $\tau \in \mathrm{Op}_{^LG}(\mathbb{P}^1)_{(z_i),\infty;(\lambda_i),\lambda_\infty}$ *is in bijection with the set of points of an open and dense subset* $(^LG/^LB)_\tau$ *of the flag variety* $^LG/^LB$.

Further, each of these solution must satisfy the equation (5.9) *for some* $w \in W$, *and the solutions which satisfy this equation with a fixed* $w \in W$ *are in bijection with an open subset* $S_w \cap (^LG/^LB)_\tau$ *of the Schubert cell* S_w.

Remark 5.9. In [MV1] (resp., [BM]), it was shown, by a method different from ours, that in the case when $\mathfrak{g}$ is of types A_n, B_n or C_n (resp., G_2) and $z_1, \ldots, z_N$ are generic, the set of solutions of the Bethe Ansatz equations satisfying (5.9) is in bijection with a disjoint union of open and dense subsets of S_w. The connection between the results of [MV1] and our results is explained in [F3]. □

In [F3] we obtained a generalization of Theorem 5.8 to the case when $\mathfrak{g}$ is an arbitrary Kac-Moody algebra.

---

[2]For example, it follows from the results of Mukhin and Varchenko in [MV2] that for fixed $\lambda_1, \ldots, \lambda_N, \lambda_\infty$ this open set may not contain the one point Schubert cell $S_1 \subset G/B$ even if we allow $z_1, \ldots, z_N$ to be generic.

5.5. Degenerate opers

By Lemma 5.6 we have a bijection between the set $\mathrm{Conn}(\Omega^{-\rho})^{\mathrm{gen}}_{(z_i),\infty;(\lambda_i),\lambda_\infty}$ and an open subset $\mathrm{MOp}^{\mathrm{gen}}_{{}^LG}(\mathbb{P}^1)_{(z_i),\infty;(\lambda_i),\lambda_\infty}$ of non-degenerate Miura opers in

$$\mathrm{MOp}_{{}^LG}(\mathbb{P}^1)_{(z_i),\infty;(\lambda_i),\lambda_\infty}.$$

Now we wish to extend this bijection to the entire set $\mathrm{MOp}_{{}^LG}(\mathbb{P}^1)_{(z_i),\infty;(\lambda_i),\lambda_\infty}$. To any point of $\mathrm{MOp}_{{}^LG}(\mathbb{P}^1)_{(z_i),\infty;(\lambda_i),\lambda_\infty}$ we can still assign, as before, a connection $\overline{\nabla}$ on $\Omega^{-\rho}$ with regular singularities at $z_1,\dots,z_N,\infty$ and some additional points $w_1,\dots,w_m$. By Proposition 3.4, the residue of $\overline{\nabla}$ at z_i (resp., ∞, w_j) must be equal to $-y_i(\check{\lambda}_i+\check{\rho})+\check{\rho}$ (resp., $-y_\infty(\check{\lambda}_\infty+\check{\rho})+\check{\rho}, -y'_j(\check{\rho})+\check{\rho}$) for some elements $y_i, y_\infty, y'_j \in W$. Hence this connection has the form

$$\partial_t + \sum_{i=1}^N \frac{y_i(\lambda_i+\rho)-\rho}{t-z_i} + \sum_{j=1}^m \frac{y'_j(\rho)-\rho}{t-w_j}. \tag{5.10}$$

Considering the expansion of this connection at ∞, we obtain the following relation

$$\sum_{i=1}^N (y_i(\check{\lambda}_i+\check{\rho})-\check{\rho}) + \sum_{j=1}^m (y'_j(\check{\rho})-\check{\rho}) = y_\infty w_0(-w_0(\check{\lambda}_\infty)+\check{\rho})-\check{\rho}. \tag{5.11}$$

Let $\mathrm{Conn}(\Omega^{-\rho})_{(z_i),\infty;(\lambda_i),\lambda_\infty}$ be the set of all connections of the form (5.10) whose Miura transformation belongs to $\mathrm{Op}_{{}^LG}(\mathbb{P}^1)_{(z_i),\infty;(\lambda_i),\lambda_\infty}$. Using Lemma 5.2, we then obtain the following generalization of Lemma 5.6 (see [F3]):

Theorem 5.10. *There is a bijection between the sets*

$$\mathrm{Conn}(\Omega^{-\rho})_{(z_i),\infty;(\lambda_i),\lambda_\infty} \qquad \text{and} \qquad \mathrm{MOp}_{{}^LG}(\mathbb{P}^1)_{(z_i),\infty;(\lambda_i),\lambda_\infty}.$$

In particular, the set of those connections in $\mathrm{Conn}(\Omega^{-\rho})_{(z_i),\infty;(\lambda_i),\lambda_\infty}$ *which correspond to a fixed* ${}^LG$*-oper* $\tau \in \mathrm{Op}_{{}^LG}(\mathbb{P}^1)_{(z_i),\infty;(\check{\lambda}_i),\check{\lambda}_\infty}$ *is isomorphic to the set of points of the flag variety* ${}^LG/{}^LB$.

Moreover, the residues of these connections must satisfy the relation (5.11) *for some* $y_\infty \in W$. *The set of those connections which satisfy this relation is in bijection with the Schubert cell* ${}^LBw_0y_\infty{}^LB$ *in* ${}^LG/{}^LB$.

Consider, in particular, the unique Miura oper corresponding to an oper

$$\tau \in \mathrm{Op}_{{}^LG}(\mathbb{P}^1)_{(z_i),\infty;(\lambda_i),\lambda_\infty}$$

in which the two Borel reductions coincide at ∞ (it corresponds to the one point Schubert cell ${}^LB$ in ${}^LG/{}^LB$). Let $\overline{\nabla}_\tau$ be the corresponding connection in $\mathrm{Conn}(\Omega^{-\rho})_{(z_i),\infty;(\lambda_i),\lambda_\infty}$. It has the form (5.10) and its residues satisfy the relation (5.11) with $y_\infty = w_0$:

$$\sum_{i=1}^N (y_i(\lambda_i+\rho)-\rho) + \sum_{j=1}^m (y'_j(\rho)-\rho) = -w_0(\lambda_\infty).$$

The oper $\tau \in \mathrm{Op}_{^L G}(\mathbb{P}^1)_{(z_i),\infty;(\lambda_i),\lambda_\infty}$ is degenerate if and only if either some of the elements y_i are not equal to 1 or some of the elements y'_j have lengths greater than 1 (i.e., are not simple reflections).

Observe that we can write $y_i(\lambda_i+\rho)-\rho$ as λ_i minus a combination of simple roots with non-negative coefficients, and likewise, $y'_j(\rho)-\rho$ is equal to minus a linear combination of the simple roots with non-negative coefficients. It is instructive to think of the connection $\overline{\nabla}_\tau$ corresponding to degenerate opers as the limit of a family of connections corresponding to non-degenerate opers, under which some of the points w_j's either collide with some of the z_i's or with each other (leading to the degeneracies of types (1) and (2), respectively). The resulting residues of the connection become the sums of the residues of the points that have coalesced together. Proposition 3.4 places severe restrictions on what kinds of residues (and hence collisions) can occur: namely, at the point z_i the residue must lie in the W-orbit of λ_i, whereas at the additional points w_j the residues must be in the W-orbit of 0 (under the ρ-shifted action of W).

We expect that if $z_1,\dots,z_N$ are generic, then for any $\tau \in \mathrm{Op}_{^L G}(\mathbb{P}^1)_{(z_i),\infty;(\lambda_i),\lambda_\infty}$ we have $y_i = 1$ for all $i = 1,\dots,N$ in the connection $\overline{\nabla}_\tau$. In other words, this Miura oper satisfies condition (1) from Section 5.2, but does not satisfy condition (2), that is at least one of the y'_j's is not a simple reflection.

Suppose that τ is such an oper, so in particular we have

$$\sum_{i=1}^{N} \lambda_i + \sum_{j=1}^{m} (y'_j(\rho)-\rho) = -w_0(\lambda_\infty). \tag{5.12}$$

Then we can still attach to the connection (5.10) an eigenvector of the generalized Gaudin hamiltonians in $V^G_{(\lambda_i),\lambda_\infty}$ with eigenvalue τ by generalizing the procedure of Section 4.4.

More precisely, suppose that the residue of $\lambda(t)$ at w_j is equal to $y'_j(\check{\rho})-\check{\rho}$ for some $y'_j \in W$. Then the expansion of this connection at w_j is $\mu_j(t-w_j)$, where

$$\mu_j(t) = \frac{y'_j(\check{\rho})-\check{\rho}}{t-w_j} + \sum_{n\geq 0} \mu_{j,n} t^n, \qquad \mu_{j,n} \in \mathfrak{h}^*.$$

The condition that the oper τ has no monodromy at w_j (i.e., that the Miura transformation of the connection $\partial_t + \mu_j(t)$ is regular at $t=0$) translates into a system of equations on the coefficients $\mu_{j,n}$ of this expansion. We denote this system by S_j. Recall from Lemma 3.5 that when $y'_j = s_{i_j}$ there is only one equation $\langle \check{\alpha}_{i_j}, \mu_{j,0}\rangle = 0$.

To see what these equations look like, we consider the example when $\mathfrak{g} = \mathfrak{sl}_n$. The PGL_n-opers may be represented by nth order differential operators of the form

$$\partial_t^n + v_1(t)\partial_t^{n-2} + \dots + v_{n-1}(t). \tag{5.13}$$

Let us identify the dual Cartan subalgebra of $\mathfrak{sl}_n$ with the hyperplane $\sum_{k=1}^n \epsilon_k = 0$ of the vector space $\mathrm{span}\{\epsilon_k\}_{k=1,\dots,n}$. Then an ${}^L\mathfrak{h}$-valued connection has the form

$\partial_t + \sum_{k=1}^n u_k(t)\epsilon_k$. The Miura transformation of this connection is given by the formula

$$\partial_t^n + v_1(t)\partial_t^{n-2} + \ldots + v_{n-1}(t) = (\partial_t - u_1(t)) \ldots (\partial_t - u_n(t)). \tag{5.14}$$

The system of equations S_j is obtained by writing $\mu_j(t)$ as $\sum_{k=1}^n u_k(t)\epsilon_k$, substituting the functions $u_k(t)$ into formula (5.14) and setting to zero all coefficients in front of the negative powers of t in the resulting nth order differential operator. For example, if $y_j' = s_a s_b, |a-b| > 1$, then we have

$$u_m(t) = -(\delta_{m,a} - \delta_{m,a+1} + \delta_{m,b} - \delta_{m,b+1})t^{-1} + \sum_{r \geq 0} u_{m,r} t^r,$$

and it is easy to write down the equations on the coefficients $u_{m,r}$ corresponding to the regularity of the operator (5.14) at $t = 0$.

The system of equations S_j also has a nice interpretation in the theory of the Wakimoto modules. Namely, consider the Wakimoto module $W_{\mu_j(z)}$. Then it contains a $\mathfrak{g}[[t]]$-invariant vector P_j if and only if the system S_j of equations on $\mu_j(z)$ is satisfied. In the case when $y_j' = s_{i_j}$ and $\mu_j(z) = -\alpha_{i_j} z^{-1} + \ldots$, this is the statement of Lemma 4.5; this vector is given by the formula $e^R_{i_j,-1}|0\rangle$ in this case. In general the formulas for these vectors are much more complicated.

It follows from the definition that for our connection $\overline{\nabla}_\tau$ the systems S_j are satisfied for all $j = 1, \ldots, m$. Therefore each module $W_{\mu_j(z)}$ contains a $\mathfrak{g}[[t]]$-invariant vector P_j. Then, in the same way as in Section 4.4, we obtain a homomorphism of $\widehat{\mathfrak{g}}_{\kappa_c}$-modules

$$\bigotimes_{i=1}^N W_{\lambda_i(z)} \otimes \mathbb{V}_0^{\otimes m} \otimes W'_{\lambda_\infty(z)} \to \bigotimes_{i=1}^N W_{\lambda_i(z)} \otimes \bigotimes_{j=1}^m W_{\mu_j(z)} \otimes W'_{\lambda_\infty(z)},$$

which sends the vacuum vector in the jth copy of $\mathbb{V}_0$ to $P_j \in W_{\mu_j(z)}$. Therefore, composing the corresponding map of the spaces of coinvariants with the functional $\tau_{(z_i),(w_j)}$ introduced at the end of Section 4.3, we obtain a linear functional

$$\widetilde{\tau}_{(z_i),(w_j)} : H((W_{\lambda_i(z)}), W'_{\lambda_\infty(z)}) \to \mathbb{C}.$$

By construction, $\widetilde{\tau}_{(z_i),(w_j)}$ is an eigenvector of the Gaudin algebra $\mathcal{Z}_{(z_i)}(\mathfrak{g})$, and the corresponding eigenvalue is given by the ${}^LG$-oper $\overline{\mathbf{b}}^*_{(z_i),\infty;(\lambda_i),\lambda_\infty}(\overline{\nabla}_\tau) = \tau$.

Continuing along the lines of the construction presented in Section 4.4, we obtain an $\mathfrak{n}_+$-invariant vector in $\otimes_{i=1}^N M_{\lambda_i}$ of weight $\sum_{i=1}^N \lambda_i + \sum_{j=1}^m (y_j'(\rho) - \rho)$, which, according to formula (5.12), is equal to $-w_0(\lambda_\infty)$ which is a dominant integral weight. The projection of this vector onto $\otimes_{i=1}^N V_{\lambda_i}$ gives rise to an eigenvector of the Gaudin algebra $\mathcal{Z}_{(z_i)}(\mathfrak{g})$ in $V^G_{(\lambda_i),\lambda_\infty}$ with the eigenvalue corresponding to our degenerate oper τ.

Thus, we have assigned to any oper τ in $\operatorname{Op}_{^LG}(\mathbb{P}^1)_{(z_i),\infty;(\lambda_i),\lambda_\infty}$ such that the corresponding special connection $\overline{\nabla}_\tau$ satisfies condition (1) that $y_i = 1$, for all $i = 1, \ldots, N$, in (5.10), an eigenvector of the Gaudin algebra in $V^G_{(\lambda_i),\lambda_\infty}$ with the eigenvalue τ.

The next question is what happens if $\overline{\nabla}_\tau$ has $y_i \neq 1$ for some i. In this case the above construction gives rise to an eigenvector of the Gaudin algebra $\mathcal{Z}_{(z_i)}(\mathfrak{g})$ in the space

$$(\otimes_{i=1}^N M_{y_i(\lambda_i+\rho)-\rho})^{\mathfrak{n}_+}_{-w_0(\lambda_\infty)}$$

of the diagonal $\mathfrak{n}_+$-invariants of $\otimes_{i=1}^N M_{y_i(\lambda_i+\rho)-\rho}$ of weight $-w_0(\lambda_\infty)$.

Now recall that for any dominant integral weight λ and $y \in W$ the Verma module $M_{y(\lambda+\rho)-\rho}$ is a $\mathfrak{g}$-submodule of M_λ. Hence $\otimes_{i=1}^N M_{y_i(\lambda_i+\rho)-\rho}$ is naturally a subspace of $\otimes_{i=1}^N M_{\lambda_i}$. It is also a $\mathfrak{g}_{\mathrm{diag}}$-submodule of $\otimes_{i=1}^N M_{\lambda_i}$ which is contained in the maximal proper $\mathfrak{g}_{\mathrm{diag}}$-submodule $I_{(\lambda_i)}$. The quotient of $\otimes_{i=1}^N M_{\lambda_i}$ by $I_{(\lambda_i)}$ is isomorphic to $\otimes_{i=1}^N V_{\lambda_i}$. We have thus associated to $\overline{\nabla}_\tau$ an eigenvector of the Gaudin algebra in $\otimes_{i=1}^N M_{\lambda_i}$ with the eigenvalue τ, but this eigenvector is inside its maximal proper $\mathfrak{g}_{\mathrm{diag}}$-submodule $I_{(\lambda_i)}$. Some sample computations that we have made in the case when $\mathfrak{g} = \mathfrak{sl}_2$ suggest that in this case the generalized eigenspace of the Gaudin algebra in $(\otimes_{i=1}^N M_{\lambda_i})^{\mathfrak{n}_+}_{-w_0(\lambda_\infty)}$ corresponding to the eigenvalue τ has dimension greater than one and that its projection onto $(\otimes_{i=1}^N V_{\lambda_i})^{\mathfrak{n}_+}_{-w_0(\lambda_\infty)} \simeq V^G_{(\lambda_i),\lambda_\infty}$ is non-zero. In other words, we expect that in this case there also exists an eigenvector of the Gaudin algebra $\mathcal{Z}_{(z_i)}(\mathfrak{g})$ in $V^G_{(\lambda_i),\lambda_\infty}$ with the same eigenvalue τ.

If true, this would explain the meaning of the degeneracy of the connection $\overline{\nabla}_\tau$ at the point z_i. Namely, for some special values of z_i it may happen that some of the eigenvalues of the Gaudin operators in $(\otimes_{i=1}^N V_{\lambda_i})^{\mathfrak{n}_+}_{-w_0(\lambda_\infty)}$ and in $(I_{(\lambda_i)})^{\mathfrak{n}_+}_{-w_0(\lambda_\infty)}$ become equal. Then they may combine into a Jordan block such that the eigenvector in $(\otimes_{i=1}^N V_{\lambda_i})^{\mathfrak{n}_+}_{-w_0(\lambda_\infty)}$ is only a generalized eigenvector in $(\otimes_{i=1}^N M_{\lambda_i})^{\mathfrak{n}_+}_{-w_0(\lambda_\infty)}$. If that happens, we can no longer obtain this eigenvector by projection from $(\otimes_{i=1}^N M_{\lambda_i})^{\mathfrak{n}_+}_{-w_0(\lambda_\infty)}$ (but we could potentially obtain it by considering the family of eigenvectors corresponding to small perturbations of the z_i's).

Let $\otimes_{i=1}^N M_{y_i(\lambda_i+\rho)-\rho}$ be the smallest $\mathfrak{g}^{\otimes N}$-submodule of $\otimes_{i=1}^N M_{\lambda_i}$ in which the corresponding "true" eigenvector in $(\otimes_{i=1}^N M_{\lambda_i})^{\mathfrak{n}_+}_{-w_0(\lambda_\infty)}$ is contained. We believe that this is precisely the situation when the connection $\overline{\nabla}\tau$ develops a singularity with residue $y_i(\lambda_i+\rho)-\rho$ at the point z_i.

Let us summarize the emerging picture. As we explained at the beginning of Section 5, we have an injective map from the spectrum of the Gaudin algebra $\mathcal{Z}_{(z_i)}(\mathfrak{g})$ in $V^G_{(\lambda_i),\lambda_\infty}$ to $\mathrm{Op}_{^LG}(\mathbb{P}^1)_{(z_i),\infty;(\lambda_i),\lambda_\infty}$. We wish to construct the inverse map, in other words, to assign to each oper $\tau \in \mathrm{Op}_{^LG}(\mathbb{P}^1)_{(z_i),\infty;(\lambda_i),\lambda_\infty}$ an eigenvector of $\mathcal{Z}_{(z_i)}(\mathfrak{g})$ with eigenvalue τ in $V^G_{(\lambda_i),\lambda_\infty}$. First, we associate to τ the Miura oper in which the horizontal Borel reduction coincides with the oper reduction with the oper reduction at the point ∞. This Miura oper gives rise to a connection $\overline{\nabla}_\tau \in \mathrm{Conn}(\Omega^{-\rho})_{(z_i),\infty;(\lambda_i),\lambda_\infty}$. It has the form (5.10), and the residues satisfy the condition (5.12).

The simplest situation occurs if all elements y_i in (5.10) are equal to 1 and all elements y'_j are simple reflections. This is the situation where Bethe Ansatz is applicable. Namely, as explained in Section 4.4, we assign to $\overline{\nabla}_\tau$ a certain coinvariant of the tensor product of the Wakimoto modules, which gives rise to an eigenvector of $\mathcal{Z}_{(z_i)}(\mathfrak{g})$ with eigenvalue τ in $V^G_{(\lambda_i),\lambda_\infty}$. This is the Bethe vector given by an explicit formula (4.21). It was believed that this was in fact a generic situation, i.e., for fixed $\lambda_1, \ldots, \lambda_N, \lambda_\infty$ and generic $z_1, \ldots, z_N$ the connection $\overline{\nabla}_\tau$ satisfies these conditions for all opers $\tau \in \mathrm{Op}_{{}^LG}(\mathbb{P}^1)_{(z_i),\infty;(\lambda_i),\lambda_\infty}$. But recent results of Mukhin and Varchenko [MV2] show that this is not the case. Hence we analize possible degeneracies.

The first type of degeneracy that may occur is the following: while all elements y_i in (5.10) are still equal to 1, some of the elements y'_j are no longer simple reflections. We expect this to be the generic situation. In this case we can still construct an eigenvector of $\mathcal{Z}_{(z_i)}(\mathfrak{g})$ with eigenvalue τ in $V^G_{(\lambda_i),\lambda_\infty}$, as explained above. A formula for this vector will be more complicated than the formula for a Bethe vector, but in principal such a vector can be obtained by an algorithmic procedure.

The most general and most difficult case is when some of the elements y_i are not equal to 1 and some of the elements y'_j are not simple reflections. In this case, as explained above, we still believe that there exists an eigenvector of $\mathcal{Z}_{(z_i)}(\mathfrak{g})$ with eigenvalue τ in $V^G_{(\lambda_i),\lambda_\infty}$. But this eigenvector cannot be constructed directly by using the Wakimoto modules, as above. Indeed, our construction using the Wakimoto modules gives eigenvectors in $\otimes_{i=1}^N M_{\lambda_i}$, but we expect that in this most degenerate case the eigenvector in $V^G_{(\lambda_i),\lambda_\infty}$ cannot be lifted to an eigenvector in $\otimes_{i=1}^N M_{\lambda_i}$ (only to a generalized eigenvector). However, one can probably obtain this eigenvector by considering families of eigenvectors corresponding to small perturbations of the z_i's.

Finally, there is a question as to whether all of the eigenvectors constructed by means of the Wakimoto module construction (including the Bethe vectors) are non-zero. It was conjectured by S. Chmutov and I. Scherbak in [CS] that the Bethe vectors are always non-zero. Hence optimistically one can hope that the more general eigenvectors that we have constructed (for y'_j not being simple reflections) are also non-zero. This optimistic view then leads to the following conjecture:

Conjecture 1. *For any set of integral dominant weights $\lambda_1, \ldots, \lambda_N, \lambda_\infty$ and an arbitrary collection of distinct complex numbers $z_1, \ldots, z_N$ there is a bijection between the set $\mathrm{Op}_{{}^LG}(\mathbb{P}^1)_{(z_i),\infty;(\lambda_i),\lambda_\infty}$ and the spectrum of the Gaudin algebra $\mathcal{Z}_{(z_i)}(\mathfrak{g})$ in $V^G_{(\lambda_i),\lambda_\infty}$ (not counting multiplicities).*

A possible approach to proving this conjecture (which may be true even if some of the ingredients in the argument suggested above do not work out) comes from the geometric Langlands correspondence. As explained in [F1], to each ${}^LG$-oper τ on $\mathbb{P}^1$ with regular singularities at $z_1, \ldots, z_N$ and ∞ one assigns a D-module on the moduli space of G-bundles on $\mathbb{P}^1$ with B-reductions at the points $z_1, \ldots, z_N, \infty$ (this moduli space is isomorphic to $(G/B)^N/B_{\mathrm{diag}}$). This D-module is a Hecke eigensheaf whose "eigenvalue" is the flat ${}^LG$-bundle on $\mathbb{P}^1 \backslash \{z_1, \ldots, z_N, \infty\}$ obtained by forgetting the ${}^LB$-reduction of τ. Now if τ is in

$\mathrm{Op}_{^LG}(\mathbb{P}^1)_{(z_i),\infty;(\lambda_i),\lambda_\infty}$, then this flat ^LG-bundle has trivial monodromy and hence is isomorphic to the trivial flat ^LG-bundle. It is natural to expect that the corresponding Hecke eigensheaf "does not depend" on the B-reductions at the points z_i and ∞, i.e., it is just the structure sheaf on $(G/B)^N/B_{\mathrm{diag}}$ (or a direct sum of copies of it). This is equivalent to the existence of a non-zero eigenvector of the Gaudin algebra with eigenvalue τ in $V^G_{(\lambda_i),\lambda_\infty}$, as explained in [F1], and hence we obtain a proof of the conjecture.

Acknowledgements

This paper reviews the results of our previous works [FFR, F1, F3], some of which were obtained jointly with B. Feigin and N. Reshetikhin.

I thank the organizers of the Workshop "Infinite-dimensional algebras and quantum integrable systems" in Faro in July of 2003 for their invitation to give a talk on this subject and for encouraging me to write this review.

References

[ATY] H. Awata, A. Tsuchiya and Y. Yamada, *Integral formulas for the WZNW correlation functions*, Nucl. Phys. **B 365** (1991) 680–698.

[BaFl] H.M. Babujian and R. Flüme, *Off-shell Bethe Ansatz equation for Gaudin magnets and solutions of Knizhnik-Zamolodchikov equations*, Mod. Phys. Lett. **A 9** (1994) 2029–2039.

[BD1] A. Beilinson and V. Drinfeld, *Quantization of Hitchin's integrable system and Hecke eigensheaves*, Preprint, available at www.math.uchicago.edu/∼benzvi.

[BD2] A. Beilinson and V. Drinfeld, *Opers*, Preprint.

[BM] L. Borisov and E. Mukhin, *Self-self-dual spaces of polynomials*, Preprint math. QA/0308128.

[CS] S. Chmutov and I. Scherbak, *Intersections of Schubert varieties and highest weight vectors in tensor products of sl_{N+1}-representations*, Preprint math.RT/0407367.

[DS] V. Drinfeld and V. Sokolov, *Lie algebras and KdV type equations*, J. Sov. Math. **30** (1985) 1975–2036.

[EH] D. Eisenbud, J. Harris, *Divisors on general curves and cuspidal rational curves*, Invent. Math. **74** (1983) 371–418.

[ER] B. Enriquez and V. Rubtsov, *Hitchin systems, higher Gaudin operators and R-matrices*, Math. Res. Lett. **3** (1996) 343–357.

[FF1] B. Feigin, E. Frenkel, *Affine Kac-Moody Algebras and semi-infinite flag manifolds*, Comm. Math. Phys. **128**, 161–189 (1990).

[FF2] B. Feigin and E. Frenkel, *Affine Kac-Moody algebras at the critical level and Gelfand-Dikii algebras*, Int. Jour. Mod. Phys. **A7**, Supplement 1A (1992) 197–215.

[FFR] B. Feigin, E. Frenkel and N. Reshetikhin, *Gaudin model, Bethe Ansatz and critical level*, Comm. Math. Phys. **166** (1994) 27–62.

[F1] E. Frenkel, *Affine algebras, Langlands duality and Bethe Ansatz*, in Proceedings of the International Congress of Mathematical Physics, Paris, 1994, ed. D. Iagolnitzer, pp. 606–642, International Press, 1995.

[F2] E. Frenkel, *Lectures on Wakimoto modules, opers and the center at the critical level*, Preprint math.QA/0210029.

[F3] E. Frenkel, *Opers on the projective line, flag manifolds and Bethe Ansatz*, Preprint math.QA/0308269, to appear in Moscow Math. Journal.

[FB] E. Frenkel and D. Ben-Zvi, *Vertex algebras and algebraic curves*, Second Edition, Mathematical Surveys and Monographs, vol. 88. AMS 2004.

[FG] E. Frenkel and D. Gaitsgory, to appear.

[K] V.G. Kac, *Infinite-dimensional Lie Algebras*, 3rd Edition, Cambridge University Press, 1990.

[M] E. Markman, *Spectral curves and integrable systems*, Compositio Math. **93** (1994) 255–290.

[MV1] E. Mukhin and A. Varchenko, *Critical points of master functions and flag varieties*, Preprint math.QA/0209017.

[MV2] E. Mukhin and A. Varchenko, *Multiple orthogonal polynomials and a counterexample to Gaudin Bethe Ansatz Conjecture*, Preprint math.QA/0501144.

[RV] N. Reshetikhin and A. Varchenko, *Quasiclassical asymptotics of solutions of the KZ equations*, in *Geometry, Topology and Physics for Raoul Bott*, pp. 293–322, International Press, 1994.

[S1] I. Scherbak, *Rational functions with prescribed critical points*, Geom. Anal. Funct. Anal. **12** (2002) 1–16.

[S2] I. Scherbak, *A theorem of Heine-Stieltjes, the Wronski map, and Bethe vectors in the sl_p Gaudin model*, Preprint math.AG/0211377.

[SV] I. Scherbak and A. Varchenko, *Critical points of functions, sl_2 representations, and Fuchsian differential equations with only univalued solutions*, Moscow Math. J. **3** (2003), no. 2, 621–645.

[Sk] E. Sklyanin, *Separation of variables in the Gaudin model*, J. Soviet Math. **47** (1989) 2473–2488.

[V] A. Varchenko, *Critical points of the product of powers of linear functions and families of bases of singular vectors*, Compositio Math. **97** (1995), 385–401.

Edward Frenkel
Department of Mathematics
University of California
Berkeley
CA 94720, USA

Progress in Mathematics, Vol. 237, 59–87

Integrable Models with Unstable Particles

Olalla Castro-Alvaredo and Andreas Fring

Abstract. We review some recent results concerning integrable quantum field theories in $1+1$ space-time dimensions which contain unstable particles in their spectrum. Recalling first the main features of analytic scattering theories associated to integrable models, we subsequently propose a new bootstrap principle which allows for the construction of particle spectra involving unstable as well as stable particles. We describe the general Lie algebraic structure which underlies theories with unstable particles and formulate a decoupling rule, which predicts the renormalization group flow in dependence of the relative ordering of the resonance parameters. We extend these ideas to theories with an infinite spectrum of unstable particles. We provide new expressions for the scattering amplitudes in the soliton-antisoliton sector of the elliptic sine-Gordon model in terms of infinite products of q-deformed gamma functions. When relaxing the usual restriction on the coupling constants, the model contains additional bound states which admit an interpretation as breathers. For that situation we compute the complete S-matrix of all sectors. We carry out various reductions of the model, one of them leading to a new type of theory, namely an elliptic version of the minimal $SO(n)$-affine Toda field theory.

Mathematics Subject Classification (2000). 81U15, 81U20, 81T10, 81T40, 81R10.

Keywords. Exactly and quasi-solvable systems, S-matrix theory, model quantum field theories, two-dimensional field theories, conformal field theories, infinite-dimensional groups and algebras motivated by physics, including Virasoro, Kac-Moody.

1. Introduction

The structure of integrable quantum field theories (IQFT) in $1+1$ space-time dimensions has been unravelled to a very large extend. Many theories can be solved even exactly, that is to all orders in perturbation theory, in this context. However,

We thank the organizers of the workshop on "Infinite-dimensional algebras and quantum integrable systems" (Faro, Portugal, July, 2003), especially Nenad Manojlovic, for their kind hospitality and untiring engagement to make things work.

the large majority of investigations concentrates on theories which involve exclusively stable particles, despite the fact that in nature most particles are unstable. Since of course one of the motivations to study IQFT is to reproduce realistic features, there is an apparent need to investigate also theories which have unstable particles in their spectrum. The aim of this talk is to review some recent results which deal with such theories.

2. Analytic scattering theory of factorizable integrable models

Since not all participants of this conference work directly on integrable quantum field theories, we briefly recall some well-known facts on analytic scattering theories in $1+1$ space-time dimensions. Having in mind to emphasize features related to unstable particles this will also be useful to the experts. As a starting point in every scattering theory one requires a complete set of asymptotic in and out states ($t \to \pm\infty$). These states consist of operators $Z_\mu(p)$ acting on the vacuum $|0\rangle$ and creating in this way a stable particle of the type μ with momentum p. Already at this point enters the fundamental difference between stable and unstable particles. Even though experimentally unstable particles with a very long lifetime can very often not be distinguished from stable ones, mathematically they are very distinct. They can never be associated to an asymptotic state, even when they have an extremely long lifetime, as by their very nature they will have decayed in the infinite future or were never produced in the infinite past. Then the scattering matrix is defined to be the operator which relates a stable n-particle in state to a stable m-particle out state

$$Z_{\mu_m}(\theta_m^{'})\ldots Z_{\mu_1}(\theta_1^{'})\,|0\rangle_{\text{out}} = S^{\mu_1\mu_2\ldots\mu_n}_{\mu_1\mu_2\ldots\mu_m}(\theta_1^{'},\ldots,\theta_n) Z_{\mu_1}(\theta_1)\ldots Z_{\mu_n}(\theta_n)\,|0\rangle_{\text{in}}\,. \tag{2.1}$$

Conveniently one parameterizes the two-momentum by the rapidity θ as $\vec{p} = m(\cosh\theta, \sinh\theta)$. Now there are some very special features happening in integrable (that means here there exists at least one non-trivial conserved charge) quantum field theories in $1+1$ dimensions [1, 2, 3, 4, 5]. There is no particle production and furthermore the incoming and outgoing momenta coincide

$$\{\theta_1^{'},\theta_2^{'},\ldots,\theta_m^{'}\} = \{\theta_1,\theta_2,\ldots,\theta_n\} \qquad \text{with } n=m\,. \tag{2.2}$$

In addition, the n-particle S-matrix factorizes into a set of two-particle S-matrices

$$S^{\mu_1\mu_2\ldots\mu_n}_{\mu_1\mu_2\ldots\mu_m}(\theta_1^{'},\ldots,\theta_n) = \prod_{1\le i<j\le n} S_{\mu_i\mu_j}(\theta_i,\theta_j)\,. \tag{2.3}$$

Obviously, this is a considerable simplification in comparison with the general situation (2.1), as it means that once we know the two-particle S-matrix, we control the entire scattering matrix. Because of this fact, we refer from now on to the two-particle scattering matrix as *the* S-matrix.

How does one construct this S-matrix? In general one is limited to the use of perturbation theory in the coupling constant. In particular in higher dimensions that is essentially the only method available. In contrast, two dimensions are very special as they miraculously allow to determine S exactly to all orders in perturbation theory. This is one of the major successes of this area of research and one of the reasons for the continued interest in such theories. The original ideas which lead to explicit expressions for S go back to what is called the bootstrap approach [6, 1, 2]. It consists of using various properties for the scattering matrix, which one motivates by some physical principles in order to set up an axiomatic system for S in the hope that it might be so constraining that it determines S completely. Indeed, these hopes are not in vain.

We recall the S-matrix properties:

i) Lorentz invariance

Dealing with relativistic scattering theories, we expect the scattering matrix to be Lorentz invariant, i.e., it should depend only on covariant combinations of the momenta. The Mandelstam variables are precisely such quantities, see, e.g., [7]. In $1+1$ dimensions only one of them is independent, usually taken to be $s_{ab} = (p_a + p_b)^2 = m_a^2 + m_b^2 + 2m_a m_b \cosh(\theta_a - \theta_b)$. Hence, Lorentz invariance is simply guaranteed when S depends either only on s_{ab} or the rapidity difference $\theta_{ab} := \theta_a - \theta_b$

$$S_{ab}(p_a, p_b) = S_{ab}(\theta_a, \theta_b) = S_{ab}(s_{ab}) = S_{ab}(\theta_{ab}). \tag{2.4}$$

Since s_{ab} admits the interpretation as the total energy in the center of mass system, θ_{ab} has to be real for a physical process, such that $s_{ab} \geq (m_a + m_b)^2$.

ii) Hermitian analyticity

As a central assumption of analytic S-matrix theory [7] one assumes that the S-matrix can be continued to the complex plane and depends on $s_{ab}, \theta_{ab} \in \mathbb{C}$. Physical scattering amplitudes are then assumed to be real boundary values of analytic functions, which can be obtained from a generalization of Feynman's $i\varepsilon$ prescription of perturbation theory

$$S_{ab}^{\text{physical}} = \lim_{\varepsilon \to 0} S_{ab}(s + i\varepsilon) = S_{ab}(\theta) \qquad s \in \mathbb{R}, \varepsilon, \theta \in \mathbb{R}^+. \tag{2.5}$$

The choice of the signs is important and relates to causality. Since a two-particle wave function, having here plane waves in mind modulated by some enveloping function, will depend on the sum of the momenta, i.e., on $\sqrt{s_{ab}}$, one has lost the single valuedness of the scattering matrix by an analytic continuation. This is remedied by branch cuts along the real axis at $s_{ab} \geq (m_a + m_b)^2$ and $s_{ab} \leq (m_a - m_b)^2$, the latter being motivated by crossing see iv). Hermitian analyticity is now a postulate which states how to continue over these cuts [8, 9]

$$\lim_{\varepsilon \to 0} S_{ab}(s + i\varepsilon) = \lim_{\varepsilon \to 0} S_{ab}(s - i\varepsilon) \quad \Leftrightarrow \quad S_{ab}(\theta) = [S_{ba}(-\theta^*)]^* \ . \tag{2.6}$$

once more for $s \in \mathbb{R}, \varepsilon, \theta \in \mathbb{R}^+$. The equivalence is due to the fact that the analytic continuation $s + i\varepsilon \leftrightarrow s - i\varepsilon$ corresponds to $\theta \leftrightarrow -\theta$. Often one merely uses real analyticity $S_{ab}(\theta) = [S_{ab}(-\theta^*)]^*$ instead of (2.6), which only coincides when $S_{ab} = S_{ba}$, that is in parity invariant theories. This difference is very important with regard to the theories consider below, which involve unstable particles as they unavoidably break parity invariance. Further support for (2.6) comes from perturbation theory [8], general considerations in analytic S-matrix theory [10, 7] and explicitly constructed examples.

iii) Unitarity

Assuming that the states in (2.1) are complete and orthogonal, the operator which maps them to each other has to be unitary

$$SS^{\dagger} = S^{\dagger}S = 1 \; . \tag{2.7}$$

The combination of (2.6) and (2.7) leads to the simpler relation $S_{ab}(\theta)S_{ba}(-\theta) = 1$, which may also be derived from applying twice the Zamolodchikov algebra $Z_a(\theta_1)Z_b(\theta_2) = S_{ab}(\theta_{12})Z_b(\theta_2)Z_a(\theta_1)$.

iv) Crossing symmetry

Crossing symmetry can be motivated by the Lehmann-Symanzik-Zimmermann (LSZ) formalism [11] and consists of the replacement of an incoming particle a by its anti-particle $\bar{a}$ with reversed momentum. A discussion of the anti-particle theorem can be found in [10]. The prescription amounts to the continuation of the Mandelstam variable s_{ab} to the variable $t_{ab} = (p_a - p_b)^2$

$$\lim_{\varepsilon \to 0} S_{ab}(s + i\varepsilon) = \lim_{\varepsilon \to 0} S_{b\bar{a}}(t - i\varepsilon) \quad \Leftrightarrow \quad S_{b\bar{a}}(\theta) = S_{ab}(i\pi - \theta) \; . \tag{2.8}$$

It is easy to check that the analytic continuation $s + i\varepsilon \leftrightarrow t - i\varepsilon$ corresponds to $\theta \leftrightarrow i\pi - \theta$.

v) Yang-Baxter equation

In (2.3) we already indicated that the conserved charge(s) of an integrable theory can be used to disentangle an n-particle scattering process into a consecutive scattering of two particles only. An additional consequence of this argument is that the order in which this takes place does not matter, such that two different orderings are taken to be equivalent. As in general the S-matrices do not commute, this leads to a new constraint. In other words this amounts to say that the operators Z in (2.1) obey an associative algebra. As a result of this one obtains the Yang-Baxter equation [12, 13]

$$S(\theta_{12}) \otimes S(\theta_{13}) \otimes S(\theta_{23}) = S(\theta_{23}) \otimes S(\theta_{13}) \otimes S(\theta_{12}) \; . \tag{2.9}$$

For diagonal theories, i.e., when backscattering is absent, we simply have $S_{ab}^{cd}(\theta) \to S_{ab}(\theta)$ such that (2.9) is trivially satisfied.

vi) Fusing bootstrap equation

By the same reasoning as in v) integrability, i.e., factorizability, yields a further constraining equation, when two particles are allowed to form a bound state (what that means is discussed in vii)). For instance, the particles a, b fuse to a third particle $\bar{c}$, i.e., $a + b \to \bar{c}$. One makes now a further assumption, sometimes referred to as nuclear democracy, namely that also the particle of type $\bar{c}$ exists asymptotically.

Then, by integrability, it is equivalent if a third particle, say l, scatters with the bound state $\bar{c}$ or consecutively with the two particles a, b. For S this reads

$$S_{l\bar{c}}(\theta) = S_{la}(\theta + i\bar{\eta}_{ac}^{b})S_{lb}(\theta - i\bar{\eta}_{bc}^{a}) \ . \tag{2.10}$$

The $\bar{\eta}_{ac}^{b} \in \mathbb{R}^{+}$ are the fusing angles specific to the individual theory considered. It is clear that the assumption of nuclear democracy does not hold if $\bar{c}$ is an unstable particle, such that (2.10) cannot be valid in the form stated for that case. We will now indicate the origin for the possibility to form bound states, which is the

vii) Pole structure

In general, the S-matrix can have a quite intricate singularity structure consisting of poles of finite order distributed all over the complex s, θ-plane. A strong further constraint is to assume that all singularities which emerge in S admit a consistent explanation. As a slightly weaker assumption one could suppose that all explainable poles form a coherent system, in the sense that the bootstrap (2.10) closes etc., and allow some redundant poles.

Single order poles are most important as they determine the particle spectrum of the theory. In the s-plane they might be on the real axis between the two branch cuts at $s = m_{\bar{c}}^2$, interpreted as an on-shell bound state particle, or in the second Riemann sheet at $s = (m_{\bar{c}} - i\Gamma_{\bar{c}}/2)^2$ corresponding to an unstable particle with finite lifetime $\tau = 1/\Gamma_{\bar{c}}$. The discussion is more conveniently carried out in the θ-plane, since $S(\theta)$ is a meromorphic function unlike $S(s)$. Near the singularity S has to be of the form

$$S_{ab}(\theta) \sim \frac{iR_{ab}^{c}}{(\theta - i\eta_{ab}^{c} + \sigma_{ab}^{c})} \ . \tag{2.11}$$

Depending on the location and signs of the residues we have the following interpretations

s-channel bound state:	$R_{ab}^{c} \in \mathbb{R}^{+}, \eta_{ab}^{c} \in \mathbb{R}^{+}, \sigma_{ab}^{c} = 0$
t-channel bound state:	$R_{ab}^{c} \in \mathbb{R}^{-}, \eta_{ab}^{c} \in \mathbb{R}^{+}, \sigma_{ab}^{c} = 0$
unstable particle:	$R_{ab}^{c} \in \mathbb{R}, \eta_{ab}^{c} \in \mathbb{R}^{-}, \sigma_{ab}^{c} \in \mathbb{R}^{-}$

The relation between the poles in the s and θ planes are the Breit-Wigner (BW) equations [14]

$$m_{\bar{c}}^2 - \frac{\Gamma_{\bar{c}}^2}{4} = m_a^2 + m_b^2 + 2m_a m_b \cosh\sigma_{ab}^{\bar{c}} \cos\eta_{ab}^{\bar{c}} \tag{2.12}$$

$$m_{\bar{c}}\Gamma_{\bar{c}} = 2m_a m_b \sinh\sigma_{ab}^{\bar{c}} \sin\eta_{ab}^{\bar{c}} \ , \tag{2.13}$$

which allow to express the mass $m_{\bar{c}}$ and decay width $\Gamma_{\bar{c}}$ of the unstable particle as functions of $m_a, m_b, \eta_{ab}^{\bar{c}}, \sigma_{ab}^{\bar{c}}$. For the stable particle formation we have the following relation between the fusing angles in (2.10) and the poles in (2.11): $\bar{\eta}_{ac}^{b} = \pi - \eta_{ab}^{c}$. Note further that for the unstable particle formation in (2.11) we made the definite choice that the unstable particle $\bar{c}$ is formed in the process

$$a + b \to \bar{c} \tag{2.14}$$

rather than in $b+a$, which is not equivalent to $a+b$. It is clear that parity has to be broken, as with the choice $\eta_{ab}^{c}, \sigma_{ab}^{c} \in \mathbb{R}^{-}$ the amplitude $S_{b\bar{a}}(\theta)$ will have a pole at $i\pi - i\eta_{ab}^{c} + \sigma_{ab}^{c}$, leaving in (2.13) the choice that either $m_{\tilde{c}} < 0$ or $\Gamma_{\tilde{c}} < 0$, which is of course both non-physical.

Below we will be particularly interested in the situation for large resonance parameters $\sigma_{ab}^{\bar{c}}$, when the mass of the unstable particles can be approximated as

$$m_{\bar{c}} \sim \sqrt{m_a m_b} e^{-\sigma_{ab}^{\bar{c}}/2} . \tag{2.15}$$

In terms of perturbation theory in a coupling constant β, the pole (2.11) would be of second order, i.e., $R_{ab}^{c}(\beta^2)$, corresponding to a tree diagram. Similarly, higher order poles admit interpretations in form of more complicated singular Feynman diagrams. In some simple theories, such as for example sine-Gordon, the highest order of the poles is two. In that context it was suggested [15] that such type of poles are of order β^4 and may be viewed as box diagrams. For quite some time higher order poles were ignored and also here we will not enter into a deeper discussion of them, which can be found for instance in [16, 17]. From what is said it is clear that such poles will not alter the particle spectrum. Nonetheless, one should be able to draw the relevant Feynman diagrams, which means one needs certain three-point couplings to be non-vanishing. Consequently this is a constraint on the existence of certain three point couplings R_{ab}^{c}.

Remarkably, *the constraints* i)–vii) *allow to determine the S-matrix exactly, that is to all orders in perturbation theory.* However, one should say that the solution constructed this way is not unique, as there exists always the possibility to multiply with so-called CDD-factors [18]. To fix them requires additional arguments beyond the scheme outlined above, such as ultraviolet limits, certain inputs from Lagrangians, etc.

2.1. A proposal for a construction principle of unstable particle spectra

We have seen in the previous section, that there exists a powerful construction principle for the spectrum of stable particles, consisting of solving the equations (axioms) i)–vii). For unstable particles we do not have yet such a construction tool, as by now they emerge rather passively as poles in the unphysical sheet as by-products in the scattering process of two stable particles. Furthermore, a description of the scattering process of an unstable particle with another stable or unstable particle is entirely missing in this context. Obviously, scattering processes involving unstable particles do occur in nature, such that the quest for a proper prescription is of physical relevance. In addition, one aims of course always at a description which has predictive power.

From what has been said, it is clear that such a description can not be a scattering theory in the usual sense, since for that one requires the particles involved to exist asymptotically. Any unstable particle will vanish in this limit rendering such formulation meaningless at first sight. Nonetheless, some particles have extremely long lifetimes, and seem to exist quasi infinitely long from an experimentalists point of view. It appears therefore natural to seek a principle closely related to the conventional bootstrap for stable particles. Inspired by this we proposed [19] the following construction principle:

Let us assume that in the time interval $0 < t < \tau_{\bar{c}}$ we can formally associate to the unstable particle some creation operator $\tilde{Z}_{\bar{c}}^{\dagger}(\theta)$, with $\lim_{t\to\infty} \tilde{Z}_{\bar{c}}^{\dagger}(\theta) = 1$ if $\tau_{\bar{c}} < \infty$. It is clear that these operators do not exist asymptotically, but for the stated time interval they can mimic an asymptotic state. Let us now further suppose that these operators satisfy a Zamolodchikov algebra

$$Z_a(\theta_1)\tilde{Z}_b(\theta_2) = \tilde{S}_{ab}(\theta_{12})\tilde{Z}_b(\theta_2)Z_a(\theta_1) \tag{2.16}$$

$$\tilde{Z}_a(\theta_1)\tilde{Z}_b(\theta_2) = \tilde{S}_{ab}(\theta_{12})\tilde{Z}_b(\theta_2)\tilde{Z}_a(\theta_1) \tag{2.17}$$

which can be used to generate an S-matrix type of amplitude $\tilde{S}_{ab}$, describing the scattering of one unstable particle with a stable one (2.16) or the scattering of two unstable particles (2.17). We may proceed as before and ask which type of properties might be satisfied for $\tilde{S}$.

i) Lorentz invariance

As already indicated in (2.16), (2.17) it is natural to expect Lorentz invariance also for this amplitude such that $\tilde{S}$ depends only the rapidity difference θ_{ab}

$$\tilde{S}_{ab}(p_a, p_b) = \tilde{S}_{ab}(\theta_a, \theta_b) = \tilde{S}_{ab}(s_{ab}) = \tilde{S}_{ab}(\theta_{ab}). \tag{2.18}$$

ii, iii) Hermitian analyticity, unitarity

We will not make any assumption on hermitian analyticity here and in fact we do not expect unitarity to hold, since the states formed with the $\tilde{Z}$ are not complete. However, applying (2.16) or (2.17) twice yields

$$\tilde{S}_{ab}(\theta)\tilde{S}_{ba}(-\theta) = 1, \tag{2.19}$$

which also holds for S, derivable from combining (2.6) and (2.7) in that case. In fact, also for the construction of S it is really only the corresponding equation to (2.19) which is employed, rather than individually (2.6) and (2.7).

iv) Crossing symmetry

The validity of crossing can also be argued as before, but now we have to continue as

$$\lim_{\varepsilon\to 0} \tilde{S}_{ab}(s - i\varepsilon) = \lim_{\varepsilon\to 0} \tilde{S}_{b\bar{a}}(t + i\varepsilon) \quad \Leftrightarrow \quad \tilde{S}_{b\bar{a}}(-\theta) = \tilde{S}_{ab}(i\pi + \theta)\ , \tag{2.20}$$

which in the θ-plane amounts to the same equation as the one for S.

v) Yang-Baxter equation

Supposing the algebra related to (2.16), (2.17) is associative we have by the same reasoning as for stable particles the Yang-Baxter equation

$$\tilde{S}(\theta_{12}) \otimes \tilde{S}(\theta_{13}) \otimes \tilde{S}(\theta_{23}) = \tilde{S}(\theta_{23}) \otimes \tilde{S}(\theta_{13}) \otimes \tilde{S}(\theta_{12})\ . \tag{2.21}$$

vi) Fusing bootstrap equation

We commence with the fusing of two stable particles to create an unstable particle as in the process (2.14). To this process we can associate bootstrap equations almost in the usual way. We scatter for this with an additional stable or unstable particle, say of type l, and obtain the $\tilde{S}$ bootstrap equations

$$\tilde{S}_{la}(\theta - \bar{\gamma}_{ca}^{\bar{b}})\, \tilde{S}_{lb}(\theta + \bar{\gamma}_{bc}^{\bar{a}}) = \tilde{S}_{l\bar{c}}(\theta), \tag{2.22}$$

where $\bar{\gamma} = \pm i\pi - \gamma$, $\gamma = i\eta - \sigma$ and also $\bar{\gamma} \to -\bar{\gamma}$ is not a symmetry. The angles should be measured anti-clockwise, which explains the signs. We also note that we do not assume parity invariance, such that in general $\bar{\gamma}_{ba}^{\bar{c}} \neq \bar{\gamma}_{ab}^{\bar{c}}$. With the help of (2.19), (2.20) one derives the bootstrap equations for the opposite parity and the ones for the crossed processes $a + c \to \bar{b}$ and $b + c \to \bar{a}$ and from (2.22)

$$\tilde{S}_{\bar{c}l}(\theta) = \tilde{S}_{al}(\theta + \bar{\gamma}_{ca}^{\bar{b}})\, \tilde{S}_{bl}(\theta - \bar{\gamma}_{bc}^{\bar{a}}), \tag{2.23}$$

$$\tilde{S}_{l\bar{j}}(\theta) = \tilde{S}_{lc}(\theta - \bar{\gamma}_{bc}^{\bar{a}})\, \tilde{S}_{la}(\theta \pm i\pi - \bar{\gamma}_{ca}^{\bar{b}} - \bar{\gamma}_{bc}^{\bar{a}}), \tag{2.24}$$

$$\tilde{S}_{l\bar{i}}(\theta) = \tilde{S}_{lc}(\theta + \bar{\gamma}_{ca}^{\bar{b}})\tilde{S}_{lb}(\theta \pm i\pi + \bar{\gamma}_{ca}^{\bar{b}} + \bar{\gamma}_{bc}^{\bar{a}}). \tag{2.25}$$

From the crossing relation for the "scattering matrix" and (2.24) or (2.25) one obtains some relations between the various fusing angles

$$\bar{\gamma}_{ab}^{\bar{c}} + \bar{\gamma}_{ca}^{\bar{b}} + \bar{\gamma}_{bc}^{\bar{a}} = \pm i\pi . \tag{2.26}$$

At first sight this looks very much like the usual bootstrap prescription, but there are some differences. As is clear from the scattering process of two stable particles producing an unstable one, the angle $\bar{\gamma}_{ab}^{\bar{c}}$ is not purely complex any longer as it is for the situation when exclusively stable particles scatter. As a consequence, this property then extends to the other angles $\bar{\gamma}_{ca}^{\bar{b}}$ and $\bar{\gamma}_{bc}^{\bar{a}}$ in (2.22), which also possess some non-vanishing real parts. Note that (2.26) implies that the real parts of the three angels involved add up to zero. At this point we do not have an entirely compelling reason for demanding that, but this formulation will turn out to work well.

Of course the above equations are only a proposal, which needs to be put on more solid ground. Nonetheless, at this point our proposal gains support from self-consistency and its predictive power, which may be double checked: a) The bootstrap closes consistently for many non-trivial examples, which we calculated. As for stable particles this is never guaranteed and by no means self-evident. b) The bootstrap yields the amount of unstable particles together with their mass. This prediction can be used to explain a mass degeneracy of some unstable particles which can not be seen in a thermodynamic Bethe ansatz (TBA) analysis for the concrete example of the homogeneous sine-Gordon (HSG) models, see below. c) The bootstrap is in agreement with a general Lie algebraic decoupling rule, which we also present below, describing the behavior when certain resonance parameters tend to infinity. d) The bootstrap yields the three-point couplings of all possible interactions, that is, involving stable as well as unstable particles.

2.2. An example: The $\mathbf{g}_k$-HSG model

The $\mathbf{g}_k$-homogeneous sine-Gordon models (HSG) [20, 21], with $\mathbf{g}$ being a simple Lie algebra of rank ℓ and level k, will be our standard example in what follows. In fact they have been the first models with a well-defined Lagrangian containing unstable particles which have been the subject of a systematic analysis [22, 23, 24,

25, 26, 27, 28, 19, 29]. They can be viewed as perturbed conformal field theories (CFTs)[1]

$$\mathcal{H}_{G_k\text{-HSG}} = \mathcal{H}_{G_k/U(1)^\ell\text{-CFT}} - \lambda \int d^2x \phi(x,t) \,. \tag{2.27}$$

The underlying ultraviolet CFT is a Wess-Zumino-Novikov-Witten-$G_k/U(1)^\ell$-coset theory [31, 32]. The corresponding Virasoro central charge c is computed with standard arguments of [32] and the perturbing operator ϕ is identified with a primary field of conformal dimensions $\Delta, \bar{\Delta}$. One finds

$$c = \ell \, \frac{k\,h - h^\vee}{k + h^\vee} \qquad \text{and} \qquad \Delta = \bar{\Delta} = \frac{h^\vee}{k + h^\vee} \,. \tag{2.28}$$

Here $(h^\vee)\, h$ is the (dual) Coxeter number of $\mathbf{g}$. For simplicity we will drop in the following the explicit mentioning of the subalgebra $U(1)^\ell$ which were indicated in (2.27).

The scattering matrix for $\mathbf{g}_k$-HSG-models with $\mathbf{g}$ simply laced algebras was constructed in [22]. For $k = 2$ it can be brought into the simple form

$$S_{ij}(\theta, \sigma_{ij}) = (-1)^{\delta_{ij}} \,\varepsilon(\sigma_{ij}) (\sigma_{ij}, 2)^{I_{ij}} \,, \qquad 1 \le i,j \le \ell \tag{2.29}$$

where I denotes the incidence matrix of $\mathbf{g}$ and $\varepsilon(x)$ is the step-function, i.e., $\varepsilon(x) = 1$ for $x \ge 0$, $\varepsilon(x) = -1$ for $x < 0$. It is convenient to use the abbreviation

$$(\sigma, x) := \tanh(\theta + \sigma - i\pi x/4)/2 \,. \tag{2.30}$$

Let us now consider the concrete case $SU(3)_2$. We can start with the known part of the scattering matrix (2.29) for the stable particles, and leave the remaining entries which involve unstable particles unknown. From this we construct consistent solutions to the bootstrap equations (2.22), (2.24) and (2.25). We can fix the imaginary parts of the fusing angles by the requirement that for vanishing resonance parameters we want to reproduce the masses predicted by the Breit-Wigner formula. When choosing the masses of the stable particles to be $m_1 = m_2 = m$, the one for the unstable results to $m_{(12)} = \sqrt{2}m$. This argument does not constrain the real parts of the fusing angles, such that they are not completely fixed and still contain a certain ambiguity. The different choices of these parameters give rise to slightly different theories. First we consider the case $\sigma_{21} > 0$.

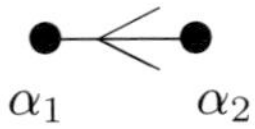

For this choice of the resonance parameter, we then find the following bootstrap equations

$$\tilde{S}_{l(12)}(\theta) = \tilde{S}_{l1}(\theta + (1-\nu)\sigma_{12} + i\pi/4)\tilde{S}_{l2}(\theta - \nu\sigma_{12} - i\pi/4) \tag{2.31}$$

[1] For the particular case of the $SU(3)_2$-HSG model it was shown [30] that it can be described alternatively as a perturbation of a tensor product of two minimal CFTs.

from which we construct

$$\tilde{S}_{SU(3)}(\theta,\sigma_{12}) = \begin{pmatrix} -1 & -(\sigma_{12},2) & -((1-\nu)\sigma_{12},3) \\ (\sigma_{21},2) & -1 & -(\nu\sigma_{21},1) \\ -((\nu-1)\sigma_{12},1) & -(\nu\sigma_{12},3) & -1 \end{pmatrix}. \tag{2.32}$$

Here we label the rows and columns in the order $\{1,2,(12)\}$. According to the principles outlined above, the S-matrix (2.32) allows for the processes

$$1+2 \to (12), \qquad 2+(12) \to 1, \qquad (12)+1 \to 2. \tag{2.33}$$

The related fusing angles are read off from (2.32) as

$$\gamma_{12}^{(12)} = -\frac{i\pi}{2} + \sigma_{21}, \quad \gamma_{(12)1}^{2} = -\frac{3i\pi}{4} + (1-\nu)\sigma_{12}, \quad \gamma_{2(12)}^{1} = -\frac{3i\pi}{4} + \nu\sigma_{12} \tag{2.34}$$

and are interrelated through equation (2.26), which still holds even though the γ's have non-vanishing real parts. We can employ these fusing angles and compute the masses and decay widths by means of the Breit-Wigner formulae (2.12) and (2.13). Taking again for simplicity $m_1 = m_2 = m$ and in addition $\nu = 1/2$, we obtain for the first process in (2.33)

$$m_{(12)} = \sqrt{2}m\cosh\sigma_{21}/2 \quad \text{and} \quad \Gamma_{(12)} = 2\sqrt{2}m\sinh\sigma_{21}/2\,. \tag{2.35}$$

Employing now also in the process $2+(12) \to 1$ the Breit-Wigner formula, we reproduce in the limit $\sigma_{12} \to 0$ the values $m_1 = m$ and $\Gamma_1 = 0$. Likewise, in the last process in (2.33) we obtain $m_2 = m$ and $\Gamma_2 = 0$.

The asymptotic limit $t \to \infty$ becomes meaningful when we operate on an energy scale at which the unstable particle has not even been created yet, i.e., $\Gamma_{(12)} \to \infty \equiv \sigma_{21} \to \infty$. In that case the theory decouples into two $SU(2)_2$-models, i.e., free fermions, with $S_{11} = S_{22} = -1$. This is a simple version of the decoupling rule (3.3).

Next we consider a different theory with $\sigma_{12} > 0$.

α_1 α_2

Taking also in this case for simplicity $\nu = 1/2$, we find the following bootstrap satisfied

$$\tilde{S}_{l(12)}(\theta) = \tilde{S}_{l2}(\theta - \sigma_{12}/2 + i\pi/4)\tilde{S}_{l1}(\theta + \sigma_{12}/2 - i\pi/4), \tag{2.36}$$

which yields the S-matrix

$$\tilde{S}_{SU(3)}(\theta,\sigma_{21}) = \begin{pmatrix} -1 & (\sigma_{12},2) & -(\sigma_{12}/2,1) \\ -(\sigma_{21},2) & -1 & -(\sigma_{21}/2,3) \\ -(\sigma_{21}/2,3) & -(\sigma_{12}/2,1) & -1 \end{pmatrix}. \tag{2.37}$$

The S-matrix (2.37) allows for the processes

$$2+1 \to (12), \qquad 1+(12) \to 2, \qquad (12)+2 \to 1, \tag{2.38}$$

instead of (2.33). Now the fusing angles are read off as

$$\gamma_{21}^{(12)} = -\frac{i\pi}{2} + \sigma_{12}, \quad \gamma_{1(12)}^{2} = -\frac{3i\pi}{4} - \frac{\sigma_{12}}{2}, \quad \gamma_{2(12)}^{1} = -\frac{3i\pi}{4} - \frac{\sigma_{12}}{2} \tag{2.39}$$

and also satisfy (2.26). The masses and decay width are obtained again from (2.12) and (2.13) with $\sigma_{12} \to \sigma_{21}$. As a whole, we can think of this theory simply as being obtained from the $\mathbb{Z}_2$-Dynkin diagram automorphism which exchanges the roles of the particles 1 and 2. However, since parity invariance is now broken this is not a symmetry any more and the two theories are different. In the asymptotic limit $\sigma_{12} \to \infty$, we obtain once again a simple version of the decoupling rule (3.3) and the theory decouples into two $SU(2)_2$-models.

The next example, $SU(4)_2$-HSG, is more intriguing as it leads to the prediction a new unstable particle. Proceeding in the way as before we construct the corresponding amplitudes $\tilde{S}$, for details see [19]. We found there the processes

$$\begin{array}{llllll} 1+2 \to & (12), & (12)+1 \to & 2, & 2+(12) \to & 1, \\ 3+2 \to & (23), & (23)+3 \to & 2, & 2+(23) \to & 3, \end{array} \tag{2.40}$$

which simply correspond to two copies of $SU(3)_2$-HSG. It is interesting to note that the amplitudes $\tilde{S}_{(12)3}$ and $\tilde{S}_{(23)1}$ contain poles at

$$\gamma_{(12)3}^{(123)} = \frac{\sigma_{21} - 2\sigma_{23}}{2} - \frac{3i\pi}{4} \quad \text{and} \quad \gamma_{(23)1}^{(123)} = \frac{\sigma_{23} - 2\sigma_{21}}{2} - \frac{3i\pi}{4}, \tag{2.41}$$

which yield the possible processes

$$\begin{array}{llllll} (12)+3 \to (123), & (123)+(12) \to & 3, & 3+(123) \to (12), \\ (23)+1 \to (123), & (123)+(23) \to & 1, & 1+(123) \to (23). \end{array} \tag{2.42}$$

An interesting prediction results from the consideration of the first two processes in (2.42). Making in the first process the particle (12) and in the second the particle (23) stable, by $\sigma_2 \to \sigma_1$ and by $\sigma_2 \to \sigma_3$, respectively, both predict the mass of the particle (123) as

$$m_{(123)} \sim m e^{|\sigma_{13}|/2}. \tag{2.43}$$

This value is precisely the one we expect from the approximation in the Breit-Wigner formula (2.15). Note that in one case we obtain σ_{13} and in the other σ_{31} as a resonance parameter. The difference results from the fact that according to the processes (2.42), the particle (123) is either formed as $(1+2)+3$ or $3+(2+1)$. Thus the different parity shows up in this process, but this has no effect on the values for the mass.

In [19] we presented more examples and remarkably we found consistency in each case. We take the closure of the bootstrap equations as a non-trivial confirmation for our proposal.

3. Lie algebraic structure for theories with unstable particles

There exist some concrete Lagrangian formulations for integrable theories with unstable particles, such as the aforementioned HSG-models (2.27). Inspired by the structure of these models, we present here a slightly more general Lie algebraic picture. We keep the discussion here abstract and supply below concrete examples. For our formulation we need an arbitrary simply laced Lie algebra $\tilde{\mathbf{g}}$ (possibly with a subalgebra $\tilde{\mathbf{h}}$) with rank $\tilde{\ell}$ together with its associated Dynkin diagram (see for instance [33]). To each node we attach a simply laced Lie algebra $\mathbf{g}_i$ with rank ℓ_i and to each link between the nodes i and j a resonance parameter $\sigma_{ij} = \sigma_i - \sigma_j$, as depicted in the following $\tilde{\mathbf{g}}/\tilde{\mathbf{h}}$-coset Dynkin diagram

Besides the usual rules for Dynkin diagrams, we adopt here the convention that we add an arrow to the link, which manifests the parity breaking and allows to identify the signs of the resonance parameters. An arrow pointing from the node i to j simply indicates that $\sigma_{ij} > 0$. Since we are dealing exclusively with simply laced Lie algebras, this should not lead to confusion. To each simple root of the algebras $\mathbf{g}_i$, we associate now a stable particle and to each positive non-simple root of $\tilde{\mathbf{g}}$ an unstable particle, such that

$$\#\text{ of stable particles} = \sum_{i=1}^{\tilde{\ell}} \ell_i, \qquad \#\text{ of unstable particles} = \frac{\tilde{\ell}\,(\tilde{h}-2)}{2}. \tag{3.1}$$

From the discussion above, we expect that the σ's will be associated to unstable particles, but we note that the

$$\#\text{ of resonance parameters} = \frac{\tilde{\ell}(\tilde{\ell}-1)}{2} \tag{3.2}$$

only agrees with the amount of unstable particles for $\tilde{h} = \tilde{\ell} + 1$, e.g., for $\tilde{\mathbf{g}} = SU(\tilde{\ell}+1)$. Since the resonance parameters govern the mass of the unstable particles, this discrepancy is interpreted as an unavoidable mass degeneracy.

Concrete examples for this formulations are the $\tilde{\mathbf{g}}_k$-homogeneous sine-Gordon models [20, 21], for which one can choose $\tilde{\mathbf{g}}$ to be simply laced and $\mathbf{g}_1 = \cdots = \mathbf{g}_{\tilde{\ell}} = SU(k)$. This is generalized [34] when taking instead $\tilde{\mathbf{g}}$ to be non-simply laced and $\mathbf{g}_i = SU(2k/\alpha_i^2)$, with α_i being the simple roots of $\tilde{\mathbf{g}}$. The choice $\mathbf{g}_1 = \cdots = \mathbf{g}_{\tilde{\ell}} = \mathbf{g}$ with $\mathbf{g}$ being any arbitrary simply laced Lie algebra gives the $\mathbf{g}|\tilde{\mathbf{g}}$-theories [35]. An example for a theory associated to a coset is the roaming sinh-Gordon model [36], which can be thought of as $\tilde{\mathbf{g}}/\tilde{\mathbf{h}} \equiv \lim_{k\to\infty} SU(k+1)/SU(k)$ with $\mathbf{g}_1 = \cdots = \mathbf{g}_{\tilde{\ell}} = SU(2)$. It is clear that the examples presented here do not exhaust yet all possible combinations and the structure mentioned above allows for more combinations of algebras, which are not yet explored. One is also not limited to Dynkin diagrams and may consider more general graphs which have multiple links, i.e., resonance parameters, between various nodes. Examples for such theories were proposed and studied in [37].

3.1. Decoupling rule

Of special interest is to investigate the behavior of previously defined systems when certain resonance parameters σ become very large or tend to infinity. The physical motivation for that is to describe a renormalization group (RG) flow, which we shall discuss in more detail below. Here we present first the mathematical set-up.

Decoupling rule: *Call the overall Dynkin diagram* $\mathcal{C}$ *and denote the associated Lie group and Lie algebra by* $\tilde{G}_{\mathcal{C}}$ *and* $\mathbf{g}_{\mathcal{C}}$*, respectively. Let* σ_{ij} *be some resonance parameter related to the link between the nodes* i *and* j*. To each node* i *attach a simply laced Lie algebra* $\mathbf{g}_i$*. Produce a reduced diagram* $\mathcal{C}_{ji}$ *containing the node* j *by cutting the link adjacent to it in the direction* i*. Likewise produce a reduced diagram* $\mathcal{C}_{ij}$ *containing the node* i *by cutting the link adjacent to it in the direction* j*. Then the* $\tilde{G}_{\mathcal{C}}$*-theory decouples according to the rule*

$$\lim_{\sigma_{ij}\to\infty} \tilde{G}_{\mathcal{C}} = \tilde{G}_{(\mathcal{C}-\mathcal{C}_{ij})} \otimes \tilde{G}_{(\mathcal{C}-\mathcal{C}_{ji})} / \tilde{G}_{(\mathcal{C}-\mathcal{C}_{ij}-\mathcal{C}_{ji})} \,. \tag{3.3}$$

We depict this rule also graphically in terms of Dynkin diagrams:

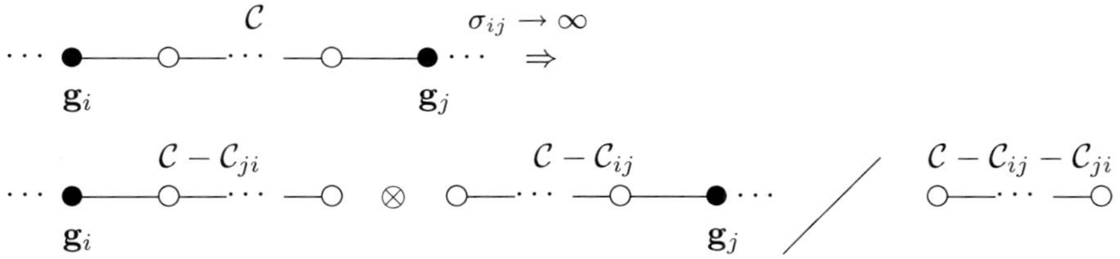

$$\underset{\mathbf{g}_i}{\cdots\bullet}\!-\!\!\circ\!-\!\cdots\overset{\mathcal{C}}{}-\!\circ\!-\!\!\underset{\mathbf{g}_j}{\bullet\cdots} \quad \overset{\sigma_{ij}\to\infty}{\Rightarrow}$$

$$\underset{\mathbf{g}_i}{\cdots\bullet}\!-\!\!\circ\!-\!\overset{\mathcal{C}-\mathcal{C}_{ji}}{\cdots}-\!\circ \;\otimes\; \circ\!-\!\overset{\mathcal{C}-\mathcal{C}_{ij}}{\cdots}-\!\circ\!-\!\!\underset{\mathbf{g}_j}{\bullet\cdots} \;\Big/\; \circ\!-\!\overset{\mathcal{C}-\mathcal{C}_{ij}-\mathcal{C}_{ji}}{\cdots}-\!\circ$$

According to the GKO-coset construction [32], this means that the Virasoro central charge flows as

$$c_{\tilde{\mathbf{g}}_{\mathcal{C}}} \to c_{\tilde{\mathbf{g}}_{\mathcal{C}-\mathcal{C}_{ij}}} + c_{\tilde{\mathbf{g}}_{\mathcal{C}-\mathcal{C}_{ji}}} - c_{\tilde{\mathbf{g}}_{\mathcal{C}-\mathcal{C}_{ij}-\mathcal{C}_{ji}}} \,. \tag{3.4}$$

The rule may be applied consecutively to each disconnected subgraph produced according to the decoupling rule (3.3). Note that this rule describes a decoupling and not a fusing, as it only predicts the flow in one direction and the limit is not reversible. From a physical point of view this is natural as analogously the RG flow is also irreversible. The rule (3.3) generalizes a rule proposed in [26], which was based on the assumption that unstable particles are associated exclusively to positive roots of height two.

More familiar in the mathematical literature is a decoupling rule found by Dynkin [38] for the construction of semi-simple[2] subalgebras $\tilde{\mathbf{h}}$ from a given algebra $\tilde{\mathbf{g}}$. For the more general diagrams which can be related to the $\tilde{\mathbf{g}}_k$-HSG models the generalized rule can be found in [39]. These rules are all based on removing some of the nodes rather than links. For our physical situation at hand this corresponds to sending the masses of all stable particles which are associated to the algebra of a particular node to infinity. As in the decoupling rule (3.3) the number of stable

[2]The subalgebras constructed in this way are not necessarily maximal and regular. A guarantee for obtaining those, except in six special cases, is only given when one manipulates adequately the extended Dynkin diagram.

particles remains preserved, it is evident that the two rules are inequivalent. Letting for instance the mass scale in $\mathbf{g}_j$ go to infinity, the generalized (in the sense that $\mathbf{g}_j$ can be different from A_ℓ) rule of Kuniba is simply depicted as

$$\begin{array}{ccccc} \mathcal{C} & m_j \to \infty & \mathcal{C}-\mathcal{C}_{ji} & & \mathcal{C}-\mathcal{C}_{jk} \\ \cdots \bullet\!-\!\!-\!\bullet\!-\!\!-\!\bullet \cdots & \Rightarrow & \cdots \bullet & \otimes & \bullet \cdots \\ \mathbf{g}_i \quad \mathbf{g}_j \quad \mathbf{g}_k & & \mathbf{g}_i & & \mathbf{g}_k \end{array}$$

Clearly this cannot be produced with (3.3).

3.2. A simple example: The $SU(4)_2$-HSG model

We illustrate the working of the rule (3.3) with a simple example. We take $\tilde{\mathbf{g}}$ to be $SU(4)$, attach to each node simply an $SU(2)$ algebra and to the links the resonance parameters $\sigma_{12}, \sigma_{13}, \sigma_{23}$. This corresponds to the $SU(4)_2$-HSG model. For the ordering $\sigma_{13} > \sigma_{12} > \sigma_{23}$ the rule (3.3) predicts the following flow

$$\begin{array}{lll} & \underset{\alpha_1}{\bullet}\!-\!\!-\!\underset{\alpha_2}{\bullet}\!-\!\!-\!\underset{\alpha_3}{\bullet} & \tilde{\mathbf{g}} = SU(4)_2 \qquad c = 2 \\ \to \sigma_{13} & \underset{\alpha_1}{\bullet}\!-\!\!-\!\underset{\alpha_2}{\bullet} \otimes \underset{\alpha_2}{\bullet}\!-\!\!-\!\underset{\alpha_3}{\bullet} \;\Big/\; \underset{\alpha_2}{\bullet} & \tilde{\mathbf{g}} = SU(3)_2^{\otimes 2}/SU(2)_2 \quad c = 1.9 \\ \to \sigma_{12} & \underset{\alpha_1}{\bullet} \otimes \underset{\alpha_2}{\bullet}\!-\!\!-\!\underset{\alpha_3}{\bullet} & \tilde{\mathbf{g}} = SU(3)_2 \otimes SU(2)_2 \quad c = 1.7 \\ \to \sigma_{23} & \underset{\alpha_1}{\bullet} \otimes \underset{\alpha_2}{\bullet} \otimes \underset{\alpha_3}{\bullet} & \tilde{\mathbf{g}} = SU(2)_2^{\otimes 3} \qquad c = 1.5 \end{array}$$

The central charges are obtained from (2.28) using (3.4). Choosing instead the ordering $\sigma_{23} > \sigma_{13} > \sigma_{12}$, we compute

$$\begin{array}{lll} & \underset{\alpha_1}{\bullet}\!-\!\!-\!\underset{\alpha_2}{\bullet}\!-\!\!-\!\underset{\alpha_3}{\bullet} & \tilde{\mathbf{g}} = SU(4)_2 \qquad c = 2 \\ \to \sigma_{23} & \underset{\alpha_1}{\bullet}\!-\!\!-\!\underset{\alpha_2}{\bullet} \otimes \underset{\alpha_3}{\bullet} & \tilde{\mathbf{g}} = SU(3)_2 \otimes SU(2)_2 \quad c = 1.7 \\ \to \sigma_{13} & \text{is already decoupled} & \\ \to \sigma_{12} & \underset{\alpha_1}{\bullet} \otimes \underset{\alpha_2}{\bullet} \otimes \underset{\alpha_3}{\bullet} & \tilde{\mathbf{g}} = SU(2)_2^{\otimes 3} \qquad c = 1.5 \end{array}$$

It is important to note the non-commutative nature of the limiting procedures. For more complicated algebras it is essential to keep track of the labels on the nodes, since only in this way one can decide whether they cancel against the subgroup diagrams or not.

3.3. A non-trivial example: The $(E_6)_2$-HSG model

As by now we do not have a rigorous proof of the decoupling rule (3.3), we take the support for its validity from the working of various examples. We will check below the analytic predictions of the rule against some alternative method. As the previous example was a simple pedagogical one, we will consider next a non-trivial one leading to an intricate prediction for the RG-flow. The confirmative double check below can hardy be accidental and we take that as very strong support for the validity of (3.3).

We consider now the $(E_6)_2$-HSG model. In this case we have $\tilde{\ell} = 6, \tilde{h} = 12$ such that we have 6 stable particles, 30 unstable particles and 15 resonance parameters. From the 5! possible orderings for the resonance parameters we present here only two concrete ones, which will predict different types of flows and mass degeneracies. Note that this degeneracy is not the unavoidable one resulting from the difference between the number of resonance parameters and non-simple positive roots that is 30 − 15, as discussed for (3.1) and (3.2). The degeneracies discussed here are a consequence of the particular choices of the resonance parameters. The conventions for the labelling of our particles are indicated in the following Dynkin diagram:

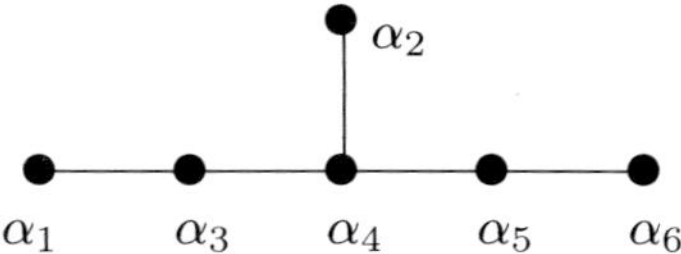

We choose first the ordering and values for resonance parameters as

$$\sigma_{13} = 100 > \sigma_{34} = 80 > \sigma_{45} = 60 > \sigma_{56} = 40 > \sigma_{24} = 20\,. \tag{3.5}$$

According to the decoupling rule (3.3), we predict therefore the flow:

	E_6	$\frac{36}{7} \sim 5.14$
$\rightarrow \sigma_{16} = 280$	$SO(10)^{\otimes 2}/SO(8)$	5
$\rightarrow \sigma_{15} = 240$	$SO(10) \otimes SU(5)/SU(4)$	$\frac{34}{7} \sim 4.86$
$\rightarrow \sigma_{14}, \sigma_{36} = 180$	$SO(8) \otimes SU(5) \otimes SU(3)/SU(4) \otimes SU(2)$	$\frac{319}{70} \sim 4.56$
$\rightarrow \sigma_{12} = 160$	is already decoupled	
$\rightarrow \sigma_{35} = 140$	$SU(5) \otimes SU(4) \otimes SU(3)/SU(3) \otimes SU(2)$	$\frac{61}{14} \sim 4.36$
$\rightarrow \sigma_{26} = 120$	$SU(4)^{\otimes 3} \otimes SU(3)/SU(3)^{\otimes 2} \otimes SU(2)$	4.3
$\rightarrow \sigma_{13}, \sigma_{46} = 100$	$SU(4)^{\otimes 2} \otimes SU(3) \otimes SU(2)/SU(3) \otimes SU(2)$	4
$\rightarrow \sigma_{25}, \sigma_{34} = 80$	$SU(3)^{\otimes 3} \otimes SU(2)^{\otimes 2}/SU(2)^{\otimes 2}$	3.6
$\rightarrow \sigma_{32}, \sigma_{45} = 60$	$SU(3)^{\otimes 2} \otimes SU(2)^{\otimes 2}$	3.4
$\rightarrow \sigma_{56} = 40$	$SU(3) \otimes SU(2)^{\otimes 4}$	3.2
$\rightarrow \sigma_{24} = 20$	$SU(2)^{\otimes 6}$	3

Note that eight particles are pairwise degenerate and we therefore expect to find $15 - 8/2 = 11$ plateaux in the flow. The first step which corresponds to one of these degeneracies occurs for instance at $\sigma_{14} = \sigma_{36}$ and we have to apply the decoupling rule twice at this point before we get a new fixed point theory.

Next we arrange the couplings as

$$\sigma_{45} = 100 > \sigma_{34} = 80 > \sigma_{13} = 60 > \sigma_{56} = 40 > \sigma_{24} = 20\,. \tag{3.6}$$

and compute from (3.3) the flow

$$\begin{array}{lll}
 & E_6 & \frac{36}{7} \sim 5.14 \\
\to \sigma_{16} = 280 & SO(10)^{\otimes 2}/SO(8) & 5 \\
\to \sigma_{15} = 240 & SO(10) \otimes SU(5)/SU(4) & \frac{34}{7} \sim 4.86 \\
\to \sigma_{36} = 220 & SU(5)^{\otimes 2} \otimes SO(8)/SU(4)^{\otimes 2} & \frac{33}{7} \sim 4.71 \\
\to \sigma_{35} = 180 & SU(5)^{\otimes 2}/SU(3) & \frac{158}{35} \sim 4.51 \\
\to \sigma_{26} = 160 & SU(4)^{\otimes 2} \otimes SU(5)/SU(3)^{\otimes 2} & \frac{156}{35} \sim 4.46 \\
\to \sigma_{14} = \sigma_{46} = 140 & SU(4)^{\otimes 2} \otimes SU(3)^{\otimes 2}/SU(2)^{\otimes 2} \otimes SU(3) & 4.2 \\
\to \sigma_{12} = \sigma_{25} = 120 & SU(4) \otimes SU(3)^{\otimes 3}/SU(2)^{\otimes 3} & 4.1 \\
\to \sigma_{45} = 100 & SU(4) \otimes SU(3)^{\otimes 2}/SU(2) & 3.9 \\
\to \sigma_{34} = 80 & SU(3)^{\otimes 3} & 3.6 \\
\to \sigma_{13} = \sigma_{32} = 60 & SU(3)^{\otimes 2} \otimes SU(2)^{\otimes 2} & 3.4 \\
\to \sigma_{56} = 40 & SU(3) \otimes SU(2)^{\otimes 4} & 3.2 \\
\to \sigma_{24} = 20 & SU(2)^{\otimes 6} & 3
\end{array}$$

In this case we have only six particles pairwise degenerate and we expect to find $15 - 6/2 = 12$ plateaux. In the next section we find that the predictions made here are confirmed even for this involved case.

4. How to detect unstable particles?

In Section 2 we described several arguments which predict the spectrum of unstable particles and now we will present some methods which allow to test these predictions. In particular with regard to the bootstrap proposal this will be important, as it is not yet rigorously supported. Computing renormalization group (RG) flows will allow to detect the unstable particles. Roughly speaking, the central idea of an RG analysis is to probe different energy scales of a theory. We can flow from an energy regime so large that the unstable particle can energetically not exist to one in which it is formed. As a consequence, the particle content of the theory will be altered, which is visible in form of a typical staircase pattern of the RG scaling function.

There are various ways to compute such scaling functions, such as the evaluation of the c-theorem [40] or an analysis by means of the thermodynamic Bethe ansatz (TBA) [41]. In the first case we have to evaluate the expression

$$c(r_0) = \frac{3}{2} \int_{r_0}^{\infty} dr\, r^3 \;\; \langle \Theta(r)\Theta(0)\rangle \;\; . \tag{4.1}$$

The main difficulty in this approach is to evaluate the two-point correlation function $\langle \Theta(r)\Theta(0)\rangle$ for the trace of the energy-momentum tensor Θ depending on the radial distance r. Most effectively, one can do this by expanding it in terms of form factors, for a general recent introduction see, e.g., [42] and references therein. It is well known that for many, even quite non-trivial, theories such form factor expansions converge extremely fast, see [27] for the computation of (4.1) for the $SU(3)_2$-HSG model.

Here we will concentrate more on the TBA, which is simpler to handle in most cases. As a prerequisite, one assumes to know all scattering matrices $S_{ij}(\theta)$ for the **stable** particles of the type i,j with masses m_i, m_j. Besides this dynamical interaction one also makes an assumption on the statistical interaction between the particles, which are chosen here to be of fermionic type. The TBA consists now of compactifying the space of this $1+1$-dimensional relativistic model into a circle of finite circumference R, such that all energies become discrete and functions of R. The function similar to (4.1), which scales now these energies takes on the form

$$c_{\text{eff}}(r) = \frac{3\,r}{\pi^2} \sum_i m_i \int_{-\infty}^{\infty} d\theta \, \cosh\theta \, \ln(1 + e^{-\varepsilon_i(\theta,r)})\,. \tag{4.2}$$

One identifies the circumference R with the inverse temperature T and introduces the scaling parameter $r = m/T$, with m being an overall mass scale. The $\varepsilon_i(\theta, r)$ are the so-called the pseudo-energies which can be obtained as solutions of the thermodynamic Bethe ansatz equations

$$r m_i \cosh\theta = \varepsilon_i(\theta, r) + \sum_j [\varphi_{ij} * \ln(1 + e^{-\varepsilon_j})](\theta, r)\,. \tag{4.3}$$

Here the $*$ denotes the convolution of two functions $(f * g)(\theta) := 1/(2\pi) \int d\theta' f(\theta - \theta')g(\theta')$ and the S (for the stable particles only!) enter via their logarithmic derivatives $\varphi_{ij}(\theta) = -i d \ln S_{ij}(\theta)/d\theta$. The main difficulty in this approach is to solve (4.3), which are coupled non-linear integral equations and therefore not solvable in a systematic analytical way.

Now it is clear, that the two functions (4.1) and (4.2) cannot be the same, but nevertheless they contain the same qualitative information. The functions will flow through various fixed points, at which the theory become effectively conformal field theories and the normalizations are chosen in such a way that the values of both functions coincide with the corresponding Virasoro central charges. When the theory is not unitary, (4.2) has to be corrected by an additive term to achieve this. Computing then a flow from the infrared to the ultraviolet, one passes now various CFT plateaux, where the changes are associated to the formation of unstable particles with mass (2.15). The challenge is of course to predict the positions, that is, the height and the on-set of the plateaux, as a function of the scaling parameter. The on-set is related to the energy scale of the unstable particles and thus simply determined by the formula (2.15). To predict the height is less trivial and the proposal made in [19] is that the decoupling rule (3.3) achieves this. It is important to note here that $\sigma \to \infty$ in (3.3), which means in the RG context $\sigma \gg$ all other resonance parameters. In the following picture we present the numerical computation for the $(E_6)_2$-HSG model, which precisely confirms our analytical predictions made by the decoupling rule in Section 3.3

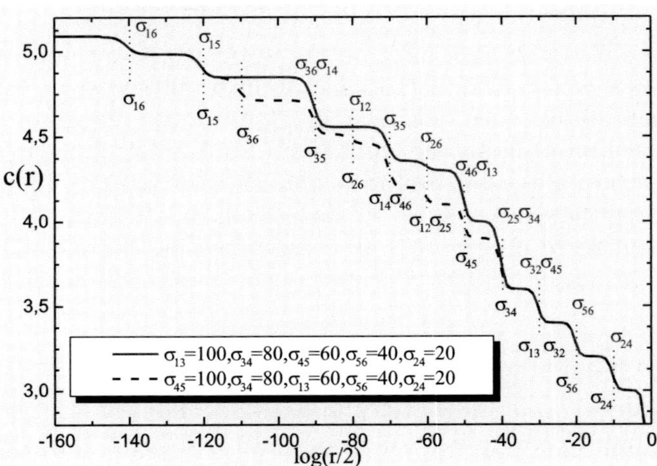

Having confirmed the predictions of our decoupling rule with a TBA-analysis, let us now discuss how the results of this analysis are compatible with our bootstrap proposal with a simple example: We consider the processes (2.40), (2.42). In order to be able to interpret the BW-formula for the production of the particle (123) from $(12)+3$ or $(23)+1$ one has to "make" (12) and (23) stable, which is achieved when σ_{12} or σ_{23} is zero. One has to do that as otherwise the BW cannot be applied, it only makes sense for stable particles. The first not obvious result here is that the resulting mass for (123) turns out to be the same from both cases (2.42) (and in all other examples!!!). Looking at the outcome of the TBA calculation (see [19] for the numerics on this case) one finds precisely the value (2.43) reproduced by the TBA at the onset $\ln(r/2) \sim -\sigma_{13}/2 = -(\sigma_{12}+\sigma_{23})/2$. Now apparently in the TBA analysis σ_{12} or σ_{23} are not zero, which seems to contradict the previous assumptions in the bootstrap analysis. To understand this, one should keep in mind the meaning of the steps in the TBA. The formation of the particle (123) takes place when its mass becomes greater than the energy scale of the RG-flow, i.e., when $m\exp(\sigma_{13}/2) > 2m/r$. Let us chose for instance $\sigma_{12} = 30$, $\sigma_{23} = 60$, then $\exp(\sigma_{13}/2) \sim \exp(45) \sim 3.49 \times 10^{19}$. To resolve the apparent contradiction, it is now important to note that the other unstable particles are formed several orders of magnitude below at $\exp(30) \sim 1.06 \times 10^{13}$ and $\exp(15) \sim 3.72 \times 10^{6}$. This means in comparison to the formation energy scale of particle (123) the parameters σ_{12}, σ_{23} can be regarded as approximately zero, which is in agreement with the assumption in the bootstrap analysis!

This is just the same picture as put forward in the decoupling rule: In the formulation we say $\sigma_{13} \to \infty$, but inside the TBA analysis this is a milder statement and just means $\sigma_{13} \gg \sigma_{12}, \sigma_{23}$. Further quite non-obvious confirmation comes from the results when choosing the parameters differently, i.e., in the example discussed here $\sigma_{12} \to -\sigma_{12}$. The two pictures completely coincide. With regard to previous studies, it is very important to note that the occurrence of the step at $\exp(\sigma_{13}/2) \sim 2/r$ had no explanation at all before. Only the onsets at $\exp(\sigma_{23}/2) \sim 2/r$ and $\exp(\sigma_{12}/2) \sim 2/r$ could be explained as they correspond to the formation of unstable particles from two stable ones. The additional step (for other algebras there are far more) was a mystery pointed out first in [29]. In [19] we provided for the first time an explanation for this feature: We predict its height and on-set, thus explaining also why it is absent when the resonance parameters are chosen differently. For all other examples studied (not even all have been presented in this proceeding, see [19] for more) this picture is completely consistent.

5. Theories with an infinite amount of unstable particles

We address now the question of how to enlarge a given finite particle spectrum of a theory to an infinite one. In general the bootstrap (2.10), which is the central construction principle for the S-matrix, is assumed to close after a finite number of steps, which means it involves a finite number of particles. However, from a physical as well as from a mathematical point of view, it appears to be natural to extend the construction in such a way that it would involve an infinite number of particles. The physical motivation for this are string theories, which admit an infinite particle spectrum. Mathematically the infinite bootstrap would be an analogy to infinite-dimensional groups, in the sense that two entries of the S-matrix are combined into a third, which is again a member of the same infinite set. It appears to us that it is impossible to construct an infinite bootstrap system involving asymptotic states, although we do not know a rigorous proof of such a no-go theorem. Instead, we will demonstrate that it is possible to introduce an infinite number of unstable particles into the spectrum.

5.1. q-deformed gamma functions and Jacobian elliptic functions

In general, the S-matrix amplitudes consist of (in)finite products of hyperbolic or/and gamma functions. Here we will argue, that to enlarge the spectrum to an infinite number one should replace these functions by q-deformed quantities or elliptic functions. Let us first recall some mathematical facts in this section. We start with some properties of q-deformed quantities, which have turned out to be very useful objects as they allow for instance to carry out elegantly (semi)-classical limits when the deformation parameter is associated to Planck's constant. Here we define a deformation parameter q and its Jacobian imaginary transformed version, i.e., $\tau \to -1/\tau$, as

$$q = \exp(i\pi\tau), \qquad \hat{q} = \exp(-i\pi/\tau), \qquad \tau = iK_{1-\ell}/K_\ell \,. \tag{5.1}$$

We introduced here the quarter periods K_ℓ of the Jacobian elliptic functions depending on the parameter $\ell \in [0,1]$, defined in the usual way through the complete elliptic integrals $K_\ell = \int_0^{\pi/2}(1-\ell\sin^2\theta)^{-1/2}d\theta$. Then

$$\lim_{\ell\to 0,\hat{q}\to 1} K_\ell = \lim_{\ell\to 1,q\to 1} K_{1-\ell} = \pi/2, \qquad \lim_{\ell\to 0,\hat{q}\to 1} K_{1-\ell} = \lim_{\ell\to 1,q\to 1} K_\ell \to \infty \ . \tag{5.2}$$

It will turn out below that quantities in $\hat{q}$ will be most relevant for our purposes and therefore we state several identities directly in $\hat{q}$, rather than q, even when they hold for generic values. The most basic q-deformed objects one defines are q-deformed integers (numbers), for which we take the convention

$$[n]_{\hat{q}} := \frac{\hat{q}^n - \hat{q}^{-n}}{\hat{q} - \hat{q}^{-1}}. \tag{5.3}$$

They have the obvious properties

$$\lim_{\ell\to 0}[n]_{\hat{q}} = n, \tag{5.4}$$

$$\lim_{\ell\to 0}\frac{[n+m\tau]_{\hat{q}}}{[n'+m'\tau]_{\hat{q}}} = \begin{cases} 1 & \text{for } m,m'\neq 0 \\ n/n' & \text{for } m=m'=0 \end{cases} \ . \tag{5.5}$$

Next we define a q-deformed version of Euler's gamma function

$$\Gamma_{\hat{q}}(x+1) := \prod_{n=1}^{\infty}\frac{[1+n]_{\hat{q}}^x[n]_{\hat{q}}}{[x+n]_{\hat{q}}[n]_{\hat{q}}^x} \ . \tag{5.6}$$

The crucial property of the function $\Gamma_{\hat{q}}$, which coins also its name, is

$$\lim_{\ell\to 0}\Gamma_{\hat{q}}(x+1) = \lim_{\hat{q}\to 1}\Gamma_{\hat{q}}(x+1) = \prod_{n=1}^{\infty}\frac{n}{n+x}\left(\frac{1+n}{n}\right)^x = \Gamma(x+1) \ . \tag{5.7}$$

We can relate deformations in q and $\hat{q}$ through

$$\frac{\hat{q}^{(x+\tau/2-1/2)^2}}{\hat{q}^{(y+\tau/2-1/2)^2}}\frac{\Gamma_{\hat{q}}(y)\Gamma_{\hat{q}}(1-y)}{\Gamma_{\hat{q}}(x)\Gamma_{\hat{q}}(1-x)} = \frac{\Gamma_q(-y/\tau)\Gamma_q(1+y/\tau)}{\Gamma_q(-x/\tau)\Gamma_q(1+x/\tau)} \ . \tag{5.8}$$

Frequently we have to shift the argument by integer values

$$\Gamma_{\hat{q}}(x+1) = \hat{q}^{x-1}[x]_{\hat{q}}\Gamma_{\hat{q}}(x) \ . \tag{5.9}$$

Relation (5.9) can be obtained directly from (5.6). As a consequence of this we also have

$$\Gamma_{\hat{q}}(x+m) = \Gamma_{\hat{q}}(x)\prod_{l=0}^{m-1}\hat{q}^{x+l-1}[x+l]_{\hat{q}} \qquad m\in\mathbb{Z} \tag{5.10}$$

$$\Gamma_{\hat{q}}(x) = \Gamma_{\hat{q}}(x-m)\prod_{l=0}^{m-1}\hat{q}^{x-l-2}[x-l-1]_{\hat{q}} \qquad m\in\mathbb{Z} \ . \tag{5.11}$$

Whereas (5.9)–(5.10) hold for generic q, the following identities are only valid for $\hat{q}$

$$\Gamma_{\hat{q}}(1/2-\tau/2)\Gamma_{\hat{q}}(1/2+\tau/2)=\ell^{1/4}\Gamma_{\hat{q}}(1/2)^2 \tag{5.12}$$

$$\frac{\Gamma_{\hat{q}}(x+2\tau)}{\Gamma_{\hat{q}}(y+2\tau)}=\frac{\Gamma_{\hat{q}}(x)}{\Gamma_{\hat{q}}(y)} \tag{5.13}$$

$$\prod_{i=1}^{p}\frac{\Gamma_{\hat{q}}(x_i)\Gamma_{\hat{q}}(x_i\pm\tau/2)}{\Gamma_{\hat{q}}(y_i)\Gamma_{\hat{q}}(y_i\pm\tau/2)}=\prod_{i=1}^{p}\frac{\Gamma_{\hat{q}^2}(x_i)}{\Gamma_{\hat{q}^2}(y_i)} \quad \text{if} \quad \sum_{i=1}^{p}x_i=\sum_{i=1}^{p}y_i \tag{5.14}$$

$$\lim_{\hat{q}\to 1}\prod_{i=1}^{p}\frac{\Gamma_{\hat{q}}(x_i\pm\tau/2)}{\Gamma_{\hat{q}}(y_i\pm\tau/2)}=1 \quad \text{if} \quad \sum_{i=1}^{p}x_i=\sum_{i=1}^{p}y_i \tag{5.15}$$

$$\lim_{\hat{q}\to 1}\frac{1}{\ell^{1/4}}\Gamma_{\hat{q}}(\frac{x}{2K_\ell}\mp\frac{\tau}{2})\Gamma_{\hat{q}}(1-\frac{x}{2K_\ell}\pm\frac{\tau}{2})=\pi \quad \text{for} \quad x\neq 0 \tag{5.16}$$

Most of these properties can be checked directly by means of the defining relation (5.6). The singularity structure will be important for the physical applications. It follows from (5.6) that the $\Gamma_{\hat{q}}$-function has no zeros, but poles

$$\lim_{\theta\to\theta_{\Gamma,p}^{nm}=m\tau-n}\Gamma_{\hat{q}}(\theta+1)\to\infty \qquad \text{for} \;\; m\in\mathbb{Z},n\in\mathbb{N}\;. \tag{5.17}$$

Next we define

$$\{x\}_\theta^\sigma:=\frac{\tanh(\theta-i\pi x+\sigma)/2}{\tanh(\theta+i\pi x+\sigma)/2},\quad \{x\}_{\theta,\ell}^\sigma:=\prod_{n=-\infty}^{\infty}\{x\}_{\theta-n\log q}^\sigma=\frac{\operatorname{sc}\theta_-\operatorname{dn}\theta_+}{\operatorname{sc}\theta_+\operatorname{dn}\theta_-} \tag{5.18}$$

with $x\in\mathbb{Q}$, $\sigma\in\mathbb{R}$ and $\theta_\pm=(\theta\pm i\pi x+\sigma)iK_\ell/\pi$. We employed here the Jacobian elliptic functions for which we use the common notation $\mathrm{pq}(z)$ with $p,q\in\{$s,c,d,n$\}$ (see, e.g., [43] for standard properties). We derive important relations between the q-deformed gamma functions and the Jacobian elliptic sn-function

$$\begin{aligned}\operatorname{sn}(x) &= \frac{1}{\ell^{\frac14}}\frac{\Gamma_{\hat{q}}(\frac{x}{2K_\ell}\mp\frac{\tau}{2})\Gamma_{\hat{q}}(1-\frac{x}{2K_\ell}\pm\frac{\tau}{2})}{\Gamma_{\hat{q}}(\frac{x}{2K_\ell})\Gamma_{\hat{q}}(1-\frac{x}{2K_\ell})}, &(5.19)\\ &= \frac{q^{\frac14-\frac{ix}{2K_{1-\ell}}}}{i\ell^{\frac14}}\frac{\Gamma_q(\frac12+\frac{ix}{2K_{1-\ell}})\Gamma_q(\frac12-\frac{ix}{2K_{1-\ell}})}{\Gamma_q(1-\frac{ix}{2K_{1-\ell}})\Gamma_q(\frac{ix}{2K_{1-\ell}})}\,. &(5.20)\end{aligned}$$

These relations can be used to obtain some of the above-mentioned expressions. For instance, recalling that $\operatorname{sn}(K_\ell)=1$, we obtain (5.12). With (5.6) we recover from this the well-known identity $\operatorname{sn}(iK_{1-\ell}/2)=i/\ell^{1/4}$. The trigonometric limits

$$\begin{aligned}\lim_{\ell\to 0}\operatorname{sn}(x) &= \lim_{\hat{q}\to 1}\operatorname{sn}(x)=\frac{\pi}{\Gamma(\frac{x}{\pi})\Gamma(1-\frac{x}{\pi})}=\sin(x) &(5.21)\\ \lim_{\ell\to 1}\operatorname{sn}(x) &= \lim_{q\to 1}\operatorname{sn}(x)=\frac{1}{i}\frac{\Gamma(\frac12+\frac{ix}{\pi})\Gamma(\frac12-\frac{ix}{\pi})}{\Gamma(1-\frac{ix}{\pi})\Gamma(\frac{ix}{\pi})}=\tanh(x). &(5.22)\end{aligned}$$

can be read off directly recalling (5.2), (5.7) and presuming that (5.16) holds. We recall the zeros and poles of the Jacobian elliptic $\operatorname{sn}(\theta)$-function, which in our conventions are located at

$$\text{zeros:} \qquad \theta^{lm}_{\text{sn},0} = 2lK_\ell + i2mK_{1-\ell} \qquad l,m \in \mathbb{Z} \tag{5.23}$$

$$\text{poles:} \qquad \theta^{lm}_{\text{sn},p} = 2lK_\ell + i(2m+1)K_{1-\ell} \qquad l,m \in \mathbb{Z}\,. \tag{5.24}$$

We have now assembled the main properties of the q-deformed functions which we shall use below.

5.2. Generalizing diagonal S-matrices

Here we follow [37] and propose a quite simple principle which introduces an infinite number of unstable particles into the spectrum. We note first, that in general many scattering matrices factorize in the following form

$$S_{ab}(\theta) = S^{\text{min}}_{ab}(\theta) S^{\text{CDD}}_{ab}(\theta)\,. \tag{5.25}$$

Here $S^{\text{min}}_{ab}(\theta)$ denotes the so-called minimal S-matrix which satisfies the consistency relations i)–vii) of Section 2. The CDD-factor [18], only satisfies i)–vi) and has its poles in the sheet $-\pi \leq \operatorname{Im}\theta \leq 0$, which is the "physical one" for resonance states. We note now that the minimal part is of the general form

$$S_{ab}(\theta) = \prod_{x\in\mathcal{A}} \{x\}^{\sigma}_{\theta}\,, \tag{5.26}$$

with $\mathcal{A}$ being a finite set specific to each theory. Then we may define a new S-matrix

$$\hat{S}_{ab}(\theta) = \prod_{x\in\mathcal{A}} \{x\}^{\sigma}_{\theta} \{x\}^{\sigma}_{\theta,\ell} \tag{5.27}$$

and note that the additional factor in (5.27) is just of CDD-type. Therefore (5.27) constitutes a solution to the consistency relations i)–vii) of Section 2, and thus a strong candidate for a scattering matrix of a proper quantum field theory. Note that whereas (5.26) was a finite product of hyperbolic functions, the new proposal (5.27) contains, according to the identity (5.18) in addition elliptic functions, which lead to the desired spectrum of infinitely many unstable particles according to the principles outlined in Section 2.

5.3. Non-diagonal S-matrices

We discuss now the elliptic sine-Gordon model, which may be related to the continuum limit of the eight-vertex model. The (anti)-soliton sector was studied many years ago in [44]. In [45] we demonstrated that it is possible to associate a consistent breather sector to this model. Let us recall the argument by recalling the Zamolodchikov algebra for the soliton sector

$$Z(\theta_1)Z(\theta_2) = a(\theta_{12})Z(\theta_2)Z(\theta_1) + d(\theta_{12})\bar{Z}(\theta_2)\bar{Z}(\theta_1)\,, \tag{5.28}$$

$$Z(\theta_1)\bar{Z}(\theta_2) = b(\theta_{12})\bar{Z}(\theta_2)Z(\theta_1) + c(\theta_{12})Z(\theta_2)\bar{Z}(\theta_1)\,. \tag{5.29}$$

In comparison with the more extensively studied sine-Gordon model the difference is the occurrence of the amplitude d in (5.28), i.e., the possibility that two solitons change into two anti-solitons and vice versa. Invoking the consistency equations i)–v) one finds [44, 45]

$$\begin{aligned} a(\theta) &= \Phi(\theta)\prod_{k=0}^{\infty}\left(\frac{\Gamma_{\hat{q}^2}[-\hat{\theta}-\frac{1+2k}{2}\lambda]\Gamma_{\hat{q}^2}[1-\hat{\theta}-\frac{1+2k}{2}\lambda]}{\Gamma_{\hat{q}^2}[\hat{\theta}-\frac{1+2k}{2}\lambda]\Gamma_{\hat{q}^2}[1+\hat{\theta}-\frac{1+2k}{2}\lambda]}\right. \\ &\qquad\left.\times\frac{\Gamma_{\hat{q}^2}[\hat{\theta}-(k+1)\lambda]\Gamma_{\hat{q}^2}[1+\hat{\theta}-k\lambda]}{\Gamma_{\hat{q}^2}[-\hat{\theta}-(k+1)\lambda]\Gamma_{\hat{q}^2}[1-\hat{\theta}-k\lambda]}\right) \end{aligned} \tag{5.30}$$

$$b(\theta) = -\frac{\mathrm{sn}(i\theta/\nu)}{\mathrm{sn}(i\theta/\nu+\pi/\nu)}a(\theta), \tag{5.31}$$

$$c(\theta) = \frac{\mathrm{sn}(\pi/\nu)}{\mathrm{sn}(i\theta/\nu+\pi/\nu)}a(\theta), \tag{5.32}$$

$$d(\theta) = -\sqrt{\ell}\,\mathrm{sn}(i\theta/\nu)\,\mathrm{sn}(\pi/\nu)a(\theta), \tag{5.33}$$

$$\Phi(\theta) = \frac{\Gamma_{\hat{q}}[1+\frac{\tau}{2}]\Gamma_{\hat{q}}[-\frac{\tau}{2}]\Gamma_{\hat{q}}[1-\hat{\theta}+\frac{\lambda}{2}+\frac{\tau}{2}]\Gamma_{\hat{q}}[\hat{\theta}-\frac{\lambda}{2}-\frac{\tau}{2}]}{\Gamma_{\hat{q}}[1+\hat{\theta}+\frac{\tau}{2}]\Gamma_{\hat{q}}[-\hat{\theta}-\frac{\tau}{2}]\Gamma_{\hat{q}}[1+\frac{\lambda}{2}+\frac{\tau}{2}]\Gamma_{\hat{q}}[-\frac{\lambda}{2}-\frac{\tau}{2}]}. \tag{5.34}$$

Here we used $\lambda = -\pi/K_\ell\nu$, $\hat{\theta} = i\theta/2K_\ell\nu$ with $\nu\in\mathbb{R}$ being the coupling constant of the model.With regard to property vii), it is clear that it is important to analyze the singularity structure of the amplitudes (5.30)–(5.33) to judge whether there exists a breather sector. For this we appeal to the relations (5.17), (5.23) and (5.24) and find the following pole structure inside the physical sheet

$$\begin{aligned} \theta^{nm}_{a_1,p} &= 2m\nu K_{1-\ell}+i2n\nu K_\ell, & \theta^{nm}_{a_2,p} &= (2m+1)\nu K_{1-\ell}+i(\pi-2n\nu K_\ell), \\ \theta^{lm}_{b_1,p} &= 2m\nu K_{1-\ell}+i(\pi-2l\nu K_\ell), & \theta^{lm}_{b_2,p} &= (2m+1)\nu K_{1-\ell}+i2l\nu K_\ell, \\ \theta^{lm}_{c_1,p} &= 2m\nu K_{1-\ell}+i2l\nu K_\ell, & \theta^{lm}_{c_2,p} &= 2m\nu K_{1-\ell}+i(\pi-2l\nu K_\ell), \\ \theta^{lm}_{d_1,p} &= (2m+1)\nu K_{1-\ell}+i2l\nu K_\ell, & \theta^{lm}_{d_2,p} &= (2m+1)\nu K_{1-\ell}+i(\pi-2n\nu K_\ell). \end{aligned}$$

We took $l,m\in\mathbb{Z}, n\in\mathbb{N}$ and associated always two sets of poles $\theta^{nm}_{a_1,p}$ and $\theta^{nm}_{a_2,p}$ to $a(\theta)$, $\theta^{nm}_{b_1,p}$ and $\theta^{nm}_{b_2,p}$ to $b(\theta)$ etc. One readily sees from this that if one restricts the parameter $\nu\geq\pi/2K_\ell$ all poles move out of the physical sheet into the non-physical one, where they can be interpreted in principle as unstable particles. This was already stated in [44], where the choice $\nu\geq\pi/2K_\ell$ was made in order to avoid the occurrence of non-physical states. This is clear from our discussion of property vii) in Section 2, as we would have poles in the physical sheet beyond the imaginary axis, which when interpreted with the Breit-Wigner formula leave the choice that either $m_{\tilde{c}}<0$ or $\Gamma_{\tilde{c}}<0$, i.e., we either violate causality or we have Tachyons. The restriction on the parameters makes the model somewhat unattractive as this limitation eliminates the analogue of the entire breather sector which is present in the sine-Gordon model, such that also in the trigonometric limit one only obtains the soliton-antisoliton sector of that model, instead of a theory with a richer particle content. For this reason we relax here the restriction on ν and note that the poles

$$\theta^{n0}_{b_1,p} = \theta^{n0}_{c_2,p} \qquad \text{for } 0<n<n_{\max}=[\pi/2\nu K_\ell], n\in\mathbb{N} \tag{5.35}$$

are located on the imaginary axis inside the physical sheet and are therefore candidates for the analogue of the nth-breather bound states in the sine-Gordon model. We indicate here the integer part of x by $[x]$. In other words, there are at most $n_{\max}-1$ breathers for fixed ν and ℓ. The price one pays for the occurrence of these new particles in the elliptic sine-Gordon model is that one unavoidably also introduces additional Tachyons into the model as the poles always emerge in "strings". It remains to be established whether the poles (5.35) may really be associated to a breather type behavior.

Let us now see if the poles on the imaginary axis inside the physical sheet can be associated consistently with breathers. We proceed similarly as for the sine-Gordon model [46], even though in the latter approach the following ansatz is inspired by the classical theory and here we do not have a classical counterpart. We define the auxiliary state

$$Z_n(\theta_1,\theta_2) := \frac{1}{\sqrt{2}}\left[Z(\theta_1)\bar{Z}(\theta_2)+(-1)^n\bar{Z}(\theta_1)Z(\theta_2)\right] \; . \tag{5.36}$$

This state has properties of the classical sine-Gordon breather being chargeless and having parity $(-1)^n$. Choosing thereafter the rapidities such that the state (5.36) is on-shell, we can speak of a breather bound state

$$\lim_{(p_1+p_2)^2\to m_{b_n}^2} Z_n(\theta_1,\theta_2) \equiv \lim_{\theta_{12}\to\theta+\theta_{12}^{b_n}} Z_n(\theta_1,\theta_2) = Z_n(\theta) \; . \tag{5.37}$$

Here $\theta_{12}^{b_n}$ is the fusing angle related to the poles in the soliton-antisoliton scattering amplitudes. We compute now with the help of (5.28) and (5.29) the exchange relation

$$Z_n(\theta_1)Z(\theta_2) = S_{b_n s}(\theta_{12})Z(\theta_2)Z_n(\theta_1) \; , \tag{5.38}$$

where

$$S_{b_n s}(\theta) = \frac{\mathrm{sn}(\frac{i\theta}{\nu}-\frac{\pi}{2\nu}+nK_\ell)}{\mathrm{sn}(\frac{i\theta}{\nu}+\frac{\pi}{2\nu}+nK_\ell)}[\ell\mathrm{sn}^2\frac{\pi}{\nu}\mathrm{sn}^2\left(\frac{i\theta}{\nu}+\frac{\pi}{2\nu}+nK_\ell\right)-1]\bar{a} \tag{5.39}$$

and

$$\bar{a} = \frac{\Gamma_{\hat{q}^2}[1+\hat{\theta}+\frac{\lambda}{4}-\frac{n}{2}]\Gamma_{\hat{q}^2}[-\hat{\theta}-\frac{\lambda}{4}-\frac{n}{2}]\Gamma_{\hat{q}^2}[-\hat{\theta}+\frac{\lambda}{4}+\frac{n}{2}]\Gamma_{\hat{q}^2}[\hat{\theta}+\frac{\lambda}{4}-\frac{n}{2}]}{\Gamma_{\hat{q}^2}[1-\hat{\theta}+\frac{\lambda}{4}-\frac{n}{2}]\Gamma_{\hat{q}^2}[\hat{\theta}-\frac{\lambda}{4}-\frac{n}{2}]\Gamma_{\hat{q}^2}[\hat{\theta}+\frac{\lambda}{4}+\frac{n}{2}]\Gamma_{\hat{q}^2}[-\hat{\theta}+\frac{\lambda}{4}-\frac{n}{2}]}$$
$$\times\Phi_{13}\Phi_{23}\prod_{l=1}^{n-1}\frac{[\hat{\theta}-\frac{n}{2}+\frac{\lambda}{4}-k\lambda+l]_{\hat{q}^2}^2[-\hat{\theta}+\frac{n}{2}-\frac{\lambda}{4}-k\lambda-l]_{\hat{q}^2}^2}{[-\hat{\theta}-\frac{n}{2}+\frac{\lambda}{4}-k\lambda+l]_{\hat{q}^2}^2[\hat{\theta}+\frac{n}{2}-\frac{\lambda}{4}-k\lambda-l]_{\hat{q}^2}^2}$$
$$\times\prod_{k=0}^{\infty}\frac{[\hat{\theta}-\frac{n}{2}+\frac{\lambda}{4}-k\lambda]_{\hat{q}^2}[-\hat{\theta}+\frac{n}{2}-\frac{\lambda}{4}-k\lambda]_{\hat{q}^2}}{[-\hat{\theta}-\frac{n}{2}+\frac{\lambda}{4}-k\lambda]_{\hat{q}^2}[\hat{\theta}+\frac{n}{2}-\frac{\lambda}{4}-k\lambda]_{\hat{q}^2}}$$
$$\times\prod_{k=0}^{\infty}\frac{[\hat{\theta}+\frac{n}{2}+\frac{\lambda}{4}-k\lambda]_{\hat{q}^2}[-\hat{\theta}-\frac{n}{2}-\frac{\lambda}{4}-k\lambda]_{\hat{q}^2}}{[-\hat{\theta}+\frac{n}{2}+\frac{\lambda}{4}-k\lambda]_{\hat{q}^2}[\hat{\theta}-\frac{n}{2}-\frac{\lambda}{4}-k\lambda]_{\hat{q}^2}} \; .$$

We abbreviate $\Phi_{ij}=\Phi(\theta_{ij})$ with θ_{ij} being the difference of the on-shell rapidities. What is remarkable here and cannot be anticipated a priori, is that all off-diagonal terms vanish, thus as (5.38) expresses in the soliton breather scattering there is no backscattering. Similarly, but more lengthy, we compute the scattering amplitude between the nth-breather and mth-breather

$$Z_n(\theta_1)Z_m(\theta_2) = S_{b_n b_m}(\theta_{12})Z_m(\theta_2)Z_n(\theta_1) \tag{5.40}$$

where

$$S_{b_n b_m}(\theta) = \left[1 - \ell \mathrm{sn}^2 \frac{\pi}{\nu} \mathrm{sn}^2\left(\frac{i\theta}{\nu} + (n+m)K_\ell + \frac{\pi}{\nu}\right)\right] \tag{5.41}$$
$$\times \left[1 - \ell \mathrm{sn}^2 \frac{\pi}{\nu} \mathrm{sn}^2\left(\frac{i\theta}{\nu} + (n+m)K_\ell\right)\right] \frac{\mathrm{sn}(i\theta/\nu - \pi/\nu + (n+m)K_\ell)}{\mathrm{sn}(i\theta/\nu + \pi/\nu + (n+m)K_\ell)}\tilde{a}$$

and

$$\begin{aligned}
\tilde{a} = {} & \Phi_{13}\Phi_{14}\Phi_{23}\Phi_{24} \frac{\Gamma_{\hat{q}^2}(1+\frac{m}{2}+\frac{n}{2}+\hat{\theta}+\frac{\lambda}{2})\Gamma_{\hat{q}^2}(\frac{-m}{2}-\frac{n}{2}-\hat{\theta}-\frac{\lambda}{2})}{\Gamma_{\hat{q}^2}(1+\frac{m}{2}+\frac{n}{2}-\hat{\theta}+\frac{\lambda}{2})\Gamma_{\hat{q}^2}(\frac{-m}{2}-\frac{n}{2}+\hat{\theta}-\frac{\lambda}{2})} \\
& \times \prod_{k=1}^{\infty}\prod_{l=1}^{n-1} \frac{[\frac{m}{2}+\frac{n}{2}-l-\hat{\theta}-k\,\lambda+\lambda]_{\hat{q}^2}[\frac{-m}{2}-\frac{n}{2}+l+\hat{\theta}-k\,\lambda]_{\hat{q}^2}}{[\frac{m}{2}+\frac{n}{2}-l+\hat{\theta}-k\,\lambda+\lambda]_{\hat{q}^2}[\frac{-m}{2}-\frac{n}{2}+l-\hat{\theta}-k\,\lambda]_{\hat{q}^2}} \\
& \times \prod_{k=0}^{\infty}\prod_{l=1}^{n-1} \frac{[\frac{m}{2}+\frac{n}{2}-l+\hat{\theta}-\frac{\lambda}{2}-k\,\lambda]_{\hat{q}^2}[-\frac{m}{2}-\frac{n}{2}+l-\hat{\theta}-\frac{\lambda}{2}-k\,\lambda]_{\hat{q}^2}}{[\frac{m}{2}+\frac{n}{2}-l-\hat{\theta}-\frac{\lambda}{2}-k\,\lambda]_{\hat{q}^2}[-\frac{m}{2}-\frac{n}{2}+l+\hat{\theta}-\frac{\lambda}{2}-k\,\lambda]_{\hat{q}^2}} \\
& \times \prod_{k=1}^{\infty}\prod_{l=0}^{m-1} \frac{[\frac{m}{2}+\frac{n}{2}-l-\hat{\theta}-k\,\lambda+\lambda]_{\hat{q}^2}[-\frac{m}{2}-\frac{n}{2}+l+\hat{\theta}-k\,\lambda]_{\hat{q}^2}}{[\frac{m}{2}+\frac{n}{2}-l+\hat{\theta}-k\,\lambda+\lambda]_{\hat{q}^2}[-\frac{m}{2}-\frac{n}{2}+l-\hat{\theta}-k\,\lambda]_{\hat{q}^2}} \\
& \times \prod_{k=0}^{\infty}\prod_{l=0}^{m-1} \frac{[\frac{m}{2}+\frac{n}{2}-l+\hat{\theta}-\frac{\lambda}{2}-k\,\lambda]_{\hat{q}^2}[-\frac{m}{2}-\frac{n}{2}+l-\hat{\theta}-\frac{\lambda}{2}-k\,\lambda]_{\hat{q}^2}}{[\frac{m}{2}+\frac{n}{2}-l-\hat{\theta}-\frac{\lambda}{2}-k\,\lambda]_{\hat{q}^2}[-\frac{m}{2}-\frac{n}{2}+l+\hat{\theta}-\frac{\lambda}{2}-k\,\lambda]_{\hat{q}^2}} .
\end{aligned} \tag{5.42}$$

The latter expression (5.42) is tailored to make contact to the expressions in the literature corresponding to the trigonometric limit. Also for this amplitude the backscattering is zero.

The matrix $S_{b_n b_m}(\theta)$ also exhibits several types of poles: a) simple and double poles inside the physical sheet beyond the imaginary axis, b) double poles located on the imaginary axis, c) simple poles in the non-physical sheet and d) one simple pole on the imaginary axis inside the physical sheet at $\theta = \theta_b = i\nu(n+m)K_\ell$ which is related to the fusing process of two breathers $b_n + b_m \to b_{n+m}$. To be really sure that this pole admits such an interpretation, we have to establish according to (2.11) that the imaginary part of the residue is strictly positive, i.e.,

$$-i \lim_{\theta \to \theta_b} (\theta - \theta_b) S_{b_n b_m}(\theta) > 0 \, . \tag{5.43}$$

The explicit computation shows that this is indeed the case, see [45].

Furthermore, it is very interesting to check if also (2.10) is satisfied for the fusing process $b_n + b_m \to b_{n+m}$. For consistency, all amplitudes have to satisfy the bootstrap equations

$$S_{lb_{n+m}}(\theta) = S_{lb_n}(\theta + i\nu m K_\ell) S_{lb_m}(\theta - i\nu n K_\ell) \, , \tag{5.44}$$

for $l \in \{b_k, s, \bar{s}\}$; $k, m+n < n_{\max}$. Indeed, we verify with some algebra that (5.44) holds for the above amplitudes (5.39) and (5.41).

Finally, we carry out various limits. Our formulation in terms of q-deformed quantities and elliptic functions is useful to make this task fairly easy. We state our results here only schematically and refer the reader for details to [45]. We find

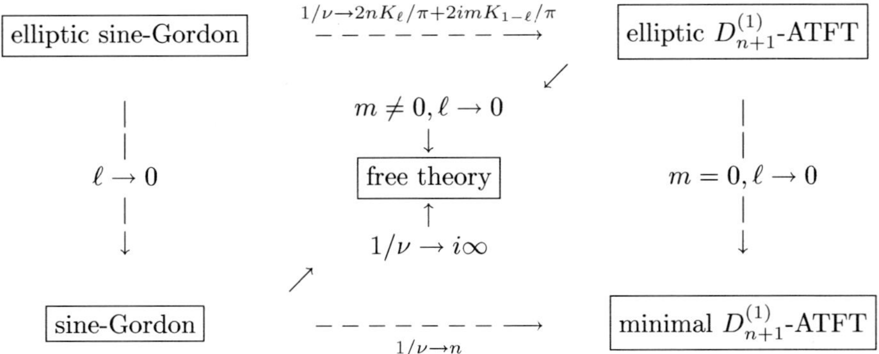

Thus we can view the elliptic sine-Gordon model as a master theory for several other models. In the limit $\ell \to 0$ we recover now all sectors, including the breathers, of the sine-Gordon model. The diagonal limit $1/\nu \to 2nK_\ell/\pi + 2imK_{1-\ell}/\pi$ is interesting as it yields a new type of theory, which we refer to as elliptic $SO(2n+2) \equiv D_{n+1}^{(1)}$-affine Toda field theory (ATFT). To coin this name for these theories seems natural as in the trigonometric limit we obtain from it the ordinary minimal $D_{n+1}^{(1)}$-ATFT.

6. Conclusions

We reviewed the general analytical scattering theory related to integrable quantum field theories in 1+1 space-time dimensions. We made a proposal for a construction principle of an S-matrix like object which describes the scattering between two unstable particles or an unstable particle and a stable one. We tested this proposal with various examples and found a remarkable agreement with the outcome of the thermodynamic Bethe ansatz in what concerns the particle content and the RG flow of the theories. We described the general Lie algebraic structure of theories with unstable particles and propose a decoupling rule which predicts the RG flow when some of the parameters in the theory become very large. Alternatively, we tested these analytical prediction with the TBA. Finally, we discussed how one can construct theories with and without backscattering which contain an infinite number of unstable particles.

Acknowledgment

We are grateful to the Deutsche Forschungsgemeinschaft (Sfb288), for financial support. This work is supported by the EU network EUCLID, *Integrable models and applications: from strings to condensed matter*, HPRN-CT-2002-00325.

References

[1] M. Karowski, H.J. Thun, T.T. Truong, and P.H. Weisz, On the uniqueness of a purely elastic S-matrix in $(1+1)$ dimensions, Phys. Lett. **B67**, 321–322 (1977).

[2] A.B. Zamolodchikov, Exact S-matrix of quantum sine-Gordon solitons, JETP Lett. **25**, 468–481 (1977).

[3] A. Zamolodchikov and A. Zamolodchikov, Factorized S-matrices in two dimensions as the exact solutions of certain relativistic quantum field models, Annals Phys. **120**, 253 (1979).

[4] R. Shankar and E. Witten, The S-matrix of the supersymmetric nonlinear sigma model, Phys. Rev. **D17**, 2134–2143 (1978).

[5] S. Parke, Absence of particle production and factorization of the S-matrix in $(1+1)$-dimensional models, Nucl. Phys. **B174**, 166–182 (1980).

[6] B. Schroer, T.T. Truong, and P. Weisz, Towards an explicit construction of the sine-Gordon theory, Phys. Lett. **B63**, 422–424 (1976).

[7] R. Eden, P. Landshoff, D.I. Olive, and J. Polkinghorne, The analytic S-matrix, Cambridge University Press (1966).

[8] D.I. Olive, Unitarity and the Evolution of Discontinuities, Nuovo Cim. **26**, 73–102 (1962).

[9] J.L. Miramontes, Hermitian analyticity versus real analyticity in two-dimensional factorised S-matrix theories, Phys. Lett. **B455**, 231–238 (1999).

[10] D.I. Olive, Exploration of S-Matrix Theory, Phys. Rev. **135**, B745–B760 (1964).

[11] H. Lehmann, K. Symanzik, and W. Zimmermann, On the formulation of quantized field theories, Nuovo Cim. **1**, 205–225 (1955).

[12] C.-N. Yang, Some exact results for the many body problems in one dimension with repulsive delta function interaction, Phys. Rev. Lett. **19**, 1312–1314 (1967).

[13] R.J. Baxter, One-dimensional anisotropic Heisenberg chain, Annals Phys. **70**, 323–327 (1972).

[14] G. Breit and E.P. Wigner, Capture of slow neutrons, Phys. Rev. **49**, 519–531 (1936).

[15] S.R. Coleman and H.J. Thun, On the prosaic origin of the double poles in the sine-Gordon S-matrix, Commun. Math. Phys. **61**, 31–51 (1978).

[16] H.W. Braden, E. Corrigan, P.E. Dorey, and R. Sasaki, Affine Toda field theory and exact S-matrices, Nucl. Phys. **B338**, 689–746 (1990).

[17] P. Christe and G. Mussardo, Integrable systems away from criticality: The Toda field theory and S-matrix of the tricritical Ising model, Nucl. Phys. **B330**, 465–487 (1990).

[18] L. Castillejo, R.H. Dalitz, and F.J. Dyson, Low's scattering equation for the charged and neutral scalar theories, Phys. Rev. **101**, 453–458 (1956).

[19] O.A. Castro-Alvaredo, J. Dreißig, and A. Fring, Integrable scattering theories with unstable particles, Euro Phys. Lett. **C35**, 393–411 (2004).

[20] Q.-H. Park, Deformed coset models from gauged WZW actions, Phys. Lett. **B328**, 329–336 (1994).

[21] C.R. Fernandez-Pousa, M.V. Gallas, T.J. Hollowood, and J.L. Miramontes, The symmetric space and homogeneous sine-Gordon theories, Nucl. Phys. **B484**, 609–630 (1997).

[22] J.L. Miramontes and C.R. Fernandez-Pousa, Integrable quantum field theories with unstable particles, Phys. Lett. **B472**, 392–401 (2000).

[23] O.A. Castro-Alvaredo, A. Fring, C. Korff, and J.L. Miramontes, Thermodynamic Bethe ansatz of the homogeneous sine-Gordon models, Nucl. Phys. **B575**, 535–560 (2000).

[24] O.A. Castro-Alvaredo, A. Fring, and C. Korff, Form factors of the homogeneous sine-Gordon models, Phys. Lett. **B484**, 167–176 (2000).

[25] O.A. Castro-Alvaredo and A. Fring, Identifying the operator content, the homogeneous sine-Gordon models, Nucl. Phys. **B604**, 367–390 (2001).

[26] O.A. Castro-Alvaredo and A. Fring, Decoupling the $SU(N)(2)$ homogeneous sine-Gordon model, Phys. Rev. **D64**, 085007 (2001).

[27] O.A. Castro-Alvaredo and A. Fring, Renormalization group flow with unstable particles, Phys. Rev. **D63**, 021701 (2001).

[28] J.L. Miramontes, Integrable quantum field theories with unstable particles, JHEP Proceedings, of the TMR conference, Nonperturbative Quantum Effects, hep-th/0010012 (Paris 2000).

[29] P. Dorey and J.L. Miramontes, Aspects of the homogeneous sine-Gordon models, JHEP Proceedings, of the workshop on Integrable Theories, Solitons and Duality, hep-th/0211174 (São Paulo 2002).

[30] P. Baseilhac, Liouville field theory coupled to a critical Ising model: Non-perturbative analysis, duality and applications, Nucl. Phys. **B636**, 465–496 (2002).

[31] E. Witten, Nonabelian bosonization in two dimensions, Commun. Math. Phys. **92**, 455–472 (1984).

[32] P. Goddard, A. Kent, and D.I. Olive, Virasoro algebras and coset space models, Phys. Lett. **B152**, 88–92 (1985).

[33] J. Humphreys, Introduction to Lie Algebras and Representation Theory, Springer, Berlin (1972).

[34] C. Korff, Color-valued scattering matrices from non simply-laced Lie algebras, Phys. Lett. **B5011**, 289–296 (2001).

[35] A. Fring and C. Korff, Color-valued scattering matrices, Phys. Lett. **B477**, 380–386 (2000).

[36] A. Zamolodchikov, Resonance factorized scattering and roaming trajectories, ENS-LPS-335-preprint.

[37] O.A. Castro-Alvaredo and A. Fring, Constructing infinite particle spectra, Phys. Rev. **D64**, 085005 (2001).

[38] E.B. Dynkin, Semisimple subalgebras of semisimple Lie algebras, Amer. Math. Soc. Trans. Ser. 2 **6**, 111–244 (1957).

[39] A. Kuniba, Thermodynamics of the $U_q(X_r^{(1)})$ Bethe ansatz system with q a root of unity, Nucl. Phys. **B389**, 209–246 (1993).

[40] A.B. Zamolodchikov, Irreversibility of the flux of the renormalization group in a 2-D field theory, JETP Lett. **43**, 730–732 (1986).

[41] A.B. Zamolodchikov, Thermodynamic Bethe ansatz in relativistic models: Scaling 3-state Potts and Lee-Yang models, Nucl. Phys. **B342**, 695–720 (1990).

[42] H. Babujian, A. Fring, M. Karowski, and A. Zapletal, Exact form factors in integrable quantum field theories: The sine-Gordon model, Nucl. Phys. **B538**, 535–586 (1999).

[43] K. Chandrasekhan, Elliptic Functions, Springer, Berlin (1985).

[44] A. Zamolodchikov, Z_4-symmetric factorised S-matrix in two space-time dimensions, Comm. Math. Phys. **69**, 165–178 (1979).

[45] O.A. Castro-Alvaredo and A. Fring, Breathers in the elliptic sine-Gordon model, J. Phys. **A36**, 10233–10249 (2003).

[46] M. Karowski and H.J. Thun, Complete S matrix of the massive Thirring model, Nucl. Phys. **B130**, 295–308 (1977).

Olalla Castro-Alvaredo
Laboratoire de Physique
Ecole Normale Supérieure de Lyon
UMR 5672 du CNRS
46 Allée d'Italie
F-69364 Lyon CEDEX, France
e-mail: ocastroa@ens-lyon.fr

Andreas Fring
Centre for Mathematical Sciences
City University
Northampton Square
London EC1V 0HB, UK
e-mail: A.Fring@city.ac.uk

Progress in Mathematics, Vol. 237, 89–131

Quantum Reduction in the Twisted Case

Victor G. Kac and Minoru Wakimoto

Mathematics Subject Classification (2000). 17B55, 17B67, 17B68, 17B69, 81R10.

Keywords. Twisted vertex algebra, quantum Hamiltonian reduction, superconformal algebra, Euler-Poincaré character, free-field realization, determinant formula.

0. Introduction

This paper is a continuation of the papers [KRW] and [KW] on structure and representation theory of vertex algebras $W_k(\mathfrak{g}, x)$ obtained by quantum Hamiltonian reduction from the affine superalgebra $\widehat{\mathfrak{g}}$. The datum one begins with is a quadruple $(\mathfrak{g}, x, f, k)$, where $\mathfrak{g}$ is a simple finite-dimensional Lie superalgebra with a non-zero invariant even supersymmetric bilinear form $(\,.\,|\,.\,)$, x is an element of $\mathfrak{g}$ such that ad x is diagonalizable with eigenvalues in $\frac{1}{2}\mathbb{Z}$, f is an even element of $\mathfrak{g}$ such that $[x, f] = -f$ and the eigenvalues of ad x on the centralizer $\mathfrak{g}^f$ of f in $\mathfrak{g}$ are non-positive, and $k \in \mathbb{C}$. Recall that a pair $\{x, f\}$ satisfying the above properties can be obtained from an $s\ell_2$-triple $\{e, x, f\}$, so that $[x, e] = e$, $[x, f] = -f$, $[e, f] = x$.

We associate to the quadruple $(\mathfrak{g}, x, f, k)$ a homology complex $\mathcal{C}(\mathfrak{g}, x, k) = (V_k(\mathfrak{g}) \otimes F_{\text{ch}} \otimes F_{\text{ne}}\ , d_0)$, where $V_k(\mathfrak{g})$ is the universal affine vertex algebra of level k associated to the affine superalgebra $\widehat{\mathfrak{g}}$, F_{ch} is the vertex algebra of free charged superfermions based on $\mathfrak{g}_+ + \mathfrak{g}_+^*$ with reversed parity, F_{ne} is the vertex algebra of free neutral superfermions based on $\mathfrak{g}_{1/2}$, and d_0 is an explicitly constructed odd derivation of the vertex algebra $\mathcal{C}(\mathfrak{g}, x, k)$ whose square is 0. Here $\mathfrak{g}_+$ (resp. $\mathfrak{g}_{1/2}$) denotes the sum of eigenspaces of ad x with positive eigenvalues (resp. with eigenvalue $1/2$), and we drop f from the notation since its different choices are conjugate. The vertex algebra $W_k(\mathfrak{g}, x)$ is the homology of the complex $(\mathcal{C}(\mathfrak{g}, x, k),\ d_0)$.

The first author was supported in part by NSF grant DMS-0201017.
The second author was supported by Grant-in-aid 13440012 for scientific research Japan.

In the present paper we begin with a diagonalizable automorphism σ of $\mathfrak{g}$ with modulus 1 eigenvalues, which leaves invariant the bilinear form $(.|.)$ and keeps the elements x and f fixed. The automorphism σ gives rise to the twisted affine superalgebra $\widehat{\mathfrak{g}}^{\mathrm{tw}}$ and the corresponding twisted vertex algebra $V_k(\mathfrak{g},\sigma)$, and to the twisted vertex algebras $F^{\mathrm{tw}}_{\mathrm{ch}}$ and $F^{\mathrm{tw}}_{\mathrm{ne}}$, and we consider the twisted complex

$$(\mathcal{C}(\mathfrak{g},\sigma,x,k) = V_k(\mathfrak{g},\sigma)\otimes F^{\mathrm{tw}}_{\mathrm{ch}}\otimes F^{\mathrm{tw}}_{\mathrm{ne}}\,,\, d_0^{\mathrm{tw}})\,.$$

Its homology is the twisted vertex algebra $W_k(\mathfrak{g},\sigma,x)$, which is the main object of our study.

In the case when $\sigma=1$ we recover the "Neveu-Schwarz sector" $W_k(\mathfrak{g},x)$ studied in our previous papers [KRW] and [KW] (and earlier in [FF, FKW, BT, FB, ST], and many other works, see [BS].) In the case when $\sigma=\sigma_R$, where $\sigma_R|_{\mathfrak{g}_{\bar 0}}=1$ and $\sigma_R|_{\mathfrak{g}_{\bar 1}}=-1$, we obtain the "Ramond sector". This terminology comes from the fact that taking the smallest simple Lie superalgebra $\mathfrak{g}=osp(1|2)$ and the only possible choice of x, we obtain as $W_k(\mathfrak{g},x)$ the vertex algebra associated to the usual Neveu-Schwarz algebra, and as $W_k(\mathfrak{g},\sigma_R,x)$ the twisted vertex algebra associated to the usual Ramond algebra. Likewise, taking $\mathfrak{g}=s\ell(2|1)$, $s\ell(2|2))/\mathbb{C}I$, $osp(3|2)$ or $D(2,1;a)$, and x the suitable multiple of the highest root θ of one of the simple components of $\mathfrak{g}_{\bar 0}$, all possible choices of σ produce all possible twists of the $N=2$, $N=4$, $N=3$ and big $N=4$ superconformal algebras.

This leads us to a unified representation theory of all twisted superconformal algebras, in particular to unified free field realizations and determinant formulas.

As in [FKW] and [KRW], we construct also a functor $M\mapsto H(M)$ from the category of restricted $\widehat{\mathfrak{g}}^{\mathrm{tw}}$-modules to the category of $\mathbb{Z}$-graded $W_k(\mathfrak{g},\sigma,x)$-modules and compute the Euler-Poincaré character of $H(M)$ in terms of the character of M. In a forthcoming paper [KW4] we shall develop a theory of characters of $W_k(\mathfrak{g},\sigma,x)$ using this functor.

1. An overview of twisted formal distributions

Let R be a Lie conformal superalgebra. Recall that this is a $\mathbb{Z}/2\mathbb{Z}$-graded $\mathbb{C}[\partial]$-module, endowed with jth products denoted by $a_{(j)}b$, $j\in\mathbb{Z}_+$, satisfying certain axioms [K4]. One associates to R a Lie superalgebra

$$\mathrm{Lie}\,(R) = R\,[t,t^{-1}]/\,\mathrm{Image}\,(\partial\otimes 1 + 1\otimes\partial_t)\,, \tag{1.1}$$

endowed with the following bracket, where $a_{(\mu)}$ stands for $a\otimes t^{\mu}\in R\,[t,t^{-1}]=R\otimes\mathbb{C}\,[t,t^{-1}]$:

$$[a_{(\mu)},b_{(\nu)}] = \sum_{j\in\mathbb{Z}_+}\binom{\mu}{j}(a_{(j)}b)_{(\mu+\nu-j)}\,. \tag{1.2}$$

Introducing formal distributions

$$a(z)=\sum_{\mu\in\mathbb{Z}} a_{(\mu)}z^{-\mu-1}\,,\quad a\in R\,, \tag{1.3}$$

one rewrites (1.2) as

$$[a(z)\,,\,b(w)]=\sum_{j\in\mathbb{Z}_+}(a_{(j)}b)(w)\partial_w^j\delta(z-w)/j!\,. \tag{1.4}$$

One also has:

$$(\partial a)(z)=\partial_z a(z)\,. \tag{1.5}$$

(The fact that Lie R is a Lie superalgebra and the distributions $\{a(z)\}_{a\in R}$ form a local system is encoded in the axioms of R.)

Let now σ be a diagonalizable automorphism of R. We shall always assume for simplicity that all eigenvalues of σ have modulus 1. We have:

$$R=\bigoplus_{\bar{\mu}\in\mathbb{R}/\mathbb{Z}}R^{\bar{\mu}},\ \text{where } R^{\bar{\mu}}=\{a\in R|\sigma(a)=e^{2\pi i\bar{\mu}}a\}\,. \tag{1.6}$$

Here and further $\bar{\mu}$ denotes the coset $\mu+\mathbb{Z}$ of $\mu\in\mathbb{R}$. We associate to the pair (R,σ) the σ-*twisted* Lie superalgebra

$$\text{Lie}\,(R,\sigma)=\bigoplus_{\mu\in\mathbb{R}}\ (R^{\bar{\mu}}\otimes t^{\mu})/\,\text{Image}\,(\partial\otimes 1+1\otimes\partial_t)\,, \tag{1.7}$$

endowed with bracket (1.2) (except that now μ and ν are not necessarily integers). Denoting by $a_{(\mu)}$ the image of $a\otimes t^{\mu}$ in $\text{Lie}\,(R,\sigma)$, and introducing the twisted formal distributions

$$a^{\text{tw}}(z)=\sum_{\mu\in\bar{\mu}}a_{(\mu)}z^{-\mu-1}\,,\quad a\in R^{\bar{\mu}}\,, \tag{1.8}$$

we get the twisted analogue of (1.4):

$$[a^{\text{tw}}(z)\,,\,b^{\text{tw}}(w)]=\sum_{j\in\mathbb{Z}_+}(a_{(j)}b)^{\text{tw}}(w)\partial_w^j\delta_{\bar{\mu}}(z-w)/j!\,, \tag{1.9}$$

where

$$\delta_{\bar{\mu}}(z-w)=z^{-1}\sum_{\mu\in\bar{\mu}}\left(\frac{w}{z}\right)^{\mu}$$

is the twisted formal δ-function.

Assuming that the $\mathbb{C}[\partial]$-module R is generated by a finite set Q, introduce a descending filtration of the Lie superalgebra $L=\text{Lie}\,(R,\sigma)$ by subspaces $F_jL=\{a_{(\mu)}|a\in Q,\mu\geqslant j\}$, and define a completion $U(L)^{\text{com}}$ of its universal enveloping algebra $U(L)$, which consists of all series $\sum_i u_i$ such that for each $N\in\mathbb{R}$ all but finitely many of the u_i's lie in $U(L)(F_NL)$. The automorphism σ of R induces one of L and of $U(L)^{\text{com}}$ in the obvious way, which we again denote by σ.

A $U(L)^{\text{com}}$-valued *twisted formal distribution* is an expression of the form

$$a(z)=\sum_{\mu\in\bar{\mu}}a_{(\mu)}z^{-\mu-1}\,,$$

where $\bar{\mu}=\mu+\mathbb{Z}$, $\sigma(a_{(\mu)})=e^{2\pi i\bar{\mu}}a_{(\mu)}$ and $a_{(\mu)}\in U(L)^{\text{com}}$ satisfy the property that for each $N\in\mathbb{R}$, $a_{(\mu)}\in U(L)(F_NL)$ for $\mu\gg 0$ and all the $a_{(\mu)}$ have the same parity, denoted by $p(a) \in \mathbb{Z}/2\mathbb{Z}$. It is clear that the derivative $\partial_z a(z)$ of a twisted formal distribution $a(z)$ is also a twisted formal distribution.

In order to define a normally ordered product of twisted formal distributions $a(z)$ and $b(z)$, we need to define a splitting $a(z) = a(z)_+ + a(z)_-$ into creation and annihilation parts $a(z)_+$ and $a(z)_-$. For that we choose s_a in the coset $\bar{\mu}$ and let

$$a(z)_+ = \sum_{\mu < s_a} a_{(\mu)} z^{-\mu-1}\,,\; a(z)_- = \sum_{\mu \geqslant s_a} a_{(\mu)} z^{-\mu-1}\,. \tag{1.10}$$

If $a(z)$ is a non-twisted formal distribution, i.e., $\bar{\mu} = \mathbb{Z}$, then one may choose $s_a = 0$, so that $\partial_z(a(z)_\pm) = (\partial_z a(z))_\pm$, but for twisted formal distributions such a choice is impossible. After making a choice of s_a, one defines the normally ordered product of twisted formal distributions in the usual way:

$$: a(z)b(z) := a(z)_+ b(z) + (-1)^{p(a)p(b)} b(z)a(z)_-\,.$$

It is easy to see that this is again a $U(L)^{\mathrm{com}}$-valued formal distribution. As usual, one defines the normally ordered product of more than two formal distributions from right to left, e.g. $: abc :=: a : bc ::$.

Denote by $V(R)$ the subspace of $U(\mathrm{Lie}\ R)^{\mathrm{com}}$ consisting of all normally ordered products of formal distributions (1.3) and 1. This is one of the constructions of the universal enveloping vertex algebra of the Lie conformal algebra R [KRW] (cf. [K4], [GMS], [BK]). The infinitesimal translation operator ∂ of $V(R)$ is defined by (1.5). The jth product $a_{(j)}b$ on $V(R)$ is defined by (1.4) for $j \in \mathbb{Z}_+$, and by $a_{(-j-1)}b =: (\partial^j a)b : /j!$ for $j \in \mathbb{Z}_+$. The automorphism σ of R induces an automorphism of $V(R)$, and we have its eigenspace decomposition:

$$V(R) = \bigoplus_{\bar{\mu} \in \mathbb{R}/\mathbb{Z}} V^{\bar{\mu}}(R)\,.$$

Likewise, denote by $V(R,\sigma)$ the subspace of $U(\mathrm{Lie}\,(R,\sigma))^{\mathrm{com}}[[z,z^{-1}]]$ consisting of all normally ordered products of twisted formal distributions (1.8) and 1. This is called a *σ-twist* of the vertex algebra $V(R)$. (It is independent of the choices of s_a used in the definition of normally ordered products.) The subspace $V(R,\sigma)$ is σ-invariant, so that we have the decomposition into its eigenspaces:

$$V(R,\sigma) = \bigoplus_{\bar{\mu} \in \mathbb{R}/\mathbb{Z}} V^{\bar{\mu}}(R,\sigma)\,.$$

The following result is well known.

Proposition 1.1. *The map $a(z) \mapsto a^{\mathrm{tw}}(z)$ $(a \in R^{\bar{\mu}}, \bar{\mu} \in \mathbb{R}/\mathbb{Z})$ extends uniquely to a σ-eigenspace preserving vector space isomorphism $V(R) \to V(R,\sigma)$, $a(z) \mapsto$*

$a^{\mathrm{tw}}(z)$, *satisfying the following properties* $(a \in V^{\bar{\mu}}(R), b \in V(R))$*:*

$$1^{\mathrm{tw}} = 1\,, \tag{1.11}$$

$$(\partial a)^{\mathrm{tw}}(z) = \partial_z a^{\mathrm{tw}}(z)\,, \tag{1.12}$$

$$[a^{\mathrm{tw}}(z), b^{\mathrm{tw}}(w)] = \sum_{j \in \mathbb{Z}_+} (a_{(j)}b)^{\mathrm{tw}}(w) \partial_w^j \delta_{\bar{\mu}}(z-w)/j!\,, \tag{1.13}$$

$$: a^{\mathrm{tw}}(z) b^{\mathrm{tw}}(z) : = \sum_{j \in \mathbb{Z}_+} \binom{s_a}{j} (a_{(j-1)}b)^{\mathrm{tw}}(z) z^{-j}\,. \tag{1.14}$$

A module M over the filtered Lie superalgebra $L = \mathrm{Lie}\,(R, \sigma)$ is called *restricted* if any vector of M is annihilated by some $F_j L$. Such an L-module can be uniquely extended to a module over the associative algebra $U(L)^{\mathrm{com}}$. Restricting this module to $V(R, \sigma)$, we obtain, in view of Proposition 1.1, what is called a *σ-twisted* module M over the vertex algebra $V(R)$.

In the examples of Lie conformal superalgebras R that follow we use the λ-bracket $[a_\lambda b] = \sum_{j \in \mathbb{Z}_+} \frac{\lambda^j}{j!} a_{(j)} b$. Due to sesquilinearity ($[\partial a_\lambda b] = -\lambda [a_\lambda b]$, $[a_\lambda \partial b] = (\partial + \lambda)[a_\lambda b]$), the λ-brackets of generators of the $\mathbb{C}[\partial]$-module R determine the λ-bracket on R. Recall also that an element K of R is called central if $[K_\lambda R] = 0 = [R_\lambda K]$.

Example 1.1. (twisted currents and Sugawara construction). Let $\mathfrak{g}$ be a simple finite-dimensional Lie superalgebra with a non-degenerate supersymmetric invariant bilinear form $(\,.\,|\,.\,)$. The associated Lie conformal superalgebra is

$$\mathrm{Cur}\mathfrak{g} = (\mathbb{C}[\partial] \otimes \mathfrak{g}) \oplus \mathbb{C}K\,,$$

where K is a central element and

$$[a_\lambda b] = [a, b] + \lambda (a|b) K\,, \quad a, b \in 1 \otimes \mathfrak{g} \equiv \mathfrak{g}\,.$$

Given a complex number k, denote by $V_k(\mathfrak{g})$ the quotient of the universal enveloping vertex algebra $V(\mathrm{Cur}\mathfrak{g})$ by the ideal generated by $K - k$. This is called the universal affine vertex algebra of level k.

Let σ be a diagonalizable automorphism of the Lie superalgebra $\mathfrak{g}$, keeping the bilinear form $(\,.\,|\,.\,)$ invariant. It extends to an automorphism of $\mathrm{Cur}\mathfrak{g}$, also denoted by σ, by letting $\sigma(P(\partial) \otimes a) = P(\partial) \otimes \sigma(a)$, $\sigma(K) = K$. Let $\mathfrak{g} = \oplus_{\bar{\mu} \in \mathbb{R}/\mathbb{Z}} \mathfrak{g}^{\bar{\mu}}$, where $\mathfrak{g}^{\bar{\mu}} = \{a \in \mathfrak{g} |\, \sigma(a) = e^{2\pi i \bar{\mu}} a\}$, be the eigenspace decomposition of $\mathfrak{g}$ for σ. Then the corresponding σ-twisted Lie superalgebra $\mathrm{Lie}\,(\mathrm{Cur}\mathfrak{g}, \sigma)$ is a twisted Kac-Moody affinization

$$\widehat{\mathfrak{g}}^{\mathrm{tw}'} = \bigoplus_{\mu \in \mathbb{R}} (\mathfrak{g}^{\bar{\mu}} \otimes t^\mu) \oplus \mathbb{C}K\,,$$

with the bracket $(a \in \mathfrak{g}^{\bar{\mu}}, b \in \mathfrak{g}^{\bar{\nu}})$:

$$[at^\mu, bt^\nu] = [a, b] t^{\mu+\nu} + \mu (a|b) \delta_{\mu, -\nu} K\,, \quad [K, \widehat{\mathfrak{g}}^{\mathrm{tw}'}] = 0\,.$$

The formal distributions

$$a^{\mathrm{tw}}(z) = \sum_{\mu \in \bar{\mu}} (at^{\mu}) z^{-\mu-1}, \quad a \in \mathfrak{g}^{\bar{\mu}},$$

are called *twisted currents.* They generate (by taking derivatives and normally ordered products) the σ-twist $V(\mathrm{Cur}\mathfrak{g}, \sigma)$ of the vertex algebra $V(\mathrm{Cur}\mathfrak{g})$. As in the non-twisted case, denote by $V_k(\mathfrak{g}, \sigma)$ the quotient by the ideal generated by $K - k$; this is the σ-twist of the vertex algebra $V_k(\mathfrak{g})$.

Choosing dual bases $\{a_i\}$ and $\{a^i\}$ of $\mathfrak{g}$, compatible with the eigenspace decomposition for σ, so that $(a_i|a^j) = \delta_{ij}$, define the twisted Sugawara field in $V_k(\mathfrak{g}, \sigma)$ (assuming that $k + h^{\vee} \neq 0$):

$$L^{\mathfrak{g},\mathrm{tw}}(z) = \frac{1}{2(k+h^{\vee})} \sum_i (-1)^{p(a_i)} :a_i a^i:^{\mathrm{tw}}(z).$$

Writing $L^{\mathfrak{g},\mathrm{tw}}(z) = \sum_{n\in\mathbb{Z}} L_n^{\mathfrak{g},\mathrm{tw}} z^{-n-2}$, and using the non-twisted Sugawara construction and formula (1.13), we obtain that the $L_n^{\mathfrak{g},\mathrm{tw}}$ satisfy the relations of the Virasoro algebra with central charge $c(k) = k\,\mathrm{sdim}\,\mathfrak{g}/(k+h^{\vee})$. Using formula (1.14), we can rewrite $L^{\mathfrak{g},\mathrm{tw}}(z)$ in terms of twisted currents and numbers $s_i = s_{a_i}$ (see (1.10)):

$$L^{\mathfrak{g},\mathrm{tw}}(z) = \frac{1}{2(k+h^{\vee})} \Bigg(\sum_i (-1)^{p(a_i)} :a_i^{\mathrm{tw}}(z) a^{i,\mathrm{tw}}(z): \tag{1.15}$$
$$- \sum_i (-1)^{p(a_i)} s_i [a_i, a^i]^{\mathrm{tw}}(z) z^{-1} - k \sum_i (-1)^{p(a_i)} \binom{s_i}{2} z^{-2} \Bigg).$$

Example 1.2. (twisted neutral free superfermions). Let A be a finite-dimensional vector superspace with a non-degenerate skew-supersymmetric bilinear form $\langle\, .\, ,\, .\,\rangle$. The associated Clifford Lie conformal superalgebra is

$$C(A) = (\mathbb{C}[\partial] \otimes A) \oplus \mathbb{C}K_A,$$

where K_A is a central element and

$$[a_\lambda b] = \langle a, b\rangle K_A.$$

Denote by $F(A)$ the quotient of the universal enveloping vertex algebra of $C(A)$ by the ideal generated by $K_A - 1$.

Let σ be a diagonalizable automorphism of the space A, keeping the bilinear form $\langle\, .\, ,\, .\,\rangle$ invariant. As above, it extends to an automorphism σ of the Lie conformal superalgebra $C(A)$. Let $A = \oplus_{\bar{\mu}\in\mathbb{R}/\mathbb{Z}} A^{\bar{\mu}}$ be the eigenspace decomposition for σ. Then the corresponding σ-twisted Lie superalgebra $\mathrm{Lie}\,(C(A), \sigma)$ is a twisted Clifford affinization

$$\widehat{A}^{\mathrm{tw}} = \oplus_{\mu\in\mathbb{R}} (A^{\bar{\mu}} \otimes t^{\mu}) \oplus \mathbb{C}K_A$$

with the bracket

$$[at^{\mu}, bt^{\nu}] = \langle a, b\rangle \delta_{\mu,-\nu-1} K_A, \quad [K_A, \widehat{A}^{\mathrm{tw}}] = 0.$$

We shall work in the Clifford algebra $U(\widehat{A}^{\mathrm{tw}})^{\mathrm{com}}/(K_A - 1)$. The formal distributions

$$\Phi^{\mathrm{tw}}(z) = \sum_{\mu \in \bar{\mu}} (\Phi t^{\mu}) z^{-\mu-1}, \quad \Phi \in A^{\bar{\mu}},$$

are called twisted neutral free superfermions. They generate the σ-twist $F(A, \sigma)$ of the vertex algebra $F(A)$.

Choosing dual bases $\{\Phi_i\}$ and $\{\Phi^i\}$ of A, compatible with the eigenspace decomposition for σ, we let

$$L^{\mathrm{ne},\mathrm{tw}}(z) = \frac{1}{2} \sum_i (-1)^{p(\Phi_i)} : \Phi_i \partial \Phi^i :^{\mathrm{tw}} (z).$$

Writing $L^{\mathrm{ne},\mathrm{tw}}(z) = \sum_{n \in \mathbb{Z}} L_n^{\mathrm{ne},\mathrm{tw}} z^{-n-2}$, we obtain a Virasoro algebra with central charge $c = -\frac{1}{2}\,\mathrm{sdim}\, A$. As in the previous example, using formula (1.14), we obtain:

$$\begin{aligned} L^{\mathrm{ne},\mathrm{tw}}(z) &= \frac{1}{2} \sum_i (-1)^{p(\Phi_i)} : \Phi_i^{\mathrm{tw}}(z) \partial \Phi^{i,\mathrm{tw}}(z) : \\ &\quad - \frac{1}{2} \sum_i (-1)^{p(\Phi_i)} \binom{s_i}{2} z^{-2}. \end{aligned} \tag{1.16}$$

Example 1.3. (twisted charged free superfermions). In notation of Example 1.2, assume that $A = A_+ \oplus A_-$, where both A_+ and A_- are isotropic and σ-invariant subspaces. Choose a basis φ_i of A_+, compatible with the eigenspace decomposition of A_+ for σ, and its dual basis φ_i^* of A_-, so that $\langle \varphi_i, \varphi_j^* \rangle = \delta_{ij}$, and define *charge* by

$$\mathrm{charge}(\varphi_i) = 1; \quad \mathrm{charge}(\varphi_i^*) = -1. \tag{1.17}$$

The formal distributions $\varphi_i^{\mathrm{tw}}(z)$ and $\varphi_i^{*\mathrm{tw}}(z)$ are called twisted charged free superfermions. Relation (1.17) gives rise to the charge decomposition:

$$F(A, \sigma) = \oplus_{m \in \mathbb{Z}} F_m(A, \sigma). \tag{1.18}$$

For a collection of complex numbers $(m_j) \in \mathbb{C}^{\dim A_+}$ we can define a Virasoro formal distribution

$$L^{\mathrm{ch},\mathrm{tw}}(z) = -\sum_i m_i : \varphi_i^* \partial \varphi_i :^{\mathrm{tw}} (z) + \sum_i (1 - m_i) : \partial \varphi_i^* \varphi_i :^{\mathrm{tw}} (z)$$

with central charge $\sum_i (-1)^{p(\varphi_i)} (12 m_i^2 - 12 m_i + 2)$. Using formula (1.14), we obtain

$$\begin{aligned} L^{\mathrm{ch},\mathrm{tw}}(z) &= -\sum_i m_i : \varphi_i^{*\mathrm{tw}}(z) \partial \varphi_i^{\mathrm{tw}}(z) : \\ &\quad + \sum_i (1 - m_i) : \partial \varphi_i^{*\mathrm{tw}}(z) \varphi_i^{\mathrm{tw}}(z) : + \sum_i (-1)^{p(\varphi_i)} \binom{s_i}{2} z^{-2}. \end{aligned} \tag{1.19}$$

2. The twisted complex

Let $\mathfrak{g}$ be a simple finite-dimensional Lie superalgebra with a non-degenerate even supersymmetric invariant bilinear form $(\,.\,|\,.\,)$. Fix an even element x of $\mathfrak{g}$ such that $\operatorname{ad} x$ is diagonalizable with half-integer eigenvalues, and let

$$\mathfrak{g} = \oplus_{j \in \frac{1}{2}\mathbb{Z}} \mathfrak{g}_j \tag{2.1}$$

be the eigenspace decomposition. Let

$$\mathfrak{g}_+ = \oplus_{j>0} \mathfrak{g}_j \,,\ \mathfrak{g}_- = \oplus_{j<0} \mathfrak{g}_j \,,\ \mathfrak{g}_\leqslant = \mathfrak{g}_0 \oplus \mathfrak{g}_- \,.$$

An even element $f \in \mathfrak{g}_{-1}$ is called *good* if its centralizer $\mathfrak{g}^f$ in $\mathfrak{g}$ lies in $\mathfrak{g}_\leqslant$, and the gradation (2.1) is called *good* if it admits a good element. We shall assume that the grading (2.1) is good and we shall fix a good element $f \in \mathfrak{g}_{-1}$ (all good elements form a Zariski dense orbit of the group $\exp \mathfrak{g}_{0,\mathrm{even}}$, hence nothing depends on the choice of f).

The most interesting good gradings come from $s\ell_2$-triples $\{e, x, f\}$, where $[x, e] = e$, $[x, f] = -f$, $[e, f] = x$, which are called Dynkin gradings. However, there are many other good gradings. In the Lie algebra case they are classified in [EK].

An important role is played by the following bilinear form $\langle\,.\,,\,.\,\rangle_{\mathrm{ne}}$ on $\mathfrak{g}_{1/2}$:

$$\langle a, b\rangle_{\mathrm{ne}} = (f|[a, b])\,, \tag{2.2}$$

which is skew-supersymmetric, even and non-degenerate.

Fix an automorphism σ of $\mathfrak{g}$ with the following three properties:

(i) $\sigma(x) = x$, $\sigma(f) = f$;
(ii) $(\sigma(a)|\sigma(b)) = (a|b)$ for all $a, b \in \mathfrak{g}$;
(iii) σ is diagonalizable and all its eigenvalues have modulus 1.

We shall construct a twisted vertex algebra $W_k(\mathfrak{g}, \sigma, x)$ depending on a complex parameter k. For $\sigma = 1$ this coincides with the vertex algebra $W_k(\mathfrak{g}, x, f)$ studied in [KRW] and [KW] (we shall drop f from the notation, since different choices of f give isomorphic algebras).

Introduce the following $\frac{1}{2}\mathbb{Z}$-graded subalgebra of $\mathfrak{g}$:

$$\mathfrak{g}(\sigma) = \oplus_{j \in \frac{1}{2}\mathbb{Z}} \mathfrak{g}_j(\sigma)\,, \ \text{where } \mathfrak{g}_j(\sigma) = \{a \in \mathfrak{g}_j | \sigma(a) = (-1)^{2j} a\}\,. \tag{2.3}$$

Choose a σ-invariant Cartan subalgebra $\mathfrak{h}$ of the even part of $\mathfrak{g}_0$, and choose a triangular decomposition of $\mathfrak{g}(\sigma)$, compatible with the gradation (2.3):

$$\mathfrak{g}(\sigma) = \mathfrak{n}(\sigma)_- \oplus \mathfrak{h}^\sigma \oplus \mathfrak{n}(\sigma)_+\,, \tag{2.4}$$

where $\mathfrak{h}^\sigma$ denotes the fixed point set of σ on $\mathfrak{h}$, such that the following properties hold:

(i) $\mathfrak{n}(\sigma)_\pm$ are isotropic with respect to $(\,.\,|\,.\,)$ nilpotent subalgebras normalized by $\mathfrak{h}^\sigma$,
(ii) $f \in \mathfrak{n}(\sigma)_+$,
(iii) $\mathfrak{n}_{1/2}(\sigma)_+ := \mathfrak{g}_{1/2}(\sigma) \cap \mathfrak{n}(\sigma)_+$ is a maximal isotropic subspace of $\mathfrak{g}_{1/2}(\sigma)$ with respect to $\langle\,.\,,\,.\,\rangle_{\mathrm{ne}}$,

(iv) $\mathfrak{n}_{1/2}(\sigma)_-$ is a direct sum of a maximal isotropic subspace $\mathfrak{n}_{1/2}(\sigma)_-'$ of $\mathfrak{g}_{1/2}(\sigma)$ with respect to $\langle\,.\,,\,.\,\rangle_{\mathrm{ne}}$ and at most 1-dimensional subspace $\mathfrak{g}_{1/2}^0(\sigma)$, normalized by $\mathfrak{h}^\sigma$.

Here and further we let $\mathfrak{n}_j(\sigma)_\pm = \mathfrak{n}(\sigma)_\pm \cap \mathfrak{g}_j(\sigma)$. We thus have the following decomposition:

$$\mathfrak{g}_{1/2}(\sigma) = \mathfrak{n}_{1/2}(\sigma)_+ + \mathfrak{g}_{1/2}(\sigma)_0 + \mathfrak{n}_{1/2}(\sigma)_-' \,, \tag{2.5}$$

where $\epsilon(\sigma) := \dim \mathfrak{g}_{1/2}(\sigma)_0 \leqslant 1$. Note that $\epsilon(\sigma) \neq 0$ iff $\dim \mathfrak{g}_{1/2}(\sigma)$ is odd.

Remark 2.1. Let $\mathfrak{g}_0(\sigma)^f$ be the centralizer in $\mathfrak{g}_0(\sigma)$ of $f \in \mathfrak{g}_{-1}(\sigma)$, and assume that there exists a semisimple element h_0 in $\mathfrak{g}_0(\sigma)^f$ such that all eigenvalues of ad h_0 on $\mathfrak{g}_{1/2}(\sigma)$ are real numbers and the multiplicity of zero is at most 1 (it follows from [EK], Theorem 1.5, that such an h_0 with all eigenvalues non-zero exists if $\mathfrak{g}(\sigma)$ is a Lie algebra). Let m denote the minimal absolute value of the non-zero eigenvalues. Let $H_0 \in \mathfrak{h}^\sigma$ be a regular element of $\mathfrak{g}(\sigma)$ such that all eigenvalues of ad H_0 are real, the eigenvalue on f is positive and their absolute values are smaller than m. Let $\mathfrak{n}(\sigma)_+$ (resp. $\mathfrak{n}(\sigma)_-$) denote the span of the eigenvectors of ad$\,(h_0 + H_0)$ in $\mathfrak{g}(\sigma)$ with positive (resp. negative) eigenvalues. This gives us a decomposition (2.4) satisfying all properties (i)–(iv). It is because the bilinear form $(\,.\,|\,.\,)$ is non-degenerate on $\mathfrak{g}(\sigma)$ and the bilinear form $\langle\,.\,,\,.\,\rangle_{\mathrm{ne}}$ is non-degenerate and invariant on $\mathfrak{g}(\sigma)_{1/2}$ with respect to $\mathfrak{g}_0(\sigma)^f$.

Let $D = -L_0^{\mathfrak{g},\mathrm{tw}}$. Recall that we have ($a \in \mathfrak{g}^{\bar{\mu}}$):

$$[D, at^\mu] = \mu(at^\mu)\,,\ [D, K] = 0\,.$$

As usual, we shall consider the extension of the Kac-Moody affinization (see Example 1.1):

$$\widehat{\mathfrak{g}}^{\mathrm{tw}} = \mathbb{C}D \ltimes \widehat{\mathfrak{g}}^{\mathrm{tw}\prime}\,.$$

The decomposition (2.4) induces a triangular decomposition of the Lie superalgebra $\widehat{\mathfrak{g}}^{\mathrm{tw}}$ (see Example 1.1):

$$\widehat{\mathfrak{g}}^{\mathrm{tw}} = \widehat{\mathfrak{n}}_- \oplus \widehat{\mathfrak{h}} \oplus \widehat{\mathfrak{n}}_+\,, \tag{2.6}$$

where

$$\widehat{\mathfrak{h}} = \mathfrak{h}^\sigma + \mathbb{C}K + \mathbb{C}D\,, \tag{2.7}$$

$$\widehat{\mathfrak{n}}_+ = \sum_{j \in \frac{1}{2}\mathbb{Z}} \Big(\mathfrak{n}_j(\sigma)_+ \otimes t^{-j} + \sum_{\substack{\mu \in \mathbb{R} \\ j+\mu > 0}} \mathfrak{g}_j^{\bar{\mu}} \otimes t^\mu\Big)\,, \tag{2.8}$$

$$\widehat{\mathfrak{n}}_- = \sum_{j \in \frac{1}{2}\mathbb{Z}} \Big(\mathfrak{n}_j(\sigma)_- \otimes t^{-j} + \sum_{\substack{\mu \in \mathbb{R} \\ j+\mu < 0}} \mathfrak{g}_j^{\bar{\mu}} \otimes t^\mu\Big)\,. \tag{2.9}$$

As usual, we extend the (non-degenerate) invariant bilinear for $(\,.\,|\,.\,)$ from $\mathfrak{h}^\sigma$ to $\widehat{\mathfrak{h}}$ by:

$$(\mathbb{C}K + \mathbb{C}D|\mathfrak{h}^\sigma) = 0\,,\ (K|K) = (D|D) = 0\,,\ (K|D) = 1\,.$$

This bilinear form is non-degenerate, and we shall identify $\widehat{\mathfrak{h}}$ and $\widehat{\mathfrak{h}}^*$ via this form.

Denote by $A_{\rm ne}$ the vector superspace $\mathfrak{g}_{1/2}$ with the bilinear form $\langle\,.\,,\,.\,\rangle_{\rm ne}$. Denote by A_+ (resp. A_-) the superspace $\mathfrak{g}_+$ (resp. its dual $\mathfrak{g}_+^*$) with reversed parity, and let $A_{\rm ch} = A_+ \oplus A_-$. Let $\langle\,.\,,\,.\,\rangle_{\rm ch}$ be the skew-supersymmetric bilinear form on $A_{\rm ch}$ defined by

$$\langle A_\pm\,,\,A_\pm\rangle_{\rm ch} = 0\,,\ \langle a\,,\,b^*\rangle_{\rm ch} = b^*(a) \text{ for } a \in A_+\,,\quad b^* \in A_-\,.$$

The automorphism σ of $\mathfrak{g}$ induces automorphisms of $A_{\rm ne}$ and $A_{\rm ch}$, which we again denote by σ, that preserve the respective bilinear forms. Finally, fix a complex number k such that $k + h^\vee \neq 0$.

We shall associate to the data $(\mathfrak{g}, x, f, k, \sigma)$ a twisted differential vertex algebra

$$(\mathcal{C}(\mathfrak{g}, \sigma, x, k),\ d_0^{\rm tw})\,.$$

Consider the twisted Kac-Moody affinization $\widehat{\mathfrak{g}}^{{\rm tw}\prime}$ and Clifford affinizations $\widehat{A}^{\rm tw}_{\rm ne}$ and $\widehat{A}^{\rm tw}_{\rm ch}$ (see Examples 1.1, 1.2 and 1.3). Let L be the direct sum of these Lie superalgebras with filtration $F_N L = F_N \widehat{\mathfrak{g}}^{{\rm tw}\prime} + F_N \widehat{A}^{\rm tw}_{\rm ne} + F_N \widehat{A}^{\rm tw}_{\rm ch}$. Let $U(L)^{\rm com}$ be the completed via this filtration universal enveloping algebra of L and let $U_k(L)^{\rm com}$ be the quotient of $U(L)^{\rm com}$ by the ideal generated by $K - k$, $K_{A_{\rm ne}} - 1$, $K_{A_{\rm ch}} - 1$. Recall that twisted currents, twisted neutral superfermions and twisted charged superfermions generate (via taking derivatives and normally ordered products) the twisted vertex algebras $V_k(\mathfrak{g}, \sigma)$, $F(A_{\rm ne}\,, \sigma)$ and $F(A_{\rm ch}, \sigma)$. We denote by $F(\mathfrak{g}, \sigma, x)$ the twisted vertex algebra generated by the last two and by $\mathcal{C}(\mathfrak{g}, \sigma, x, k)$ the one generated by all three types of formal distributions. We have:

$$F(\mathfrak{g}, \sigma, x) = F(A_{\rm ch}, \sigma) \otimes F(A_{\rm ne}\,, \sigma)\,, \mathcal{C}(\mathfrak{g}, \sigma, x, k) = V_k(\mathfrak{g}, \sigma) \otimes F(\mathfrak{g}, \sigma, x)\,.$$

By letting charge$(V_k(\mathfrak{g}, \sigma))$ = charge$(F(A_{\rm ne}\,, \sigma)) = 0$ and using (1.17), one has the induced charge decompositions:

$$F(\mathfrak{g}, \sigma, x) = \oplus_{m\in\mathbb{Z}} F(\mathfrak{g}, \sigma, x)^{\rm tw}_m\,,\ \mathcal{C}(\mathfrak{g}, \sigma, x, k) = \oplus_{m\in\mathbb{Z}} \mathcal{C}^{\rm tw}_m\,.$$

In order to define the differential $d_0^{\rm tw}$, and for further use, choose a basis $\{u_i\}_{i\in S}$ of $\mathfrak{g}$ compatible with the gradation (1.1), the σ-eigenspace decomposition and the root space decomposition with respect to $\mathfrak{h}^\sigma$. A part of this basis is a basis of $\mathfrak{g}_m$ ($m \in \frac{1}{2}\mathbb{Z}$), and of $\mathfrak{g}_+$. As in [KW], we denote the corresponding subsets of indices of S by S_m and S_+ respectively. Define the structure constants c^ℓ_{ij} by $[u_i, u_j] = \sum_\ell c^\ell_{ij} u_\ell$. Denote by $\{\varphi_i\}_{i\in S_+}$, $\{\varphi^*_i\}_{i\in S_+}$ the corresponding basis of A_+ and its dual basis of A_-, so that $\langle \varphi_i, \varphi^*_j\rangle_{\rm ch} = \delta_{ij}$, and by $\{\Phi_i\}_{i\in S_{1/2}}$ the corresponding basis of $A_{\rm ne}$. We shall denote by $\{u^i\}$ the dual basis of $\mathfrak{g}$ with respect to the form $(.|.)$ and by $\{\Phi^i\}_{i\in S_{1/2}}$ the dual basis of $A_{\rm ne}$ with respect to the form $\langle .,.\rangle_{\rm ne}$.

Recall that in the non-twisted case, i.e., when $\sigma = 1$, we defined $d_0 = \mathrm{Res}_z\, d(z)$, where $d(z)$ is the following formal distribution of the vertex algebra $\mathcal{C}(\mathfrak{g}, 1, x, k)$ [KRW], [KW]:

$$\begin{aligned} d(z) \quad &= \quad \sum_{i \in S_+} (-1)^{p(u_i)} u_i(z) \otimes \varphi_i^*(z) \otimes 1 \\ &\quad -\frac{1}{2} \sum_{i,j,\ell \in S_+} (-1)^{p(u_i)p(u_\ell)} c_{ij}^\ell \otimes : \varphi_\ell(z) \varphi_i^*(z) \varphi_j^*(z) : \otimes 1 \\ &\quad + \sum_{i \in S_+} (f|u_i) \otimes \varphi_i^*(z) \otimes 1 + \sum_{i \in S_{1/2}} 1 \otimes \varphi_i^*(z) \otimes \Phi_i(z) . \end{aligned}$$

Further on, for simplicity of notation, we shall omit the tensor sign. Note that in each summand of $d(z)$, factors are commuting formal distributions. Hence the corresponding (via Proposition 1.1) twisted formal distribution $d^{\mathrm{tw}}(z)$ of $\mathcal{C}(\mathfrak{g}, \sigma, x, k)$ is given by the same expression as $d(z)$, where all factors $u_i(z)$, $\varphi_i^*(z)$, etc.are replaced by $u_i^{\mathrm{tw}}(z)$, $\varphi_i^{*\mathrm{tw}}(z)$, etc. (it is because $: a^{\mathrm{tw}}(z) b^{\mathrm{tw}}(z) := (a_{(-1)}b)^{\mathrm{tw}}(z)$ if $a_{(j)}b = 0$ for $j \in \mathbb{Z}_+$, by (1.14)). Since $[d(z), d(w)] = 0$, it follows that $[d^{\mathrm{tw}}(z), d^{\mathrm{tw}}(w)] = 0$. Hence the odd element $d_0^{\mathrm{tw}} = \mathrm{Res}_z\, d^{tw}(z)$ has the property that $(d_0^{\mathrm{tw}})^2 = 0$. Note also that d_0^{tw} is a derivation of all products of the twisted vertex algebra $\mathcal{C}(\mathfrak{g}, \sigma, x, k)$ and $d_0^{\mathrm{tw}}(\mathcal{C}_m^{\mathrm{tw}}) \subset \mathcal{C}_{m-1}^{\mathrm{tw}}$.

Denote the homology of the complex $(\mathcal{C}(\mathfrak{g}, \sigma, x, k), d_0^{\mathrm{tw}})$ by $W_k(\mathfrak{g}, \sigma, x)$. This σ-twisted vertex algebra is called the *σ-twisted quantum reduction for the triple* $(\mathfrak{g}, \sigma, x)$. The automorphism σ of $\mathfrak{g}$ obviously induces a diagonalizable automorphism of the vertex algebra $\mathcal{C}(\mathfrak{g}, 1, x, k)$, commuting with the operator d_0. Hence it induces a diagonalizable automorphism, also denoted by σ, of the vertex algebra $W_k(\mathfrak{g}, 1, x)$.

The most important formal distribution of $W_k(\mathfrak{g}, \sigma, x)$ is the σ-twist $L^{\mathrm{tw}}(z)$ of the Virasoro formal distribution $L(z)$, defined by (2.2) of [KW]:

$$L^{\mathrm{tw}}(z) = L^{\mathfrak{g},\mathrm{tw}}(z) + L^{\mathrm{ne}\,,\mathrm{tw}}(z) + L^{\mathrm{ch},\mathrm{tw}}(z) + \partial_z x^{\mathrm{tw}}(z) ,$$

where the m_i in (1.19) are defined by $u_i \in \mathfrak{g}_{m_i}$.

Recall that the building blocks of the vertex algebra $W_k(\mathfrak{g}, x)$ are the following formal distributions [KW]:

$$J^{(v)}(z) = v(z) + \sum_{i,j \in S_+} (-1)^{p(u_i)} c_{ij}(v) : \varphi_i(z) \varphi_j^*(z) : ,$$

where $v \in \mathfrak{g}$ and $c_{ij}(v)$ is the matrix of ad v in the basis $\{u_i\}$, i.e., $[v, u_j] = \sum_i c_{ij}(v) u_i$. Using (1.14), we obtain the following formula for the corresponding twisted formal distribution ($v \in \mathfrak{g}$):

$$\begin{aligned} J^{(v)\mathrm{tw}}(z) \quad &= \quad v^{tw}(z) + \sum_{i,j \in S_+} (-1)^{p(u_i)} c_{ij}(v) : \varphi_i^{\mathrm{tw}}(z) \varphi_j^{*\mathrm{tw}}(z) : \\ &\quad - \sum_{i \in S_+} (-1)^{p(u_i)} s_i c_{ii}(v) z^{-1} . \end{aligned} \tag{2.10}$$

Theorem 4.1 of [KW] implies the following result.

Theorem 2.1.

(a) *For each* $a \in (\mathfrak{g}^{\bar{\mu}}_{-j})^f$, $j \geqslant 0$, *there exists a* d_0^{tw}*-closed twisted formal distribution* $J^{\{a\},\mathrm{tw}}(z)$ *in* $\mathcal{C}(\mathfrak{g},\sigma,x,k)$ *of conformal weight* $1+j$ *(with respect to* $L^{\mathrm{tw}}(z)$*) such that* $J^{\{a\},\mathrm{tw}}(z) - J^{(a),\mathrm{tw}}(z)$ *is a linear combination of normally ordered products of the twisted formal distributions* $J^{(b)\mathrm{tw}}(z)$*, where* $b \in \mathfrak{g}_{-s}$*,* $0 \leqslant s < j$*, the twisted formal distributions* $\Phi_i^{\mathrm{tw}}(z)$*, where* $i \in S_{1/2}$*, and their derivatives.*
(b) *The homology classes of the formal distributions* $J^{\{a_i\},\mathrm{tw}}(z)$*, where* $\{a_i\}$ *is a basis of* $\mathfrak{g}^f_{\leqslant 0}$ *compatible with the* $\frac{1}{2}\mathbb{Z}$*-gradation and* σ*-eigenspace decomposition, strongly and freely generate* $W_k(\mathfrak{g},\sigma,x)$*.*
(c) $W_k(\mathfrak{g},\sigma,x)$ *is a* σ*-twist of the vertex algebra* $W_k(\mathfrak{g},1,x)$*.*
(d) $W_k(\mathfrak{g},\sigma,x)$ *coincides with the* 0*th homology of the complex* $(\mathcal{C}(\mathfrak{g},\sigma,x,k),\, d_0^{\mathrm{tw}})$*.*

3. Modules over $W_k(\mathfrak{g},\sigma,x)$

Denote by $S' \subset S$ the subset of indices of the part of the basis $\{u_i\}_{i\in S}$ of $\mathfrak{g}$, which is a basis of $\mathfrak{g} \mod \mathfrak{h}^\sigma$, and let $S'_0 = S_0 \cap S'$. In the case when $\mathfrak{h} = \mathfrak{h}^\sigma$, S' can be identified with the set of roots of $\mathfrak{g}$ with respect $\mathfrak{h}$, but it is larger otherwise.

Recall that, given a diagonalizable automorphism σ of a vertex algebra V, so that $V = \oplus_{\bar{\mu}\in\mathbb{R}/\mathbb{Z}} V^{\bar{\mu}}$ is its eigenspace decomposition, a σ-twisted module M over V is a linear map $a \to a^{M,\mathrm{tw}}(z) = \sum_{n\in\bar{\mu}} a^M_{(n)} z^{-n-1}$ $(a \in V^{\bar{\mu}})$ satisfying equations (1.12)–(1.14), where $a^M_{(n)} \in \operatorname{End} M$ and for any $v \in M$, $a^M_{(n)}v = 0$ if $n \gg 0$. In other words, the collection of fields $a^{M,\mathrm{tw}}(z)$ forms a σ-twist of the vertex algebra V. In this section we shall discuss the properties of σ-twisted modules over $W_k(\mathfrak{g},x)$ (= modules over $W_k(\mathfrak{g},\sigma,x)$) obtained by the σ-twisted quantum reduction from restricted $\widehat{\mathfrak{g}}^{\mathrm{tw}}$-modules.

We shall embed $\mathfrak{h}^{\sigma *}$ in $\widehat{\mathfrak{h}}^*$ by letting $\lambda \in \mathfrak{h}^{\sigma *}$ be zero on K, and we define $\delta \in \widehat{\mathfrak{h}}^*$ by $\delta|_{\mathfrak{h}^\sigma + \mathbb{C}K} = 0$, $\delta(D) = 1$. Recall that, given a triangular decomposition (2.6), a *highest weight module* over the Lie superalgebra $\widehat{\mathfrak{g}}^{\mathrm{tw}}$ of level k and with highest weight $\Lambda \in \mathfrak{h}^*$ is a $\widehat{\mathfrak{g}}^{\mathrm{tw}}$-module M which admits a non-zero vector $v_{\widehat{\Lambda}}$, where $\widehat{\Lambda} = \Lambda + kD$, with the properties:

(i) $hv_{\widehat{\Lambda}} = \widehat{\Lambda}(h)v_{\widehat{\Lambda}}$, $h \in \widehat{\mathfrak{h}}$,
(ii) $\widehat{\mathfrak{n}}_+ v_{\widehat{\Lambda}} = 0$,
(iii) $U(\widehat{\mathfrak{n}}_-)v_{\widehat{\Lambda}} = M$.

For this reason, in the definition (1.10) of the annihilation part of the twisted current $u_i(z)$ $(i \in S')$, we choose

$$s_{u_i} = \min\{n |\, u_i \otimes t^n \text{ is non-zero and lies in } \widehat{\mathfrak{n}}_+\}\,, s_h = 1 \,\text{for}\, h \in \mathfrak{h}^\sigma. \tag{3.1}$$

Since each summand $\mathfrak{g}_j$ of the gradation (2.1) is σ-invariant, we have its σ-eigenspace decomposition:

$$\mathfrak{g}_j = \oplus_{\bar{\mu}\in\mathbb{R}/\mathbb{Z}}\mathfrak{g}_j^{\bar{\mu}}\,, \text{where}\, \mathfrak{g}_j^{\bar{\mu}} = \{a \in \mathfrak{g}_j | \sigma(a) = e^{2\pi i\bar{\mu}}a\}.$$

Hence for a basis element $u_i \in \mathfrak{g}_{m_i}^{\bar{\mu}_i}$ we can rewrite formula (3.1) for $s_i = s_{u_i}$ ($i \in S'$) as follows:

$$s_i = \begin{cases} \min\{n \in \bar{\mu}_i | \, n > -m_i\} \text{ if } & u_i \notin \mathfrak{n}(\sigma)_+\,, \\ -m_i \text{ if } u_i \in \mathfrak{n}(\sigma)_+\,. \end{cases} \tag{3.2}$$

It is easy to see that for a dual basis element $u^i \in \mathfrak{g}_{-m_i}^{-\bar{\mu}}$ we have for $s^i = s_{u^i}$:

$$s^i = 1 - s_i \text{ for all } i \in S'\,. \tag{3.3}$$

We extend this definition of annihilation operators to $\widehat{A}_{\text{ne}}^{\text{tw}}$ and $\widehat{A}_{\text{ch}}^{\text{tw}}$ as follows:

$$s_{\Phi_i} = s_i \,(i \in S_{1/2})\,,\; s_{\varphi_i} = s_i\,,\; s_{\varphi_i^*} = 1 - s_i \;(i \in S_+)\,. \tag{3.4}$$

It is easy to see that we have

$$s_{\Phi_i} = \mp 1/2 \text{ if } \Phi_i \in \mathfrak{n}_{1/2}(\sigma)_\pm\,,\; |s_{\Phi_i}| < 1/2 \text{ otherwise.} \tag{3.5}$$

$$s_{\Phi_i} + s_{\Phi^i} = \delta_{i,i_0}\,, \text{ where } \langle\Phi_{i_0}, \Phi_{i_0}\rangle_{\text{ne}} \neq 0. \tag{3.6}$$

We write the generating fields in the form:

$$\begin{aligned} u_i(z) &= \sum_{n\in\bar{s}_i} u_{i,n}z^{-n-1}\,,\; \Phi_i(z) = \sum_{n\in\bar{s}_i+1/2} \Phi_{i,n}z^{-n-1/2}\,, \\ \varphi_i(z) &= \sum_{n\in\bar{s}_i} \varphi_{i,n}z^{-n-1}\,,\; \varphi_i^*(z) = \sum_{n\in-\bar{s}_i} \varphi_{i,n}^* z^{-n}\,. \end{aligned}$$

Each of the Clifford affinizations $\widehat{A}_{\text{ne}}^{\text{tw}}$ and $\widehat{A}_{\text{ch}}^{\text{tw}}$ has a unique irreducible module, denoted by $F_{\text{ne}}^{\text{tw}}$ and $F_{\text{ch}}^{\text{tw}}$, respectively, admitting a non-zero vector $|0\rangle_{\text{ne}}$ and $|0\rangle_{\text{ch}}$, respectively, killed by all annihilation operators:

$$\Phi_{i,n}|0\rangle_{\text{ne}} = 0 \text{ for } n \geqslant s_i + 1/2\,, \tag{3.7}$$

$$\varphi_{i,n}|0\rangle_{\text{ch}} = 0 \text{ for } n \geqslant s_i\,,\; \varphi_{i,n}^*|0\rangle = 0 \text{ for } n \geqslant 1 - s_i\,. \tag{3.8}$$

Since these modules are restricted, they extend to the modules over $F(A_{\text{ne}}, \sigma)$ and $F(A_{ch},\sigma)$ (= twisted modules over the vertex algebras $F(A_{\text{ne}},1)$ and $F(A_{\text{ch}},1)$), respectively), hence

$$F^{\text{tw}} = F_{\text{ne}}^{\text{tw}} \otimes F_{\text{ch}}^{\text{tw}}$$

is a module over $F(\mathfrak{g}, \sigma, x)$ (= twisted module over the tensor product of these vector algebras, $F(\mathfrak{g}, 1, x)$). We let

$$|0\rangle = |0\rangle_{\text{ne}} \otimes |0\rangle_{\text{ch}} \in F^{\text{tw}}\,.$$

Thus, given a restricted $\widehat{\mathfrak{g}}^{\text{tw}}$-module M with $K = kI$, we extend it to a module over $V_k(\mathfrak{g}, \sigma)$ (= twisted module over the vertex algebra $V_k(\mathfrak{g}, 1)$), then $M \otimes F^{\text{tw}}$ becomes a module over $\mathcal{C}(\mathfrak{g}, \sigma, x, k)$ (= twisted module over the vertex algebra $\mathcal{C}(\mathfrak{g}, x, k)$). Passing to the homology of the complex $\mathcal{C}^{\text{tw}}(M) = (M \otimes F^{\text{tw}}, d_0^{\text{tw}})$, we obtain a $W_k(\mathfrak{g}, \sigma, x)$-module (= twisted $W_k(\mathfrak{g}, x)$-module) $H^{\text{tw}}(M)$. One has the charge decomposition of $C^{\rm tw}(M)$ induced by that of $F(\mathfrak{g},\sigma,x)$ by setting the charge of M to be zero. This induces a decomposition as $W_k(\mathfrak{g},\sigma,x)$-modules: $H^{\rm tw}(M)=\sum_{j\in\mathbb{Z}}H^{\rm tw}_j(M)$.

Let $\Delta^\sigma\subset\mathfrak{h}^{\sigma *}$ be the set of non-zero roots of $\mathfrak{g}$ with respect to $\mathfrak{h}^\sigma$, counted with their multiplicities. We may identify Δ^σ with a subset of S', which indexes root vectors attached to non-zero roots. (Then the remaining elements of S' index a basis of $\mathfrak{h}$ mod $\mathfrak{h}^\sigma$.) Given one of the above basis root vectors e_α, attached to $\alpha\in\Delta^\sigma$, we let $s_\alpha=s_{e_\alpha}$. One should keep in mind that the s_α corresponding to root vectors with the same α may be different (in the case $\mathfrak{h}^\sigma\neq\mathfrak{h}$).

Recall that the set of roots $\widehat{\Delta}\subset\widehat{\mathfrak{h}}^*$ of the twisted affine Lie superalgebra $\widehat{\mathfrak{g}}^{\rm tw}$ is $\widehat{\Delta}=\widehat{\Delta}^{\rm re}\cup\widehat{\Delta}^{\rm im}$, where:

$$\widehat{\Delta}^{\rm re}=\{\alpha+(m+s_\alpha)\delta|\, m\in\mathbb{Z}\,,\,\alpha\in\Delta^\sigma\}\,,\;\widehat{\Delta}^{\rm im}=\{m\delta|\, m\in E_0\backslash\{0\}\}\,,$$

where $E_0=\{\mu\in\mathbb{R}|\, e^{2\pi i\mu}$ is an eigenvalue of σ on $\mathfrak{h}\}$, and the roots are considered with their multiplicities. Then we have a subset $\widehat{\Delta}_+=\widehat{\Delta}^{\rm re}_+\cup\widehat{\Delta}^{\rm im}_+$ of positive roots in $\widehat{\Delta}$, corresponding to $\widehat{\mathfrak{n}}_+$ (see (2.8)), where

$$\widehat{\Delta}^{\rm re}_+=\{\alpha+(m+s_\alpha)\delta|\, m\in\mathbb{Z}_+\,,\,\alpha\in\Delta^\sigma\}\,,\;\widehat{\Delta}^{\rm im}_+=\{m\delta|\, m\in E_0\,,\,m>0\}\,.$$

Introduce the following subset of $\widehat{\Delta}^{\rm re}_+$:

$$\widehat{\Delta}^{\rm re}_{++}=\{\alpha+(m+s_\alpha)\delta|\,\alpha\in\Delta^\sigma\,,\,\alpha(x)\geqslant 0\,,\,m\in\mathbb{Z}_+\}\,.$$

Proposition 3.1.

(a) *If M is a restricted $\widehat{\mathfrak{g}}^{\rm tw}$-module and $v\in M$ is a singular vector, i.e., $\widehat{\mathfrak{n}}^{\rm tw}_+v=0$, then*

$$d^{\rm tw}_0(v\otimes|0\rangle)=0\,.$$

(b) *If M is a Verma module over $\widehat{\mathfrak{g}}^{\rm tw}$ with the highest weight vector $v_{\widehat{\Lambda}}$ and $v\in M$ is a singular vector with highest weight $\widehat{\Lambda}-n\alpha$, where $\alpha\in\widehat{\Delta}^{\rm re}_{++}$, then the homology class of $v\otimes|0\rangle$ in $H^{\rm tw}_0(M)$ is non-zero.*

Proof. We have: $d^{\rm tw}_0=A+B+C+D$, where

$$A=\sum_{i\in S_+}\sum_{\substack{p\in\bar{\mu}_i\\ q\in-\bar{\mu}_i\\ p+q=0}}(-1)^{p(u_i)}u_{i,p}\varphi^*_{i,q}\,,$$

$$B=-\frac{1}{2}\sum_{i,j,k\in S_+}\sum_{\substack{p\in\bar{\mu}_k\\ q\in-\bar{\mu}_i\\ r\in-\bar{\mu}_j\\ p+q+r=0}}(-1)^{p(u_i)p(u_k)}c^k_{ij}\varphi_{k,p}\varphi^*_{i,q}\varphi^*_{j,r}\,,$$

$$C=\sum_{i\in S_+}(f|u_i)\varphi^*_{i,1}\,,\;D=\sum_{i\in S_{1/2}}\sum_{\substack{p\in-\bar{\mu}_i\\ q\in\bar{\mu}_i+1/2\\ p+q=0}}\varphi^*_{i,p}\Phi_{i,q+1/2}\,.$$

A summand of A does not annihilate $v \otimes |0\rangle$ only if $p \leqslant s_i - 1$, $q \leqslant s_i$, hence there are no such summands since $p + q = 0$.

A summand of B does not annihilate $v \otimes |0\rangle$ only if $p \leqslant s_k - 1$, $q \leqslant -s_i$, $r \leqslant s_j$, which happens only if $p+q+r \leqslant s_k - s_i - s_j - 1 \leqslant -1$, since $s_k \leqslant s_i + s_j$ when $c_{ij}^k \neq 0$. Hence there are no such summands.

If $(f|u_i) \neq 0$, then $(ft|u_i t^{-1}) \neq 0$, and since $ft \in \widehat{\mathfrak{n}}_+$, we obtain that $u_i t^{-1} \in \mathfrak{n}_-$ and therefore $s_i \geqslant 0$, by definition of s_i. Hence $\varphi_{i,1}^*$ is an annihilation operator (see (3.8)) and $C(v \otimes |0\rangle) = 0$.

Finally, if a summand of D does not annihilate the $v \otimes |0\rangle$, then $p \leqslant s_i$ and $q + 1/2 \leqslant s_i - 1/2$ and therefore $p + q = -1$, which is impossible since $p + q = 0$.

This proves (a). The proof of (b) is the same as in the non-twisted case, see [KW], Lemma 7.3. □

Next, we study the formal distribution $L^{\mathrm{tw}}(z)$ of $W_k(\mathfrak{g}, \sigma, x)$. Using formulas (1.15), (1.16) and (1.19) for the first three summands, we obtain an explicit expression for $L^{\mathrm{tw}}(z) = \sum_{n \in \mathbb{Z}} L_n^{\mathrm{tw}} z^{-n-2}$. Note that the L_n^{tw} form a Virasoro algebra with the same central charge as in the non-twisted case. Examples 1.1, 1.2 and 1.3 give the following important formulas.

Proposition 3.2. *Introducing the constants*

$$s_{\mathfrak{g}} = -\frac{k}{2(k+h^\vee)} \sum_{\alpha \in S'} (-1)^{p(\alpha)} \binom{s_\alpha}{2}, \tag{3.9}$$

$$s_{\mathrm{ne}} = \frac{1}{8}\epsilon(\sigma) - \frac{1}{2} \sum_{\alpha \in S_{1/2}} (-1)^{p(\alpha)} \binom{s_\alpha}{2}, \tag{3.10}$$

$$s_{\mathrm{ch}} = \sum_{\alpha \in S_+} (-1)^{p(\alpha)} \left(\binom{s_\alpha}{2} + m_\alpha s_\alpha \right), \tag{3.11}$$

we have

$$\begin{aligned} L_0^{\mathfrak{g},\mathrm{tw}} &= \frac{1}{2(k+h^\vee)} \left(\sum_i h_i h^i - \sum_{i \in S'} (-1)^{p(\alpha)} s_\alpha \alpha \right) + s_{\mathfrak{g}} + \mathrm{ann}\ ; \\ L_0^{\mathrm{ne},\mathrm{tw}} |0\rangle_{\mathrm{ne}} &= (s_{\mathrm{ne}} + \mathrm{ann})|0\rangle_{\mathrm{ne}}\ ,\ L_0^{\mathrm{ch},\mathrm{tw}}|0\rangle_{\mathrm{ch}} = (s_{\mathrm{ch}} + \mathrm{ann})|0\rangle_{\mathrm{ch}}\ ; \\ L_0^{\mathrm{tw}} &= L_0^{\mathfrak{g},\mathrm{tw}} + L_0^{\mathrm{ne},\mathrm{tw}} + L_0^{\mathrm{ch},\mathrm{tw}} - x\,, \end{aligned}$$

where ann *(resp.* ann $|0\rangle$*) denotes the sum of terms which annihilate any singular vector in a* $\widehat{\mathfrak{g}}^{\mathrm{tw}}$*-module M of level k (resp. annihilate the vacuum vector), and $\{h_i\}$ and $\{h^i\}$ are dual bases of $\mathfrak{h}^\sigma$.*

Proof. We have: $\sum_{\alpha \in S'} (-1)^{p(\alpha)} s_\alpha [u_\alpha, u^\alpha] = \sum_i a_i h_i$, where $a_i \in \mathbb{C}$. Hence

$$a_i = \sum_{\alpha \in S'} (-1)^{p(\alpha)} s_\alpha ([u_\alpha, u^\alpha] |\, h^i) = \sum_{\alpha \in S'} (-1)^{p(\alpha)} s_\alpha \alpha(h^i).$$

Hence

$$\sum_{\alpha\in S'}(-1)^{p(\alpha)}s_\alpha[u_\alpha,u^\alpha]=\sum_{\alpha\in S'}(-1)^{p(\alpha)}s_\alpha\alpha.$$

The rest of the calculation is straightforward. □

Corollary 3.1. *Let v be a singular vector of a $\widehat{\mathfrak{g}}^{\mathrm{tw}}$ -module M of level k such that $av=\Lambda(a)v$, $a\in\mathfrak{h}^\sigma$, for some $\Lambda\in\mathfrak{h}^{\sigma *}$. Then $L_0^{\mathrm{tw}}(v\otimes|0\rangle)=h\ v\otimes|0\rangle$, where*

$$h=\frac{1}{2(k+h^\vee)}((\Lambda|\Lambda)-\sum_{\alpha\in S'}(-1)^{p(\alpha)}s_\alpha(\Lambda|\alpha))-\Lambda(x)+s_{\mathfrak{g}}+s_{\mathrm{ne}}+s_{\mathrm{ch}}\,.$$

Corollary 3.2. *Let $\gamma'=\frac{1}{2}\sum_{\alpha\in S'}(-1)^{p(\alpha)}\alpha\in\mathfrak{h}^{\sigma *}$, and let $\widehat{\rho}^{\mathrm{tw}}$ be the Weyl vector for $\widehat{\mathfrak{g}}^{\mathrm{tw}}$ (i.e., $(\widehat{\rho}^{\mathrm{tw}}|\alpha_i)=\frac{1}{2}(\alpha_i|\alpha_i)$ for all simple roots α_i of $\widehat{\mathfrak{g}}^{\mathrm{tw}}$). Then $\widehat{\rho}^{\mathrm{tw}}|_{\mathfrak{h}^\sigma}=-\gamma'$.*

Proof. By Proposition 3.2 we have in any highest weight $\widehat{\mathfrak{g}}^{\mathrm{tw}}$-module of level k with highest weight $\Lambda\in\mathfrak{h}^{\sigma *}$:

$$L_0^{\mathfrak{g},\mathrm{tw}}v_\Lambda=\frac{1}{2(k+h^\vee)}\left(((\Lambda|\Lambda)-2(\Lambda|\gamma'))+c_1\right)v_\Lambda\,, \tag{3.12}$$

where $c_1\in\mathbb{C}$ is independent of Λ. On the other hand, the operator $L_0^{\mathfrak{g},\mathrm{tw}}+D$ commutes with $\widehat{\mathfrak{g}}^{\mathrm{tw}}$, hence equals $c_2\Omega^{\mathrm{tw}}+c_3$, where $c_2,c_3\in\mathbb{C}$ are independent of Λ and Ω^{tw} is the Casimir operator of $\widehat{\mathfrak{g}}^{\mathrm{tw}}$. But $\Omega^{\mathrm{tw}}v_\Lambda=(\Lambda|\Lambda+2\widehat{\rho}^{\mathrm{tw}})v_\Lambda$ (see [K3]) and $Dv_\Lambda=0$, hence, comparing with (3.12) we obtain for any $\Lambda\in\mathfrak{h}^{\sigma *}$:

$$\frac{1}{2(k+h^\vee)}((\Lambda|\Lambda)-2(\Lambda|\gamma'))+c_1=c_2\left((\Lambda|\Lambda)+2(\widehat{\rho}^{\mathrm{tw}}|\Lambda)\right)+c_3.$$

Comparing quadratic terms in Λ we obtain $c_2=(2k+2h^\vee)^{-1}$. Comparing linear terms in Λ, we get $\widehat{\rho}^{\mathrm{tw}}|_{\mathfrak{h}^\sigma}=-\gamma'$. □

Recall that the conformal weight 1 formal distributions of the vertex algebra $W_k(\mathfrak{g},x)$ are [KW]:

$$J^{\{v\}}(z)=J^{(v)}(z)-\frac{1}{2}\sum_{i,j\in S_{1/2}}(-1)^{p(u_i)}c_{ij}(v):\Phi_i(z)\Phi^j(z):\ (v\in\mathfrak{g}_0^f).$$

Hence, by Equation (1.14), the corresponding twisted formal distributions of $W_k(\mathfrak{g},\sigma,x)$ can be explicitly expressed via twisted currents and twisted ghosts. In the sequel we shall need the following formula, in the case when $a\in\mathfrak{h}^{\sigma f}$:

$$J_0^{\{a\}\mathrm{tw}}=a-\sum_{i\in S_+}(-1)^{p(u_i)}s_ic_{ii}(a)+\frac{1}{2}\sum_{i\in S_{1/2}}(-1)^{p(u_i)}s_ic_{ii}(a)+\mathrm{ann}\,, \tag{3.13}$$

where ann denotes an operator which annihilates any vector of the form $v\otimes|0\rangle\in M\otimes F^{\mathrm{tw}}$. Formula (3.13) implies the following corollary.

Corollary 3.3. *Under the conditions of Corollary* 3.1, *the eigenvalue of* $J_0^{\{H\},\mathrm{tw}}$ ($H \in \mathfrak{h}^{\sigma f}$) *on the vector* $v \otimes |0\rangle$ *is equal to*

$$\Lambda(H) + \frac{1}{2} \sum_{\alpha \in S_{1/2}} (-1)^{p(\alpha)} s_\alpha \alpha(H) - \sum_{\alpha \in S_+} (-1)^{p(\alpha)} s_\alpha \alpha(H) .$$

As in [KRW], define the Euler-Poincaré character of $H^{\mathrm{tw}}(M)$ by the following formula, where $h \in \mathfrak{h}^{\sigma f}$ and $\tau \in \mathbb{C}$, $\mathrm{Im}\ \tau > 0$:

$$\mathrm{ch}_{H^{\mathrm{tw}}(M)}(\tau, h) = \sum_{j \in \mathbb{Z}} (-1)^j \mathrm{tr}_{\,\mathrm{H}_j^{\mathrm{tw}}(\mathrm{M})} \mathrm{e}^{2\pi \mathrm{i} \tau \mathrm{L}_0^{\mathrm{tw}}} \mathrm{e}^{2\pi \mathrm{i} \mathrm{J}_0^{\{\mathrm{h}\}}} .$$

The same argument as in [KRW] gives an explicit formula in terms of the character

$$\mathrm{ch}_M(\tau, z) = \mathrm{tr}_{\,\mathrm{M}} \mathrm{e}^{2\pi \mathrm{i}(\mathrm{z} + \tau \mathrm{L}_0^{\mathfrak{g},\mathrm{tw}})} , \mathrm{z} \in \mathfrak{h}^\sigma ,$$

of the $\widehat{\mathfrak{g}}^{\mathrm{tw}}$-module M:

$$\mathrm{ch}_{H^{\mathrm{tw}}(M)}(\tau, h) = e^{2\pi i \tau (s_{\mathrm{ne}} + s_{\mathrm{ch}})} \times \left(\mathrm{ch}_M \prod_{\substack{\alpha \in \widehat{\Delta}_+ \\ \alpha(x) \neq 0, -1/2}} (1 - \tilde{p}(\alpha) e^{-\alpha})^{\tilde{p}(\alpha)\, \mathrm{mult}\, \alpha} \right) (\tau, -\tau x + h). \tag{3.14}$$

Here and further, in order to simplify notation, we let $\tilde{p}(\alpha) = (-1)^{p(\alpha)}$, and for $\alpha \in \widehat{\mathfrak{g}}^*$, we define $\alpha(\tau, z) = 2\pi i \alpha(z - \tau D)$.

The conditions of non-vanishing of $\mathrm{ch}_{H^{\mathrm{tw}}(M)}$ are similar to those in the non-twisted case [KRW]. Namely, the same argument as in [KRW], Theorem 3.2, gives the following result.

Proposition 3.3. *Let* M *be a restricted* $\widehat{\mathfrak{g}}^{\mathrm{tw}}$*-module of level* $k \neq -h^\vee$ *and assume that* $\mathrm{ch}_M(\tau, h)$ *extends to a meromorphic function on the upper half space* $\mathrm{Im}\ \tau > 0$, $h \in \mathfrak{h}^\sigma$, *with at most simple poles at the hyperplanes* $\alpha = 0$, *where* α *are real even roots. Then* $\mathrm{ch}_{H^{\mathrm{tw}}(M)}(\tau, h)$ *is not identically zero iff the* $\widehat{\mathfrak{g}}^{\mathrm{tw}}$*-module* M *is not locally nilpotent with respect to all root spaces* $\widehat{\mathfrak{g}}_{-\alpha}$, *such that* α *are positive even real roots satisfying the following three properties:*

$$\textit{(i)}\ \alpha(D + x) = 0 , \quad \textit{(ii)}\ \alpha|_{(\mathfrak{h}^\sigma)^f} = 0 , \quad \textit{(iii)}\ \alpha(x) \neq 0 , -1/2 .$$

We shall use formula (3.14) and [KW2, KW3] to compute the characters of $W_k(\mathfrak{g}, \sigma, x)$-modules in a subsequent paper [KW4] (cf. [FKW, KRW]).

Remark 3.1. A slightly more explicit form of (3.14) is as follows:

$$\mathrm{ch}_{H^{\mathrm{tw}}(M)}(\tau, h) = e^{2\pi i \tau (s_{\mathrm{ne}} + s_{\mathrm{ch}})}$$

$$\times \Bigg(ch_M \prod_{\substack{\alpha \in S' \\ \alpha(x) > 0}} \prod_{n=1}^{\infty} \left((1 - \tilde{p}(\alpha) e^{-(n - s_\alpha)\delta + \alpha})(1 - \tilde{p}(\alpha) e^{-(n-1+s_\alpha)\delta - \alpha}) \right)^{\tilde{p}(\alpha)\, \mathrm{mult}\, \alpha}$$

$$\times \prod_{\substack{\alpha \in S' \\ \alpha(x) = 1/2}} \prod_{n=1}^{\infty} (1 - \tilde{p}(\alpha) e^{-(n - s_\alpha)\delta + \alpha})^{-\tilde{p}(\alpha)\, \mathrm{mult}\, \alpha} \Bigg) (\tau, -\tau x + h) .$$

Let $a \in (\mathfrak{g}^{\bar{\mu}}_{-j})^f$, $j \geqslant 0$, and let $J^{\{a\},\mathrm{tw}}(z)$ be the corresponding formal distribution of $W_k(\mathfrak{g},\sigma,x)$ (see Theorem 2.1). As in the non-twisted case [KW], its conformal weight with respect to $L^{\mathrm{tw}}(z)$ equals $\Delta_a = j+1$. We therefore write

$$J^{\{a\},\mathrm{tw}}(z) = \sum_{n \in \bar{\mu} - \Delta_a} J^{\{a\},\mathrm{tw}}_n z^{-n-\Delta_a}\,. \tag{3.15}$$

Recall the isomorphism as $\mathfrak{g}^f_0$-modules $\mathfrak{g}^f \cong \mathfrak{g}_0 + \mathfrak{g}_{1/2}$ given by [KW], (1.12). We shall identify $\mathfrak{g}_0$ (and its subspace $\mathfrak{h}^\sigma$) with a σ-invariant subspace of $\mathfrak{g}^f$, using this isomorphism. A $W_k(\mathfrak{g},\sigma,x)$-module M is called a *highest weight module* with highest weight $\lambda \in (\mathfrak{h}^\sigma)^*$ if there exists a non-zero vector $v_\lambda \in M$ such that:

$$\text{polynomials in the operators } J^{\{a\},\mathrm{tw}}_n \text{ applied to } v_\lambda \text{ span } M\,, \tag{3.16}$$

$$J^{\{a\},\mathrm{tw}}_0 v_\lambda = \lambda(a) v_\lambda \text{ if } a \in \mathfrak{h}^\sigma\,, \tag{3.17}$$

$$J^{\{a\},\mathrm{tw}}_m v_\lambda = 0 \text{ if } m > 0 \text{ or } m = 0 \text{ and } a \in \mathfrak{n}_0(\sigma)_+\,. \tag{3.18}$$

The Verma module is defined in the same way as in [KW], and we have the following twisted analogue of Theorem 6.3 from [KW].

Theorem 3.1. *If P is a Verma module over the Lie superalgebra $\widehat{\mathfrak{g}}^{\mathrm{tw}}$, then $H^{\mathrm{tw}}(P) = H^{\mathrm{tw}}_0(P)$, and it is a Verma module over $W_k(\mathfrak{g},\sigma,x)$.*

4. Modules over $W_k(\mathfrak{g},\sigma,\theta/2)$, the free field realizations and determinant formulas

Of particular interest are the vertex algebras $W_k(\mathfrak{g},\theta/2)$ associated to a minimal gradation of $\mathfrak{g}$ [KRW, KW] (cf. [FL]). In this case $\mathfrak{g}$ is one of the simple Lie superalgebras $s\ell(m|n)/\delta_{m,n}\mathbb{C}I$, $osp(m|n)$ $(= spo(n|m))$, $D(2,1\,;a)$, $F(4)$, $G(3)$ or one of the five exceptional Lie algebras, θ is the highest root of one of the simple components of the even part of $\mathfrak{g}$, the bilinear form $(\,.\,|\,.\,)$ is normalized by the condition $(\theta|\theta) = 2$, and $x = \theta/2$. The corresponding $\frac{1}{2}\mathbb{Z}$-gradation (2.1) looks as follows:

$$\mathfrak{g} = \mathbb{C}f + \mathfrak{g}_{-1/2} + \mathfrak{g}_0 + \mathfrak{g}_{1/2} + \mathbb{C}e\,, \tag{4.1}$$

where $\{e,x,f\}$ form an $s\ell_2$ triple , and

$$\mathfrak{g}^f_0 = \{a \in \mathfrak{g}_0 | (x|a) = 0\}\,, \quad \mathfrak{g}^f = \mathbb{C}f + \mathfrak{g}_{-1/2} + \mathfrak{g}^f_0\,.$$

Then

$$\mathfrak{g}(\sigma) = \mathbb{C}f + \mathfrak{g}^{-\sigma}_{-1/2} + \mathfrak{g}^\sigma_0 + \mathfrak{g}^{-\sigma}_{1/2} + \mathbb{C}e$$

is a minimal gradation of $\mathfrak{g}(\sigma)$. Since $\mathfrak{g}^\sigma_0 = (\mathfrak{g}^\sigma_0)^f + \mathbb{C}x$, it follows that there exists an element $h_0 \in \mathfrak{h}^{\sigma f}$ of the Lie superalgebra $\mathfrak{g}(\sigma)$ such that the eigenvalues of ad h_0 are real, h_0 is a regular element of $\mathfrak{g}^\sigma_0$, and the 0th eigenspace of ad h_0 on $\mathfrak{g}^{-\sigma}_{1/2}$ (resp. $\mathfrak{g}^{-\sigma}_{-1/2}$) is $\mathbb{C}e_{\theta/2}$ (resp. $\mathbb{C}e_{-\theta/2}$) if $e_{\theta/2}$ is a root vector of $\mathfrak{g}(\sigma)$. (Here $\theta/2$ stands for the restriction of $\theta/2$ to $\mathfrak{h}^\sigma$.) Letting $\mathfrak{n}_+(\sigma)$ (resp. $\mathfrak{n}_-(\sigma)$) be the span of all eigenvectors of ad h_0 with positive (resp. negative) eigenvalues and the vectors

$f = e_{-\theta}$ and $e_{-\theta/2}$ (resp. $e = e_\theta$ and $e = e_{\theta/2}$), we obtain the decomposition (2.4), satisfying the properties (i)–(iv). Note also that in the decomposition (2.5), $\mathfrak{h}_{1/2}(\sigma)$ (resp. $\mathfrak{n}_{1/2}(\sigma)'$) is the span of all eigenvectors of ad h_0 with positive (resp. negative) eigenvalues, and $\mathfrak{g}_{1/2}(\sigma)_0 = \mathbb{C}e_{\theta/2} \in \mathfrak{g}(\sigma)$. Thus, $\epsilon(\sigma) \neq 0$ iff $\theta/2$ is a root of $\mathfrak{g}$ with respect to $\mathfrak{h}^\sigma$ and $\sigma(e_{\theta/2}) = -e_{\theta/2}$.

Example 4.1. For the minimal gradation the numbers $s_\alpha(\alpha \in \Delta \subset \mathfrak{h}^*)$ are as follows (cf. (3.2)):

(a) If $\sigma = 1$, then $s_\alpha = 0$ (resp. 1) for $\alpha \in \Delta_+$ (resp. $-\alpha \in \Delta_+$).
(b) If $\sigma|_{\mathfrak{g}_j} = (-1)^{2j}$, then $s_\alpha = 0$ (resp. 1) if $\alpha(x) = 0$ and $\alpha(h_0) > 0$ (resp. $\alpha(h_0) < 0$), $s_\alpha = -1/2$ (resp. $\frac{1}{2}$) if $\alpha(x) = \frac{1}{2}$ and $\alpha(h_0) > 0$ (resp. $\leqslant 0$), $s_\theta = 0$ and $s_\alpha + s_{-\alpha} = 1$, $\alpha \in \Delta$.

Recall that the (Virasoro) central charge of $W_k(\mathfrak{g}, \theta/2)$ is [KW]:

$$c(k) = \frac{k \operatorname{sdim} \mathfrak{g}}{k + h^\vee} - 6k + h^\vee - 4\,,$$

and it is, of course, the same for the twisted vertex algebras $W_k(\mathfrak{g}, \sigma, \theta/2)$.

Introduce the following vectors in $\mathfrak{h}^{\sigma *}$:

$$\gamma' = \frac{1}{2} \sum_{\alpha \in S'} \tilde{p}(\alpha) s_\alpha \alpha\,, \quad \gamma_{1/2} = \frac{1}{2} \sum_{\alpha \in S_{1/2}} \tilde{p}(\alpha) s_\alpha \alpha\,.$$

Corollaries 3.1 and 3.3 give the following result, which will be used in the calculation of the determinant formula.

Proposition 4.1. *Let M be a restricted $\widehat{\mathfrak{g}}^{\rm tw}$-module of level k. Let $v \in M$ be a singular vector of M with weight $\Lambda \in \mathfrak{h}^{\sigma *}$. Then we have in the case of $W_k(\mathfrak{g}, \sigma\,, \theta/2)$:*

(a) *The eigenvalue of $L_0^{\rm tw}$ on $v \otimes |0\rangle$ is equal to*

$$h = \frac{1}{2(k + h^\vee)} ((\Lambda|\Lambda) - 2(\Lambda|\gamma')) - \Lambda(x) + s_{\mathfrak{g}} + s_{\rm gh}\,,$$

where

$$s_{\mathfrak{g}} = -\frac{k}{4(k + h^\vee)} \sum_{\alpha \in S'} \tilde{p}(\alpha) s_\alpha (s_\alpha - 1)\,, \quad s_{\rm gh} = \frac{1}{4} \sum_{\alpha \in S_{1/2}} \tilde{p}(\alpha) s_\alpha^2\,.$$

(b) *The eigenvalue of $J^{\{H\},{\rm tw}}$ for $H \in \mathfrak{h}^{\sigma f}$ on $v \otimes |0\rangle$ is equal to*

$$(\Lambda - \gamma_{1/2})(H)\,.$$

Proof. Letting $s_{\rm gh} := s_{\rm ch} + s_{\rm ne}$, we have (see (3.10) and (3.11)): $s_{\rm gh} = \frac{1}{8}\epsilon(\sigma) + \frac{1}{4}\sum_{\alpha \in S_{1/2}} \tilde{p}(\alpha)(s_\alpha^2 + s_\alpha)$. Since $\alpha \in S_{1/2}$ iff $\theta - \alpha \in S_{1/2}$, we obtain that

$$\sum_{\alpha \in S_{1/2}} \tilde{p}(\alpha) s_\alpha = -\frac{1}{2}\epsilon(\sigma)\,. \tag{4.2}$$

It is because $s_\alpha + s_{\theta - \alpha} = \delta_{\alpha, \theta/2}$, which holds due to (3.6). This proves the formula for $s_{\rm gh}$. The rest is straightforward. □

Proposition 4.2. *For the* $\frac{1}{2}\mathbb{Z}$*-gradation of* $\mathfrak{g}$ *defined by* ad x *one has:*

(a) $2\gamma'(x) = 1 - h^\vee - \frac{1}{2}\epsilon(\sigma)$.

(b) $\gamma' = 2\gamma_{1/2} + \gamma_0' - \frac{1}{2}(h^\vee - 1)\theta$, *where* $\gamma_0' = \frac{1}{2}\sum_{\substack{\alpha\in S' \\ \alpha(x)=0}} \tilde{p}(\alpha)s_\alpha\alpha$.

(c) $\gamma_{1/2}^\natural = \frac{1}{2}(\gamma'^\natural - \gamma_0'^\natural)$.

Proof. We have:

$$2\gamma'(x) = \frac{1}{2}\sum_{\alpha\in S_{1/2}} \tilde{p}(\alpha)s_\alpha - \frac{1}{2}\sum_{\alpha\in S_{-1/2}} \tilde{p}(\alpha)s_\alpha - 1 = \sum_{\alpha\in S_{1/2}} \tilde{p}(\alpha)s_\alpha - \frac{1}{2}\,\mathrm{sdim}\,\mathfrak{g}_{1/2} - 1.$$

Since sdim $\mathfrak{g}_{1/2} = 2h^\vee - 4$ (see [KW], (5.6)), formula (4.2) completes the proof of (a).

Similar calculations establish (b), and (c) is immediate by (b). □

In [KW], Theorem 5.2, we gave a realization of the vertex algebra $W_k(\mathfrak{g},\theta/2)$ as a subalgebra of $V_{\alpha_k}(\mathfrak{g}_0)\otimes F(A_{\mathrm{ne}})$, where $\mathfrak{g}_0$ is the 0th grading component in (2.1) and α_k is the "shifted" 2-cocycle: $\alpha_k(at^m, bt^n) = ((k+h^\vee)(a|b) - \frac{1}{2}\kappa_{\mathfrak{g}_0}(a,b))m\delta_{m,-n}$, where $\kappa_{\mathfrak{g}_0}$ is the Killing form on $\mathfrak{g}_0$. The twisted version of this result is derived from [KW], Theorem 5.2, by making use of (1.14), Theorem 2.1, and the following identity for formal distributions a,b,c such that $[a_\lambda b] = \langle a,b\rangle \in \mathbb{C}$, $[a_\lambda c] = \langle a,c\rangle \in \mathbb{C}$, $[b_\lambda c] = \langle b,c\rangle \in \mathbb{C}$:

$$\begin{aligned} :abc:^{\mathrm{tw}}(z) &= \ :a^{\mathrm{tw}}(z)b^{\mathrm{tw}}(z)c^{\mathrm{tw}}(z): - z^{-1}\big(s_b\langle b,c\rangle a^{\mathrm{tw}}(z) \\ &\quad + s_a\langle a,b\rangle c^{\mathrm{tw}}(z) + s_a(-1)^{p(a)p(b)}\langle a,c\rangle b^{\mathrm{tw}}(z)\big)\,. \end{aligned} \tag{4.3}$$

As in [KW], we keep the notation $J^{\{a\},\mathrm{tw}}$ if $a\in\mathfrak{g}_0^f$, but let $G^{\{v\},\mathrm{tw}} = J^{\{v\},\mathrm{tw}}$ if $v\in\mathfrak{g}_{-1/2}$. Due to Theorem 2.1, the formal distributions $J^{\{a\},\mathrm{tw}}$, $G^{\{v\},\mathrm{tw}}$ and L^{tw} strongly and freely generate the twisted vertex algebra $W_k(\mathfrak{g},\sigma,\theta/2)$.

Theorem 4.1. *The following formulas define an injective vertex algebra homomorphism of* $W_k(\mathfrak{g},\sigma,\theta/2)$ *to* $V_{\alpha_k}(\mathfrak{g}_0,\sigma)\otimes F(A_{\mathrm{ne}},\sigma)$*:*

$$\begin{aligned} J^{\{a\},\mathrm{tw}}(z) &\mapsto a^{\mathrm{tw}}(z) + \frac{(-1)^{p(a)}}{2}\sum_{\alpha\in S_{1/2}} :\Phi^{\alpha,\mathrm{tw}}(z)\Phi^{\mathrm{tw}}_{[u_\alpha,a]}(z): \\ &\quad - \frac{(-1)^{p(a)}}{2}\sum_{\alpha\in S_{1/2}} s_{\Phi^\alpha}\langle\Phi^\alpha,\Phi_{[u_\alpha,a]}\rangle_{\mathrm{ne}}\, z^{-1}\,(a\in\mathfrak{g}_0^f)\,, \\ G^{\{v\},\mathrm{tw}}(z) &\mapsto \sum_{\alpha\in S_{1/2}} :[v,u_\alpha]^{\mathrm{tw}}(z)\Phi^{\alpha,\mathrm{tw}}(z): - (k+1)\sum_{\alpha\in S_{1/2}} (v|u_\alpha)\partial\Phi^{\alpha,\mathrm{tw}}(z) \\ &\quad - \frac{(-1)^{p(v)}}{3}\sum_{\alpha,\beta\in S_{1/2}} :\Phi^{\alpha,\mathrm{tw}}(z)\Phi^{\beta,\mathrm{tw}}(z)\Phi^{\mathrm{tw}}_{[u_\beta,[u_\alpha,v]]}(z): \\ &\quad + \frac{(-1)^{p(v)}}{3}\sum_{\alpha,\beta\in S_{1/2}} \big(s_{\Phi^\beta}\langle\Phi^\beta,\Phi_{[u_\beta,[u_\alpha,v]]}\rangle_{\mathrm{ne}}\,\Phi^{\alpha,\mathrm{tw}}(z) \\ &\quad + (-1)^{p(\alpha)p(\beta)} s_{\Phi^\alpha}\langle\Phi^\alpha,\Phi_{[u_\beta,[u_\alpha,v]]}\rangle_{\mathrm{ne}}\,\Phi^{\beta,\mathrm{tw}}(z) \\ &\quad + s_{\Phi^\alpha}\langle\Phi^\alpha,\Phi^\beta\rangle_{\mathrm{ne}}\,\Phi^{\mathrm{tw}}_{[u_\beta,[u_\alpha,v]]}(z)\big)z^{-1}\ (v\in\mathfrak{g}_{-1/2})\,, \end{aligned}$$

$$L^{\mathrm{tw}}(z) \mapsto \frac{1}{2(k+h^\vee)} \sum_{\alpha \in S_0} (-1)^{p(\alpha)} : u_\alpha^{\mathrm{tw}}(z) u^{\alpha,\mathrm{tw}}(z) : + \frac{k+1}{k+h^\vee} \partial x(z)$$

$$+ \frac{1}{2} \sum_{\alpha \in S_{1/2}} (-1)^{p(\alpha)} : \Phi_\alpha^{\mathrm{tw}}(z) \partial \Phi^{\alpha,\mathrm{tw}}(z) : - \frac{1}{2(k+h^\vee)} \sum_{\alpha \in S_0'} (-1)^{p(\alpha)} s_\alpha \alpha(z) z^{-1}$$

$$+ \Big(\frac{1}{4(k+h^\vee)} \sum_{\alpha \in S_0} (-1)^{p(\alpha)} \binom{s_\alpha}{2} \kappa_{\mathfrak{g}_0}(u_\alpha, u^\alpha) - \frac{1}{2} \sum_{\alpha \in S_+ \cup S_0} (-1)^{p(\alpha)} \binom{s_\alpha}{2} \Big) z^{-2}.$$

(For $\mathfrak{g}_0^f$ simple, $\kappa_{\mathfrak{g}_0}(u_\alpha, u^\alpha) = 2h_0^\vee$, where $h_0^\vee$ is the dual Coxeter number of $\mathfrak{g}_0^f$ with respect to $(\,.\,|\,.\,)$.)

In the case of $W_k(\mathfrak{g}, \sigma, \theta/2)$, Proposition 3.1 gives the following result.

Proposition 4.3. *Let M be a $\widehat{\mathfrak{g}}^{\mathrm{tw}}$-module satisfying the conditions of Proposition 3.1. Then the Euler-Poincaré character of the $W_k(\mathfrak{g}, \sigma, \theta/2)$-module $H^{\mathrm{tw}}(M)$ is not identically zero iff $e_\theta t^{-1}$ is not locally nilpotent on M.*

Now we turn to the determinant formula for the Verma modules over $W_k(\mathfrak{g}, \sigma, \theta/2)$. To simplify notation, we let $\mathfrak{g}^\natural = \mathfrak{g}_0^f$ (resp. $\mathfrak{h}^\natural = (\mathfrak{h}^\sigma)^f$), the centralizer of f in $\mathfrak{g}_0$ (resp. in $\mathfrak{h}^\sigma$). Let $\lambda \mapsto \lambda^\natural$ denote the restriction map $\mathfrak{h}^\sigma \to \mathfrak{h}^\natural$. Let S_0 (resp. $S_{-1/2}$) $= \{\alpha \in S' |\, \alpha(x) = 0$ (resp. $\alpha(x) = -1/2)\}$, and let $\Delta_{W,\sigma}^\natural = \{\alpha^\natural |\, \alpha \in S_0 \cup S_{-1/2}\} \subset \mathfrak{h}^{\natural *}$, the multiplicity of $\alpha^\natural$ being the multiplicity of $\alpha \in S'$. Note that $\Delta_{W,\sigma}^\natural$ may contain 0 (this happens iff $\theta/2 \in \Delta^\sigma$).

Define the set of roots $\Delta_{W,\sigma}$ of $W_k(\mathfrak{g}, \sigma, \theta/2)$ as a subset of the dual of its Cartan algebra

$$\mathfrak{h}_{W,\sigma} = \mathfrak{h}^\natural \bigoplus \mathbb{C} L_0^{\mathrm{tw}},$$

defined as follows. We embed $\mathfrak{h}^{\natural *}$ in $\mathfrak{h}_{W,\sigma}^*$ by letting $\alpha \in \mathfrak{h}^{\natural *}$ to be zero on L_0^{tw}, and define $\delta' \in \mathfrak{h}_{W,\sigma}^*$ by

$$\delta'|_{\mathfrak{h}^\natural} = 0, \quad \delta'(L_0^{\mathrm{tw}}) = -1.$$

Then $\Delta_{W,\sigma} = \Delta_{W,\sigma}^{\mathrm{re}} \cup \Delta_{W,\sigma}^{\mathrm{im}}$, where

$$\begin{aligned} \Delta_{W,\sigma}^{\mathrm{re}} &= \{(n + s_\alpha + \alpha(x))\delta' + \alpha \,|\, \alpha \in \Delta_{W,\sigma}^\natural, \quad n \in \mathbb{Z}\}, \\ \Delta_{W,\sigma}^{\mathrm{im}} &= \{n\delta' |\, n \in E_0, \quad n \neq 0\}, \end{aligned}$$

where the multiplicity of a root $(n + s_\alpha + \alpha(x))\delta' + \alpha$ is equal to the multiplicity of $\alpha \in \Delta_{W,\sigma}^\natural$ with given s_α, and the multiplicity of $n\delta'$ is equal to the multiplicity of the root $n\delta$ of $\widehat{\mathfrak{g}}^{\mathrm{tw}}$. Note that 0 is a root in $\Delta_{W,\sigma}^{\mathrm{re}}$ of multiplicity $\epsilon(\sigma)(\leqslant 1)$.

We denote by $\Delta_{W,\sigma}^+$ the subset of positive roots, which consists of the subset $\Delta_{W,\sigma}^{\mathrm{im},+}$ of elements of $\Delta_{W,\sigma}^{\mathrm{im}}$ for which $n > 0$, and the subset $\Delta_{W,\sigma}^{re,+}$ of elements of $\Delta_{W,\sigma}^{re}$ for which $n \in \mathbb{Z}_+$.

Define the corresponding partition function $P_{W,\sigma}(\eta)$ on $\mathfrak{h}^*_{W,\sigma}$ as the number of ways η can be represented in the form (counting root multiplicities):

$$\eta = \sum_{\alpha\in\Delta^+_{W,\sigma}} k_\alpha \alpha\,, \text{ where } k_\alpha \in \mathbb{Z}_+ \text{ and } k_\alpha \leqslant 1 \text{ if } \alpha \text{ is odd.}$$

Remark 4.1. Denote by $P'_{W,\sigma}(\eta)$ the partition function for the set $\Delta^+_{W,\sigma}\backslash\{0\}$. Of course, $P'_{W,\sigma}(\eta) = P_{W,\sigma}(\eta)$ if $\epsilon(\sigma) = 0$, but $P'_{W,\sigma}(\eta) = \frac{1}{2}P_{W,\sigma}(\eta)$ if $\epsilon(\sigma) = 1$ and $\eta \neq 0$.

The definition (3.16)–(3.18) of a highest weight module M over the vertex algebra $W_k(\mathfrak{g}, \sigma, \theta/2)$ can be made a bit more explicit: the highest weight λ is an element of $\mathfrak{h}^*_W$, and condition (3.17) can be replaced by

$$J_0^{\{H\}} v_\lambda = \lambda^\natural(H) v_\lambda\,, H \in \mathfrak{h}^\natural\,, \text{ and } L_0^{\mathrm{tw}} v_\lambda = h v_\lambda\,, \tag{4.4}$$

where $\lambda^\natural$ denotes the restriction of λ to $\mathfrak{h}^\natural$ and h is the minimal eigenvalue of L_0^{tw} on M. We have the weight space decomposition of M:

$$M = \bigoplus_{\mu\in\mathfrak{h}^*_{W,\sigma}} M_\mu\,,\ M_\mu = \{v \in M | J_0^{\{H\}} v = \mu^\natural(H) v\,,\ H \in \mathfrak{h}^\natural\,,\ L_0^{\mathrm{tw}} v = \mu(L_0^{\mathrm{tw}}) v\}\,.$$

The Verma module $M(\lambda)$ over $W_k(\mathfrak{g}, \sigma, x)$ is a highest weight module for which

$$\dim M(\lambda)_\mu = P_{W,\sigma}(\lambda - \mu)\,.$$

In the case when $\epsilon(\sigma)(= \dim \mathfrak{g}_{1/2}(\sigma)_0) = 1$ choose a non-zero vector $e' \in \mathfrak{g}_{1/2}(\sigma)_0$ and let $f' = [f, e']$. Rescaling e', if necessary, we may assume that $[f', f'] = f$. The vector f' is a weight vector for $\mathfrak{h}^\natural$ in $\mathfrak{g}_{-1/2}$ with weight zero. Due to Theorem 2.1, we have the corresponding formal distribution in $W_k(\mathfrak{g}, \sigma, \theta/2)$:

$$G(z) := (-k - h^\vee)^{-1/2} G^{\{f'\},\mathrm{tw}}(z) = \sum_{n\in\mathbb{Z}} G_n z^{-n-3/2}\,. \tag{4.5}$$

We have the following description of the highest weight subspace of $M(\lambda)$:

$$M(\lambda)_\lambda = \begin{cases} \mathbb{C}v_\lambda \text{ if } \epsilon(\sigma) = 0\,, \\ \mathbb{C}v_\lambda + \mathbb{C}G_0 v_\lambda \text{ if } \epsilon(\sigma) = 1\,. \end{cases}$$

We shall need an explicit formula for the eigenvalue of $[G_0, G_0]$ on $v_\lambda \in M(\lambda)$, which we shall denote by $\varphi_0(k, h, \lambda^\natural)$. In order to compute the function φ_0, recall that Theorem 5.1(e) from [KW] provided an explicit expression for $[G^{\{u\}}{}_\lambda G^{\{v\}}]$, $u, v \in \mathfrak{g}_{-1/2}$, in $W_k(\mathfrak{g}, \theta/2)$. Unfortunately, the coefficient of $\frac{\lambda^2}{6}$ in this expression is correct only when $\mathfrak{g}^\natural = \mathfrak{g}_0^f$ is simple. Here is a correct expression for this coefficient, which we shall denote by γ_k:

$$\gamma_k(u, v) = -(k + h^\vee) g(u, v) c(k) + g(u, v) \sum_{\alpha\in S^\natural} \beta_k(u^\alpha, u_\alpha)$$
$$+2 \sum_{j\in S_{1/2}} \beta_k([u, u^j]^\natural\,, [u_j, v]^\natural)\,,$$

where $\beta_k(a,b) = (k + \frac{1}{2}h^\vee)(a|b) - \frac{1}{4}\kappa_{\mathfrak{g}_0}(a,b), a, b \in \mathfrak{g}_0$, $g(u,v) \in \mathbb{C}$ is defined by $[u,v] = g(u,v)f$, $a^\natural$ stands for the orthogonal projection of $a \in \mathfrak{g}_0$ on $\mathfrak{g}^\natural$, and $S^\natural$ indexes a basis of $\mathfrak{g}^\natural$. If $\mathfrak{g}^\natural$ is simple or, more generally, if $\kappa_{\mathfrak{g}_0}(a,b) = 2h_0^\vee(a|b), a, b \in \mathfrak{g}^\natural$, we have a much simpler formula:

$$\gamma_k(u,v) = -g(u,v)\Big(\big(k+h^\vee\big)c(k) - \big(k + \frac{1}{2}(h^\vee - h_0^\vee)\big)(\text{sdim}\ \mathfrak{g}_0 + \text{sdim}\ \mathfrak{g}_{1/2})\Big)\,.$$

In the case when $u = v = f'$, so that $g(u,v) = 1$, we obtain from [KW], Theorem 5.1(e):

$$\begin{aligned}
[G^{\{f'\}}{}_\lambda G^{\{f'\}}] &= -(k+h^\vee)L \\
&+\frac{1}{2}\Big(\sum_i : J^{\{h^i\}}J^{\{h_i\}} : + \sum_{\alpha\in S_0''} : J^{\{u^\alpha\}}J^{\{u_\alpha\}} :\Big) + \frac{\lambda^2}{6}\gamma_k(f',f')\,.
\end{aligned}$$

Here $\{h_i\}$ and $\{h^i\}$ are dual bases of $\mathfrak{h}^\natural$, and S_0'' is a basis of the kernel of the map ad $f' : \sum_{\alpha\in S_0'} \mathbb{C}u_\alpha \to \mathfrak{g}_{-1/2}$.

This formula is used to obtain:

$$\begin{aligned}
\varphi_0(k,h,\lambda^\natural) = h &- \frac{1}{2(k+h^\vee)}\Big(|\lambda^\natural + \gamma_{1/2}^\natural - \gamma'^\natural|^2 \\
&- |\gamma_{1/2}^\natural - \gamma'^\natural|^2 - \sum_{\alpha\in S_0''}(-1)^{p(\alpha)}\binom{s_\alpha}{2}\beta_k(u_\alpha,u^\alpha)\Big) \\
&- \frac{1}{24(k+h^\vee)}\Big(\sum_i \beta_k(h_i,h^i) + \sum_{\alpha\in S_0''}(-1)^{p(\alpha)}\beta_k(u_\alpha,u^\alpha)\Big) - \frac{c(k)}{24}\,.
\end{aligned} \tag{4.6}$$

Note that in the case when $\kappa_{\mathfrak{g}_0}(a,b) = 2h_0^\vee(a|b)$, this formula can be simplified, using $\beta_k(u_\alpha,u^\alpha) = \beta_k(h_i,h^i) = k + \frac{1}{2}(h^\vee - h_0^\vee)$. Then (4.6) becomes:

$$\begin{aligned}
\varphi_0(k,h,\lambda^\natural) = h &- \frac{1}{2(k+h^\vee)}\Big(|\lambda^\natural + \gamma_{1/2}^\natural - \gamma'^\natural|^2 \\
&- |\gamma_{1/2}^\natural - \gamma'^\natural|^2 + \frac{h^\vee + h_0^\vee}{2}\sum_{\alpha\in S_0''}(-1)^{p(\alpha)}\binom{s_\alpha}{2} - \frac{1}{2}(k+\frac{1}{2})^2 \\
&+ \frac{h^\vee(h^\vee-1)}{3} + \frac{1}{8} - \frac{h^\vee - h_0^\vee}{24}(\text{sdim}\ \mathfrak{g}_0 + \text{sdim}\ \mathfrak{g}_{1/2})\Big) \\
&+ \frac{1}{2}\sum_{\alpha\in S_0''}(-1)^{p(\alpha)}\binom{s_\alpha}{2} + \frac{h^\vee}{8}\,.
\end{aligned} \tag{4.7}$$

In order to define the contravariant bilinear form on a Verma module $M(\lambda)$ over $W_k(\mathfrak{g},\sigma,x)$, we use the anti-involution ω of $\mathfrak{g}$ introduced in [KW]; we shall assume that it commutes with σ. As in [KW], we have the following anti-involution of the associative algebra $\mathcal{A}$ generated by coefficients $J_n^{\{a\},\text{tw}}$ of formal distributions $J^{\{a\},\text{tw}}(z)$, where $a \in \mathfrak{g}^\natural \oplus \mathfrak{g}_{-1/2}$ (see (3.14)), and the L_n^{tw}:

$$\omega(L_n^{\text{tw}}) = L_{-n}^{\text{tw}}\,,\ \omega(J_n^{\{a\},\text{tw}}) = J_{-n}^{\{\omega(a)\},\text{tw}}\,.$$

The contravariant bilinear form $B(\,.\,,\,.\,)$ on a Verma module $M(\lambda)$ over $W_k(\mathfrak{g},\sigma,x)$ with highest weight vector v_λ is defined in the usual way:

$$B(av_\lambda, bv_\lambda) = \langle v_\lambda^*\,,\, \omega(a)bv_\lambda\rangle\,, \quad a,b\in\mathcal{A}\,,$$

where v_λ^* is the linear function on $M(\lambda)$, equal to 1 on v_λ and 0 on G_0v_λ and all weight spaces $M(\lambda)_\mu$, $\mu\neq\lambda$. This is a supersymmetric bilinear form, which is contravariant, i.e., $B(au,v)=B(u,\omega(a)v)$, $u,v\in M(\lambda)$, $a\in\mathcal{A}$, and $B(v_\lambda,v_\lambda)=1$, and these properties determine $B(\,.\,,\,.\,)$ uniquely. Different weight spaces are orthogonal with respect to this form and its kernel is the maximal submodule of $M(\lambda)$.

Denote by $\det_\eta(k,h,\lambda^\natural)$ the determinant of the bilinear form $B(\,.\,,\,.\,)$ restricted to the weight space $M(\lambda)_{\lambda-\eta}$, $\eta\in\mathfrak{h}^*_{W,\sigma}$. This is a function in k, h and $\lambda^\natural$ (see (4.4)) and it depends on the choice of a basis of $M(\lambda)_{\lambda-\eta}$ only up to a constant factor.

Consider the map $\pi:\widehat{\Delta}\to\mathfrak{h}^*_{W,\sigma}$, defined by

$$\pi(\alpha+m\delta)=\alpha^\natural+(m+\alpha(x))\delta'\,,\ \pi(m\delta)=m\delta'\,.$$

It is easy to see that, counting root multiplicities, π induces a bijective map:

$$\pi(\widehat{\Delta}^{\mathrm{re}}_{++}\cup\widehat{\Delta}^{\mathrm{im}}_{+})\overset{\sim}{\to}\widehat{\Delta}^{+}_{W,\sigma}\backslash\{0\}.$$

Denote by $\mathrm{mult}_{\,0}\,m\delta$ the multiplicity of the root $m\delta$ in $\widehat{\mathfrak{g}^\natural}^{\mathrm{tw}}$ $(\subset\widehat{\mathfrak{g}}^{\mathrm{tw}})$.

Theorem 4.2. *Up to a non-zero constant factor, the determinant* $\det_\eta(k,h,\lambda^\natural)$ *is given by the following formula:*

$$\varphi_0(k,h,\lambda^\natural)^{\epsilon(\sigma)P'_{W,\sigma}(\eta)}\prod_{\substack{m\in E_0^+\\ n\in\mathbb{N}}}(k+h^\vee)^{(\mathrm{mult}_{\,0}\,m\delta)P_{W,\sigma}(\eta-mn\delta')}$$

$$\times\prod_{\substack{\alpha\in\widehat{\Delta}^{\mathrm{re}}_{++}\\ n\in\mathbb{N}}}\varphi_{\alpha,n}(k,h,\lambda^\natural)^{\tilde{p}(\alpha)^{n+1}P_{W,\sigma}(\eta-n\pi(\alpha))}\,,$$

where the factor φ_0 *(occurring only when* $\epsilon(\sigma)=1$*) is given by* (4.6)*, and all the remaining factors are as follows* $(\alpha\in\Delta^\sigma, m\in E_0)$*:*

$$\varphi_{m\delta+\alpha,n}=m(k+h^\vee)+(\lambda^\natural+\gamma^\natural_{1/2}-\gamma'^\natural|\,\alpha)-\frac{n}{2}|\alpha|^2\ \text{if}\ \alpha(x)=0, \tag{4.8}$$

$$\varphi_{m\delta+\alpha,n}= \tag{4.9}$$

$$h-\frac{1}{k+h^\vee}\Big(\big(\frac{n}{2}|\alpha|^2-(m+\frac{1}{2})(k+h^\vee)-(\lambda^\natural+\gamma^\natural_{1/2}-\gamma'^\natural|\alpha)\big)^2$$

$$+\frac{1}{2}|\lambda^\natural+\gamma^\natural_{1/2}-\gamma'^\natural|^2-\frac{1}{4}(k+1-\frac{1}{2}\epsilon(\sigma))^2-\frac{1}{2}|\gamma'^\natural|^2\Big)-s_{\mathfrak{g}}-s_{\mathrm{gh}}$$

if $\alpha(x) = \frac{1}{2}$,

$$\varphi_{m\delta+\theta,n} = h - \frac{1}{4(k+h^\vee)}\Big(\big(n-(m+1)(k+h^\vee)\big)^2 \tag{4.10}$$
$$+2|\lambda^\natural + \gamma^\natural_{1/2} - \gamma'^\natural|^2 - (k+1-\frac{1}{2}\epsilon(\sigma))^2 - 2|\gamma'^\natural|^2\Big) - s_{\mathfrak{g}} - s_{\mathrm{gh}}\,.$$

(Formulas for $s_{\mathfrak{g}}$ and s_{gh} are given in Proposition 4.1*(a).)*

Proof. The proof follows the traditional lines, as in [KW]. First, let M be a Verma module over $\widehat{\mathfrak{g}}^{\mathrm{tw}}$ with highest weight $\widehat{\Lambda} = \Lambda + kD$, where $\Lambda \in \mathfrak{h}^{\sigma *}$. Then for each $\widehat{\alpha} \in \widehat{\Delta}_+$ and a positive integer n such that

$$2(\widehat{\Lambda} + \widehat{\rho}^{\mathrm{tw}}|\widehat{\alpha}) = n(\widehat{\alpha}|\widehat{\alpha}) \tag{4.11}$$

under certain conditions (stated in Lemma 7.1 of [KW]), $\widehat{\Lambda} - n\widehat{\alpha}$ is a singular weight, of multiplicity at least mult α, of M. This follows from the determinant formula for $\widehat{\mathfrak{g}}^{\mathrm{tw}}$ in [K2] (as corrected in Remark 7.1 of [KW]).

Let $h^{\vee\mathrm{tw}} = (\widehat{\rho}^{\mathrm{tw}}|\delta)$ be the dual Coxeter number of $\widehat{\mathfrak{g}}^{\mathrm{tw}}$. We have:

$$h^{\vee\mathrm{tw}} = h^\vee\,. \tag{4.12}$$

Indeed, the L^{tw}_n can be constructed for all $k \neq -h^{\vee\mathrm{tw}}$ (see [K3], Exercise 12.20). But the central charge of $L^{\mathfrak{g},\mathrm{tw}}$ is independent of σ and has singularity only at $k = -h^\vee$. This implies (4.12). Hence for $\widehat{\alpha} = \alpha + m\delta$, where $\alpha \in \Delta^\sigma$, $m \in E_0$, (4.11) can be rewritten, using also Corollary 3.2, as follows:

$$2(\Lambda|\alpha) - 2(\gamma'|\alpha) + 2m(k+h^\vee) = n(\alpha|\alpha)\,. \tag{4.13}$$

We decompose $\alpha \in \mathfrak{h}^\sigma (= \mathfrak{h}^{\sigma *})$ with respect to the orthogonal direct sum decomposition $\mathfrak{h}^\sigma = \mathbb{C}x + \mathfrak{h}^\natural$:

$$\alpha = 2\alpha(x)x + \alpha^\natural\,. \tag{4.14}$$

Next, by Theorem 3.1, the $W_k(\mathfrak{g},\sigma,\theta/2)$-module $H^{\mathrm{tw}}(M)$ is a Verma module, and its highest weight is $\lambda = h\delta' + \lambda^\natural$, where h is given by Proposition 4.1(a), and (see Proposition 4.1(b)):

$$\lambda^\natural = \Lambda^\natural - \gamma^\natural_{1/2}\,. \tag{4.15}$$

By Proposition 3.1, each singular weight $\widehat{\Lambda} - n\widehat{\alpha}$ of M satisfying (4.11) and such that $\widehat{\alpha} \in \widehat{\Delta}^{\mathrm{re}}_{++}$, gives rise to a singular weight of $H^{\mathrm{tw}}(M)$ (which is a Verma module over $W_k(\mathfrak{g},\sigma,\theta/2)$ with highest weight λ). This gives rise to a factor of $\det_\eta$. We now rewrite (4.13) in terms of k, h and $\lambda^\natural$.

In the case $\alpha(x) = 0$, substituting (4.15) in (4.13), we obtain (4.8). In the case $\alpha(x) \neq 0$, we substitute $\Lambda = 2\Lambda(x)x + \lambda^\natural + \gamma^\natural_{1/2}$ (obtained from (4.14) and

(4.15)) in the formula for h given by Proposition 4.1(a) to obtain:

$$h = \frac{1}{k+h^\vee}\Big((\Lambda(x) - \gamma'(x) - \frac{1}{2}(k+h^\vee))^2 + \frac{1}{2}|\lambda^\natural + \gamma^\natural_{1/2} - \gamma'^\natural|^2 \qquad (4.16)$$
$$- \frac{1}{4}(k+h^\vee+2\gamma'(x))^2 - \frac{1}{2}|\gamma'^\natural|^2 \Big) + s_{\mathfrak{g}} + s_{\mathrm{gh}}\,.$$

Substituting (4.14) and (4.15) in (4.13), we obtain:

$$2\alpha(x)\Lambda(x) = \frac{n}{2}|\alpha|^2 - (\lambda^\natural + \gamma^\natural_{1/2} - \gamma'|\alpha) - m(k+h^\vee)\,.$$

Finally, substituting the obtained expression for $\Lambda(x)$ in (4.16) and using Proposition 4.2(a), we get (4.9) and (4.10).

The rest of the proof is the same as in [KW]. □

Remark 4.2.

(a) If $\alpha + m\delta \in \widehat{\Delta}^{\mathrm{re}}_{++}$ is such that $(\alpha|\alpha) = 0$, the condition (4.1) becomes $(\widehat{\Lambda} + \widehat{\rho}^{\mathrm{tw}}|\alpha + m\delta) = 0$. Hence in this case the function $\varphi_{\alpha+m\delta,n}(k,h,\lambda^\natural)$ is independent of n. Since $\alpha + m\delta$ is an odd root, we therefore can simplify the corresponding factor in the formula for $\det_\eta$ (cf. [KW], Remark 7.2):

$$\prod_{n\in\mathbb{N}} \varphi_{\alpha+m\delta,n}^{\tilde{p}(\alpha)^{n+1}P_{W,\sigma}(\eta - n\pi(\alpha+m\delta))} = \varphi_{\alpha+m\delta,1}^{P_{W,\sigma\,;\,\pi(\alpha+m\delta)}(\eta-\pi(\alpha+m\delta))}\,.$$

Here $P_{W,\sigma\,;\,\widehat{\alpha}}$ stands for the partition function of the set $\Delta^+_{W,\sigma}\backslash\{\widehat{\alpha}\}$ (i.e., we reduce by 1 the multiplicity of $\widehat{\alpha}$).

(b) If $\alpha + m\delta \in \widehat{\Delta}^{\mathrm{re}}_{++}$ is such that $2(\alpha+m\delta) \in \widehat{\Delta}^{\mathrm{re}}_{++}$, and $n \in \mathbb{N}$, then condition (4.11) for the pair $\{2(\alpha+m\delta), n\}$ is the same as that for the pair $\{\alpha + m\delta, 2n\}$, hence in this case we have

$$\varphi_{\alpha+m\delta,2n} = \varphi_{2(\alpha+m\delta),n}\,,$$

and the corresponding factors in $\det_\eta$ cancel as in [KW], Remark 7.2.

(c) In all examples we have: $\varphi_0 = \varphi_{(-\delta+\theta)/2,0}$, but we do not know how to prove this in general. We conjecture that this is always the case, i.e.,

$$\varphi_0 = h - \frac{1}{2(k+h^\vee)}\Big(|\lambda^\natural + \gamma^\natural_{1/2} - \gamma'^\natural|^2 - \frac{1}{2}(k+\frac{1}{2})^2 - |\gamma'^\natural|^2 \Big) - s_{\mathfrak{g}} - s_{\mathrm{gh}}\,.$$

5. Examples

5.1. Ramond $N = 1$ algebra

Recall that the Neveu-Schwarz vertex algebra is $W_k(spo(2|1), \theta/2)$ [KW]. It corresponds to the minimal gradation of $\mathfrak{g} = spo(2|1)$, which looks as follows:

$$\mathfrak{g} = \mathbb{C}e_{-\theta} \oplus \mathbb{C}e_{-\theta/2} \oplus \mathbb{C}x \oplus \mathbb{C}e_{\theta/2} \oplus \mathbb{C}e_\theta\,,$$

where $e_{-\theta} = \frac{1}{2}E_{21}$, $e_{-\theta/2} = \frac{1}{2}(E_{31} - E_{23})$, $x = \frac{1}{2}(E_{11} - E_{22})$, $e_{\theta/2} = E_{13} + E_{32}$, $e_\theta = 2E_{12}$, $\mathfrak{h} = \mathbb{C}x$, and $\theta \in \mathfrak{h}^*$ is defined by $\theta(x) = 1$. Then $\Delta_+ = \{\theta/2, \theta\}$. Choose the invariant bilinear form $(a|b) = \mathrm{str}ab$. Then $h^\vee = 3/2$ and $(e_{\theta/2}|e_{-\theta/2}) = (e_\theta|e_{-\theta}) = 1$, $(x|x) = 1/2$. We have $x = \theta/2$ under the identification of $\mathfrak{h}$ with $\mathfrak{h}^*$.

We take $f = e_{-\theta}$. The only non-trivial automorphism σ that fixes f and x, also fixes $e = e_\theta/2$ and $\sigma(e_{\pm\theta/2}) = -e_{\pm\theta/2}$. Then we have:

$$\epsilon(\sigma) = 1\,,\ s_{\theta/2} = 1/2\,,\ s_{-\theta/2} = 1/2\,,\ s_\theta = 0\,,\ s_{-\theta} = 1\,.$$

Hence we have: $\gamma' = -\theta/2$, $\gamma_{1/2} = -\theta/8$, $s_{\mathfrak{g}} = -k/4(2k+3)$,$s_{\mathrm{gh}} = -1/16$.

In this case we have one twisted neutral free fermion $\Phi^{\mathrm{tw}}(z) = \sum_{n\in\mathbb{Z}} \Phi_n z^{-n-1/2}$, where $[\Phi_m, \Phi_n] = \delta_{m,-n}$ and $\Phi^{\mathrm{tw}}(z)_- = \sum_{n>0} \Phi_n z^{-n-1/2}$.

The twisted vertex algebra $W_k(\mathfrak{g}, \sigma, \theta/2)$ is strongly generated by the Virasoro field

$$L^{\mathrm{tw}}(z) = \sum_{n\in\mathbb{Z}} L_n^{\mathrm{tw}} z^{-n-2}$$

and the odd Ramond field

$$G^{\mathrm{tw}}(z) = \sum_{n\in\mathbb{Z}} G_n^{\mathrm{tw}} z^{-n-3/2},$$

so that the L_n and G_n satisfy the relations of the Ramond ($N = 1$) superalgebra [R] with central charge

$$c(k) = 3/2 - 12\gamma^2\,,\ \text{where } \gamma^2 = (k+1)^2/(2k+3)\,.$$

In particular, we have:

$$[G_0^{\mathrm{tw}}, G_0^{\mathrm{tw}}] = 2L_0^{\mathrm{tw}} - c(k)/12\,. \tag{5.1}$$

The free field realization, provided by Theorem 4.1, of this algebra is given in terms of a free boson $b(z) = \sum_{n\in\mathbb{Z}} b_n z^{-n-1}$, where $[b_m, b_n] = m\delta_{m,-n}$ and $b(z)_- = \sum_{n\geqslant 0} b_n z^{-n-1}$, and the twisted fermion $\Phi^{\mathrm{tw}}(z)$. We have:

$$\begin{aligned} L^{\mathrm{tw}}(z) &= \frac{1}{2} : b(z)^2 : + \gamma\partial b(z) - \frac{1}{2} : \Phi^{\mathrm{tw}}(z)\partial\Phi^{\mathrm{tw}}(z) : - \frac{1}{16} z^{-2}\,, \\ G^{\mathrm{tw}}(z) &= \frac{1}{\sqrt{2}} : \Phi^{\mathrm{tw}}(z) b(z) : + \sqrt{2}\gamma\partial\Phi^{\mathrm{tw}}(z)\,. \end{aligned}$$

In order to compute the determinant formula for the Ramond algebra we need the σ-twisted affinization $\widehat{\mathfrak{g}}^{\mathrm{tw}} = \sum_{m\in\mathbb{Z}} \mathfrak{g}_{\bar{0}} t^m + \sum_{m\in 1/2+\mathbb{Z}} \mathfrak{g}_{\bar{1}} t^m + \mathbb{C}K + \mathbb{C}D$, where $\mathfrak{g}_{\bar{0}} = \mathbb{C}e_\theta + \mathbb{C}x + \mathbb{C}e_{-\theta}$, $\mathfrak{g}_{\bar{1}} = \mathbb{C}e_{\theta/2} + \mathbb{C}e_{-\theta/2}$. The set $\widehat{\Delta}_{++}^{\mathrm{re}}$ is a union of two subsets:

$$\{m\delta + \theta/2 |\, m \in \frac{1}{2} + \mathbb{Z}_+\} \text{ and } \{m\delta + \theta |\, m \in \mathbb{Z}_+\}\,.$$

From (4.10) and Remark 4.2(b) we obtain that $\varphi_{m\delta+\theta/2,n}(k,h) = h - h_{n,2m+1}^R(k)$ and $\varphi_{m\delta+\theta,n} = h - h_{2n,m+1}^R(k)$, where

$$h_{n,m}^R(k) = \frac{1}{4(k+\frac{3}{2})}\left(\left(\frac{n}{2} - m\left(k + \frac{3}{2}\right)\right)^2 - (k+1)^2\right) + \frac{1}{16}\,. \tag{5.2}$$

It follows from (5.1) that the extra factor is equal to $h - c(k)/24$.

The set of positive even (resp. odd) roots for $W_k(spo(2|1),\sigma,\theta/2)$ is $\mathbb{N}\delta'$ (resp. $\mathbb{Z}_+\delta'$). Hence $P'_{W,\sigma}(\eta) = p^R(\eta)$, where $p^R(\eta)$ is defined by the generating series $\sum_{\eta\in\mathbb{Z}_+} p^R(\eta)q^\eta = \prod_{n=1}^\infty \frac{1+q^n}{1-q^n}$. Hence, by Theorem 4.2, we obtain the following determinant formula for the Ramond algebra, where $\eta\in\mathbb{N}$ (cf. [KW]):

$$\det_\eta(k,h) = (h-\frac{c(k)}{24})^{p^R(\eta)} \prod_{\substack{m,n\in\mathbb{N}\\ m+n\,\mathrm{odd}}} (h-h^R_{n,m}(k))^{2p^R(\eta-\frac{1}{2}mn)}.$$

5.2. $N=2$ Ramond type sector

Recall that the $N=2$ vertex algebra is $W_k(s\ell(2|1),\theta/2)$. In this section the Lie superalgebra $\mathfrak{g}=s\ell(2|1)$ consists of supertraceless matrices in the superspace $\mathbb{C}^{2|1}$, whose even part is $\mathbb{C}\epsilon_1+\mathbb{C}\epsilon_3$ and odd part is $\mathbb{C}\epsilon_2$, where $C\epsilon_1$, ϵ_2, ϵ_3 is the standard basis. We shall work in the following basis of $\mathfrak{g}$:

$$e_1=E_{12}\,,\ e_2=E_{23}\,,\ -[e_1,e_2]\,,\ f_1=E_{21}\,,\ f_2=-E_{32}\,,\ [f_1,f_2]\,,$$
$$h_1=E_{11}+E_{22}\,,\ h_2=-E_{22}-E_{33}\,.$$

The elements e_i, f_i, h_i $(i=1,2)$ are the Chevalley generators of $\mathfrak{g}$ and $\mathfrak{h}=\mathbb{C}h_1+\mathbb{C}h_2$. The elements e_i, f_i $(i=1,2)$ are all odd elements of $\mathfrak{g}$, both simple roots α_i $(i=1,2)$, attached to e_i, are odd, and $\Delta_+=\{\alpha_1,\alpha_2,\theta=\alpha_1+\alpha_2\}$. Since $\mathfrak{g}_{\bar 0}=\mathbb{C}[e_1,e_2]+\mathbb{C}[f_1,f_2]+\mathfrak{h}\simeq g\ell_2$, there is only one, up to conjugacy, nilpotent element $f=[f_1,f_2]$, which embeds in the following $s\ell_2$-triple: $\{e=-\frac{1}{2}[e_1,e_2]\,,\ x=\frac{1}{2}(h_1+h_2)\,,\ f\}$. The minimal gradation of $\mathfrak{g}$, defined by ad x, looks as follows:

$$\mathfrak{g}=\mathbb{C}f\oplus(\mathbb{C}f_1+\mathbb{C}f_2)\oplus\mathfrak{h}\oplus(\mathbb{C}e_1+\mathbb{C}e_2)\oplus\mathbb{C}e,.$$

The invariant bilinear form on $\mathfrak{g}$ is $(a|b)=\mathrm{str}ab$, and $h^\vee=1$.

First consider the Ramond type automorphisms σ_a $(-1/2<a\leqslant 1/2)$, defined by $\sigma_a(e_1)=e^{2\pi ia}e_1$, $\sigma_a(f_1)=e^{-2\pi ia}f_1$, $\sigma_a(e_2)=e^{-2\pi ia}e_2$, $\sigma_a(f_2)=e^{2\pi ia}f_2$. Then $\mathfrak{g}(\sigma_a)=\mathfrak{g}$ if $a=1/2$ (resp. $=\mathfrak{g}_{\bar 0}$ if $a<1/2$), and we choose $\mathfrak{n}(\sigma_a)_+=\mathbb{C}e_2+\mathbb{C}f_1+\mathbb{C}f$, $\mathfrak{n}(\sigma_a)_-=\mathbb{C}f_2+\mathbb{C}e_1+\mathbb{C}e$ (resp. $\mathbb{C}f$ and $\mathbb{C}e$), so that in all cases $\epsilon(\sigma_a)=0$, and

$$s_{\alpha_1}=a\,,\ s_{\alpha_2}=-a\,,\ s_\theta=0\,,\ s_{-\alpha_1}=1-a\,,\ s_{-\alpha_2}=1+a\,,\ s_{-\theta}=1\,.$$

In this case we have two twisted neutral free fermions

$$\Phi_1^{\mathrm{tw}}(z)=\sum_{n\in 1/2+a+\mathbb{Z}}\Phi^1_n z^{-n-1/2},\quad \Phi_2^{\mathrm{tw}}(z)=\sum_{n\in 1/2-a+\mathbb{Z}}\Phi^2_n z^{-n-1/2},$$

where

$$[\Phi^i_m\,,\ \Phi^j_n]=(\delta_{ij}-1)\,\delta_{m,-n},\ \Phi_1^{\mathrm{tw}}(z)_-=\sum_{n\in 1/2+a+\mathbb{Z}_+}\Phi^1_n z^{-n-1/2},\ \Phi_2^{\mathrm{tw}}(z)_-$$
$$=\sum_{n\in 1/2-a+\mathbb{Z}_+}\Phi^2_n z^{-n-1/2}.$$

The twisted vertex algebra $W_k(\mathfrak{g}, \sigma_a, \theta/2)$ is strongly generated by the Virasoro field $L^{\rm tw}(z) = \sum_{n\in\mathbb{Z}} L_n^{\rm tw} z^{-n-2}$, the current $J^{\rm tw}(z) = \sum_{n\in\mathbb{Z}} J_n^{\rm tw}\ z^{-n-1}$ and two odd fields $G^{\pm,{\rm tw}}(z) = \sum_{n\in 1/2\mp a+\mathbb{Z}} G_n^{\pm,{\rm tw}} z^{-n-3/2}$ so that L_n, J_n $(n\in\mathbb{Z})$ and $G_n^{\pm}$ $(n\in 1/2\mp a+\mathbb{Z})$ satisfy the relations of $N=2$ Ramond type superconformal algebra with central charge $c(k) = -3(2k+1)$.

The free field realization, provided by Theorem 4.1, of this algebra is given in terms of free bosons $h_i(z) = \sum_{n\in\mathbb{Z}} h_n^i z^{-n-1}$ $(i=1,2)$, where $[h_m^i, h_n^j] = (k+1)m(1-\delta_{ij})\delta_{m,-n}$, and the twisted neutral free fermions $\Phi_i^{\rm tw}(z)(i=1,2)$:

$$\begin{aligned} L^{\rm tw}(z) &= \frac{1}{k+1} : h_1(z)h_2(z) : + \frac{1}{2}(: \Phi_1^{\rm tw}(z)\partial\Phi_2^{\rm tw}(z) : \\ &\quad + : \Phi_2^{\rm tw}(z)\partial\Phi_1^{\rm tw}(z) : + \partial(h_1(z)+h_2(z))) + \frac{a^2}{2}z^{-2}, \\ J^{\rm tw}(z) &= h_1(z) - h_2(z) + : \Phi_1^{\rm tw}(z)\Phi_2^{\rm tw}(z) : + az^{-1}, \\ G^{+,{\rm tw}}(z) &= (-k-1)^{-1/2}(: \Phi_2^{\rm tw}(z)h_1(z) : + (k+1)\partial\Phi_2^{\rm tw}(z)) \\ G^{-,{\rm tw}}(z) &= (-k-1)^{-1/2}(: \Phi_1^{\rm tw}(z)h_2(z) : + (k+1)\partial\Phi_1^{\rm tw}(z)). \end{aligned}$$

The set $\widehat{\Delta}_+ = \widehat{\Delta}_+^{\rm re} \cup \widehat{\Delta}_+^{\rm im}$ of positive roots of $\widehat{\mathfrak{g}}^{\rm tw}$ is as follows: $\widehat{\Delta}_+^{\rm re} = \{(m+a)\delta+\alpha_1\,, (m-a+1)\delta-\alpha_1\,, (m-a)\delta+\alpha_2\,, (m+a+1)\delta-\alpha_2\,, m\delta+\theta\,, (m+1)\delta-\theta \,|\, m\in\mathbb{Z}_+\}$, where all roots have multiplicity 1, and $\widehat{\Delta}_+^{\rm im} = \{m\delta \,|\, m\in\mathbb{N}\}$, all having multiplicity 2. Next, $\gamma' = -aH$, $\gamma_{1/2} = -aH/2$, where $H = h_1 - h_2$, $s_{\mathfrak{g}} = ka^2/(k+1)$, $s_{\rm gh} = -a^2/2$, and the set of roots $\widehat{\Delta}_{++}^{\rm re}$ is as follows:

$$\{m\delta+\theta \,|\, m\in\mathbb{Z}_+\} \cup \{m\delta+\alpha_1 \,|\, m\in a+\mathbb{Z}_+\} \cup \{m\delta+\alpha_2 \,|\, m\in -a+\mathbb{Z}_+\}.$$

We have: $\mathfrak{h}_{W,\sigma} = \mathbb{C}H + \mathbb{C}L_0^{\rm tw}$, where $H = h_1 - h_2$. Define $\alpha\in\mathfrak{h}_{W,\sigma}^*$ by $\alpha(H) = 1$, $\alpha(L_0^{\rm tw}) = 0$. Then $\Delta_{W,\sigma}^{{\rm re},+} = \{m\delta'-\alpha \,|\, m\in 1/2+a+\mathbb{Z}_+\} \cup \{m\delta'+\alpha \,|\, m\in 1/2-a+\mathbb{Z}_+\}$, all of multiplicity 1, and $\Delta_{W,\sigma}^{{\rm im},+} = \{m\delta' \,|\, m\in\mathbb{N}\}$ of multiplicity 2. Let $P_{N=2}^a(\eta)$, $\eta\in\mathfrak{h}_{W,\sigma}^*$, be the corresponding partition function. Let h and j be the eigenvalues of $L_0^{\rm tw}$ and of $J_0^{\rm tw}$ respectively on the highest weight vector v_λ.

By Theorem 4.2 and Remark 4.2(a), we obtain the following formula for $\det_\eta(k,h,j)$, conjectured in [BFK] (cf. [KM]):

$$\begin{aligned} &\prod_{m,n\in\mathbb{N}} \Big((k+1)(h-h_{m,n}(k,j))\Big)^{P_{N=2}^a(\eta-mn\delta')} \\ \times &\prod_{m\in 1/2+a+\mathbb{Z}_+} \varphi_{m,-}(k,h,j)^{P_{N=2;m\delta'-\alpha}^a(\eta-(m\delta'-\alpha))} \\ \times &\prod_{m\in 1/2-a+\mathbb{Z}_+} \varphi_{m,+}(k,h,j)^{P_{N=2;m\delta'+\alpha}^a(\eta-(m\delta'+\alpha))}, \end{aligned}$$

where

$$h_{m,n}(k,j) = \frac{1}{4(k+1)}\Big((n-m(k+1))^2-(j-a)^2-(k+1)^2\Big)+\frac{a^2}{2},$$
$$\varphi_{m,\pm}(k,h,j) = h-(m^2+m)(k+1)\mp(m+\frac{1}{2})(j-a)-\frac{a^2}{2}.$$

5.3. $N=2$ twisted sector

In this subsection we consider the involution $\sigma=\sigma_{\rm tw}$ of $\mathfrak{g}=s\ell(2|1)$ defined by:

$$\sigma_{\rm tw}(e_1)=e_2\,,\ \sigma_{\rm tw}(f_1)=f_2\,,\ \sigma_{\rm tw}(h_1)=h_2\,.$$

Let $e^{(1)}=(e_1+e_2)/\sqrt{2}$, $e^{(2)}=(e_1-e_2)/\sqrt{2}$, $f^{(1)}=(f_1+f_2)/\sqrt{2}$, $f^{(2)}=(f_1-f_2)/\sqrt{2}$, $H=h_1-h_2$. Then

$$\mathfrak{g}(\sigma_{\rm tw})=\mathbb{C}f\oplus\mathbb{C}f^{(2)}\oplus\mathbb{C}x\oplus\mathbb{C}e^{(2)}\oplus\mathbb{C}e\,,$$

and the only possible choice for $\mathfrak{n}(\sigma_{\rm tw})_\pm$ is as follows:

$$\mathfrak{n}(\sigma_{\rm tw})_+=\mathbb{C}f^{(2)}+\mathbb{C}f\,,\ \mathfrak{n}(\sigma_{\rm tw})_-=\mathbb{C}e^{(2)}+\mathbb{C}e\,.$$

Note that $\mathfrak{n}_{1/2}(\sigma)_+=\mathfrak{n}_{1/2}(\sigma)'_-=0$ and $\mathfrak{g}_{1/2}(\sigma)_0=\mathbb{C}e^{(2)}$ (see (2.5)), so that $\epsilon(\sigma_{\rm tw})=1$. Note also that $\mathfrak{h}^\sigma=\mathbb{C}x$, so that the set Δ^σ ($\subset\mathfrak{h}^{\sigma*}$) of non-zero roots of $\mathfrak{h}^\sigma$ in $\mathfrak{g}$ is $\Delta^\sigma=\{\pm\theta,\,\pm\theta/2\}$, where $\theta(x)=1$, the roots $\pm\theta$ (resp. $\pm\theta/2$) being of multiplicity 1 (resp. 2). Thus the s_a are as follows:

$$s_H=s_{e^{(2)}}=s_{f^{(2)}}=1/2\,,\ s_{e^{(1)}}=s_e=0\,,\ s_{f^{(1)}}=s_f=1\,.$$

(Note that here s_a depends not only on the root, but also on the root vector.)

The free field realization of the twisted vertex algebra $W_k(\mathfrak{g},\sigma_{\rm tw},\theta/2)$, provided by Theorem 4.1, is given in terms of free neutral fermions

$$\Phi^{(1)}(z)=\sum_{n\in 1/2+\mathbb{Z}}\Phi^{(1)}_n z^{-n-1/2},\quad \Phi^{(2){\rm tw}}(z)=\sum_{n\in\mathbb{Z}}\Phi^{(2)}_n z^{-n-1/2},$$

where

$$[\Phi^{(i)}_m,\Phi^{(j)}_n]=(-1)^j\delta_{ij}\delta_{m,-n},$$
$$\Phi^{(1)}(z)_-=\sum_{n>0}\Phi^{(1)}_n z^{-n-1/2},\quad \Phi^{(2){\rm tw}}(z)_-=\sum_{n>0}\Phi^{(2)}_n z^{-n-1/2},$$

and free commuting bosons

$$x(z)=\sum_{n\in\mathbb{Z}}x_n z^{-n-1},\quad H^{\rm tw}(z)=\sum_{n\in 1/2+\mathbb{Z}}H_n z^{-n-1},$$

where

$$[x_m,x_n]=\frac{1}{2}(k+1)m\delta_{m,-n},$$
$$[H_m,H_n]=-2(k+1)m\delta_{m,-n},\ x(z)_-=\sum_{n\geqslant 0}x_n z^{-n-1}$$

and $H^{\rm tw}(z)_- = \sum_{n>0} H_n z^{-n-1}$:

$$\begin{aligned} L^{\rm tw}(z) &= \frac{1}{k+1}\Big(:x(z)^2: -\frac{1}{4}:H^{\rm tw}(z)^2:\Big) + \frac{1}{2}\Big(:\Phi^{(1)}(z)\partial\Phi^{(1)}(z): \\ &\quad - :\Phi^{(2),\rm tw}(z)\partial\Phi^{(2),\rm tw}(z):\Big) + \partial x(z)\,, \\ J^{\rm tw}(z) &= H^{\rm tw}(z) - :\Phi^{(1)}(z)\Phi^{(2),\rm tw}(z):\,, \\ G^{(1),\rm tw}(z) &= (-k-1)^{-1/2}\Big(:\Phi^{(1)}(z)x(z): -\frac{1}{2}:\Phi^{(2),\rm tw}(z)H^{\rm tw}(z): \\ &\quad +(k+1)\partial\Phi^{(1)}(z)\Big)\,, \\ G^{(2),\rm tw}(z) &= (-k-1)^{-1/2}\Big(:\Phi^{(2),\rm tw}(z)x(z): +\frac{1}{2}:\Phi^{(1)}(z)H^{\rm tw}(z): \\ &\quad -(k+1)\partial\Phi^{(2),\rm tw}(z)\Big)\,, \end{aligned}$$

where $G^{(1),\rm tw} = \frac{1}{\sqrt{2}}(G^{+,\rm tw} + G^{-,\rm tw})$, $G^{(2),\rm tw} = \frac{1}{\sqrt{2}}(G^{+,\rm tw} - G^{-,\rm tw})$.

Furthermore, in this case the set $\widehat{\Delta}_+ = \widehat{\Delta}_+^{\rm re} \cup \widehat{\Delta}_+^{\rm im}$ of positive roots of $\widehat{\mathfrak{g}}^{\rm tw}$ is as follows: $\widehat{\Delta}_+^{\rm re} = \{m\delta + \theta/2,\ (m+1)\delta - \theta/2,\ m\delta + \theta,\ (m+1)\delta - \theta \,|\, m \in \mathbb{Z}_+\} \cup \{m\delta \pm \theta/2 \,|\, m \in \frac{1}{2} + \mathbb{Z}_+\}$, $\widehat{\Delta}_+^{\rm im} = \{m\delta \,|\, m \in \frac{1}{2}\mathbb{N}\}$, all having multiplicity 1. Note also that the roots $m\delta \pm \theta/2$ are odd and all the other roots are even. We have: $\mathfrak{h}^\natural = 0$, $s_{\mathfrak{g}} = -k/16(k+1)$, $s_{\rm gh} = -1/16$, and

$$\widehat{\Delta}_{++}^{\rm re} = \{m\delta + \theta \,|\, m \in \mathbb{Z}_+\} \cup \{m\delta + \theta/2 \,|\, m \in \frac{1}{2}\mathbb{Z}_+\}\,,$$

all of multiplicity 1. From (4.10) and Remark 4.2(b) we obtain that

$$\varphi_{m\delta+\theta/2,n}(k,h) = h - h^{\rm tw}_{n,2m+1}(k) \quad \text{and} \quad \varphi_{m\delta+\theta,n}(k,h) = h - h^{\rm tw}_{2n,m+1}(k),$$

where

$$h^{\rm tw}_{n,m}(k) = \frac{1}{4(k+1)}\left(\left(\frac{n}{2} - m(k+1)\right)^2 - (k+1)^2\right) + \frac{1}{8}\,. \tag{5.3}$$

It is easy to compute that $[G_0^{(2)}, G_0^{(2)}] = -(L_0^{\rm tw} - c(k)/24)$, hence the extra factor equals $\varphi_{(\theta-\delta)/2,0}(h,k) = h - c(k)/24 = h + (2k+1)/8$.

The set of positive even (resp. odd) roots for $W_k(s\ell(2|1), \sigma_{\rm tw}, \theta/2)$ is $\frac{1}{2}\mathbb{N}\delta'$ (resp. $\frac{1}{2}\mathbb{Z}_+\delta'$), all of multiplicity 1. Hence $P'_{W,\sigma_{\rm tw}}(\eta) = p^{\rm tw}(\eta)$, where $p^{\rm tw}(\eta)$ is defined by the generating series

$$\sum_{\eta \in \frac{1}{2}\mathbb{Z}_+} p^{\rm tw}(\eta) q^\eta = \prod_{n=1}^{\infty} \frac{1+q^{n/2}}{1-q^{n/2}}.$$

Hence by Theorem 4.2, we obtain the following determinant formula for the $N=2$ twisted superconformal algebra, conjectured in [BFK]:

$$\det_\eta(k,h)=(h+\frac{2k+1}{8})^{p^{\mathrm{tw}}(\eta)}\prod_{\substack{m,n\in\mathbb{N}\\ n\,\mathrm{odd}}}(h-h^{\mathrm{tw}}_{n,m}(k))^{2p^{\mathrm{tw}}(\eta-\frac{1}{2}mn)}.$$

5.4. $N=4$ **Ramond type sector**

Recall that the $N=4$ vertex algebra is $W_k(\mathfrak{g},\theta/2)$, where $\mathfrak{g}=s\ell(2|2)/\mathbb{C}I$. We shall use the same basis of $\mathfrak{g}$ and keep the same notation as in [KW], Section 8.4. In particular, the simple roots are α_1, α_2, α_3, where α_1 and α_3 are odd and α_2 is even, all the non-zero scalar products between them being $(\alpha_1|\alpha_2)=(\alpha_2|\alpha_3)=1$, $(\alpha_2|\alpha_2)=-2$. The dual Coxeter number $h^\vee=0$.

Consider the Ramond type automorphisms $\sigma=\sigma_{a,b}$ of $\mathfrak{g}$, where $-1/2<a$, $b\leqslant 1/2$, defined by $\sigma(e_1)=e^{2\pi ia}e_1$, $\sigma(e_2)=e^{-2\pi i(a+b)}e_2$, $\sigma(e_3)=e^{2\pi ib}e_3$, $\sigma(h_i)=h_i$. Note that $\epsilon(\sigma_{a,b})=0$. We consider first the case when $a+b>0$. Then we have the following possibilities for $\mathfrak{n}(\sigma)_\pm$:

(i) $a,b\neq 1/2$: $\mathfrak{n}(\sigma)_-=\mathbb{C}e$, where $e=e_{123}$, $\mathfrak{n}(\sigma)_+=\mathbb{C}f$, where $f=f_{123}$;
(ii) $a=1/2$, $b\neq 1/2$: $\mathfrak{n}(\sigma)_-=\mathbb{C}e+\mathbb{C}e_1+\mathbb{C}f_{23}$, $\mathfrak{n}(\sigma)_+=\mathbb{C}f+\mathbb{C}e_{23}$; $+\mathbb{C}f_1$;
(iii) $a\neq 1/2$, $b=1/2$: $\mathfrak{n}(\sigma)_-=\mathbb{C}e+\mathbb{C}e_3+\mathbb{C}f_{12}$, $\mathfrak{n}(\sigma)_+=\mathbb{C}f+\mathbb{C}e_{12}+\mathbb{C}f_3$;
(iv) $a=b=1/2$: $\mathfrak{n}(\sigma)_-=\mathrm{span}\,\{e,e_1,e_3,f_{12},f_{23},f_2\}$,
$\mathfrak{n}(\sigma)_+=\mathrm{span}\,\{f,e_{12},e_{23},e_2,f_1,f_3\}$.

In these four cases the s_α are as follows:

$$s_{\alpha_1}=a,s_{\alpha_2}=1-a-b,s_{\alpha_3}=b,s_{\alpha_1+\alpha_2}=-b,s_{\alpha_2+\alpha_3}=-a,s_\theta=0\,,$$

where $\theta=\alpha_1+\alpha_2+\alpha_3$, and, as usual, $s_{-\alpha}=1-s_\alpha$.

Consequently, we have:

$$\begin{aligned}\widehat{\Delta}^{\mathrm{re}}_{++}\;=\;&\{(m+a)\delta+\alpha_1\,,\,(m+b)\delta+\alpha_3\,,\,(m-b)\delta+\alpha_1+\alpha_2\,,\\&(m-a)\delta+\alpha_2+\alpha_3\,,\,(m+1-a-b)\delta+\alpha_2\,,\,(m+a+b)\delta-\alpha_2\,,\\&m\delta+\theta|\,m\in\mathbb{Z}_+\}\,.\end{aligned}$$

Next $\gamma'^\natural=\frac{1}{2}\alpha_2$, $\gamma^\natural_{1/2}=\frac{a+b}{2}\alpha_2$, $s_{\mathfrak{g}}+s_{\mathrm{gh}}=-ab+(a+b)/2$.

We have: $\mathfrak{h}^\natural=\mathbb{C}h_2$, hence $\mathfrak{h}_{W,\sigma}=\mathbb{C}h_2+\mathbb{C}L^{\mathrm{tw}}_0$. Define $\alpha\in\mathfrak{h}^*_{W,\sigma}$ by $\alpha(h_2)=2$, $\alpha(L^{\mathrm{tw}}_0)=0$. Then

$$\Delta^+_{W,\sigma}=\{(m+\frac{1}{2}+a)\delta'-\frac{\alpha}{2},(m+\frac{1}{2}+b)\delta'-\frac{\alpha}{2},(m+\frac{1}{2}-b)\delta'+\frac{\alpha}{2},$$
$$(m+\frac{1}{2}-a)\delta'+\frac{\alpha}{2},(m+1-a-b)\delta'+\alpha,(m+a+b)\delta'-\alpha,(m+1)\delta'|\,m\in\mathbb{Z}_+\}$$

is the set of positive roots of $W(\mathfrak{g},\sigma_{a,b},\theta/2)$, all having multiplicity 1, except for $m\delta'$ which have multiplicity 2. We have: $\alpha_1^\natural=\alpha_3^\natural=-\alpha/2$, $\alpha_2^\natural=\alpha$.

Let $P^{a,b}_{N=4}(\eta)$ be the corresponding partition function. Let h and j be the eigenvalues of L^{tw}_0 and $J^{\{h_2\},\mathrm{tw}}_0$ on v_λ, so that $\lambda^\natural=\frac{j}{2}\alpha$. Formulas (4.8)–(4.10) give the following factors of the determinant (we introduce a simplifying notation consistent with the map π): $\varphi_{m,n} := k\varphi_{(m-1)\delta+\theta,n}$, $\varphi_{m,-\alpha/2} := \varphi_{m\delta+\alpha_1,1} = \varphi_{m\delta+\alpha_3,1}$, $\varphi_{m,\alpha/2} := \varphi_{m\delta+\alpha_1+\alpha_2,1} = \varphi_{m\delta+\alpha_2+\alpha_3,1}$, $\varphi_{m,n,\pm\alpha} = \varphi_{m\delta\pm\alpha_2,n}$, where:

$$\varphi_{m,n}(k,h,j) = 4kh-(n-mk)^2+(a+b+j-1)^2+k(k+1)+k(2a-1)(2b-1),$$

$$\varphi_{m,\pm\alpha/2}(k,h,j) = h-\Big(m+\frac{1}{2}\Big)^2 k \pm \Big(m+\frac{1}{2}\Big)(a+b+j-1)+\frac{k+1}{4}$$
$$+\Big(a-\frac{1}{2}\Big)\Big(b-\frac{1}{2}\Big),$$

$$\varphi_{m,n,\pm\alpha}(k,j) = mk \mp (a+b+j-1)+n\,.$$

By Theorem 4.2 and Remark 4.2(a), we obtain the following formula for $\det_\eta(k,h,j)$ in the case $a+b>0$:

$$\prod_{m,n\in\mathbb{N}} \varphi_{m,n}(k,h,j)^{P^{a,b}_{N=4}(\eta-mn\delta')}$$
$$\times \prod_{m\in\frac{1}{2}+\{a,b\}+\mathbb{Z}_+} \varphi_{m,-\alpha/2}(k,h,j)^{P^{a,b}_{N=4;m\delta'-\alpha/2}(\eta-(m\delta'-\alpha/2))}$$
$$\times \prod_{m\in\frac{1}{2}-\{a,b\}+\mathbb{Z}_+} \varphi_{m,\alpha/2}(k,h,j)^{P^{a,b}_{N=4;m\delta'+\alpha/2}(\eta-(m\delta'+\alpha/2))}$$
$$\times \prod_{\substack{m\in a+b+\mathbb{Z}_+\\ n\in\mathbb{N}}} \varphi_{m,n,-\alpha}(k,j)^{P^{a,b}_{N=4}(\eta-n(m\delta'-\alpha))}$$
$$\times \prod_{\substack{m\in -a-b+\mathbb{N}\\ n\in\mathbb{N}}} \varphi_{m,n,\alpha}(k,j)^{P^{a,b}_{N=4}(\eta-n(m\delta'+\alpha))}\,.$$

The case $a+b \leqslant 0$ is treated in the same fashion. The s_α's in this case are the same as in the case $a+b>0$, except for $s_{\alpha_2} = -a-b$. After the calculation, it turns out that the determinant formula in this case can be obtained from the above determinant formula by replacing a by $a+1$ and b by $b+1$ in all factors and by changing the range of m in the last two factors by exchanging $\mathbb{Z}_+$ and $\mathbb{N}$. Some cases of this determinant formula were conjectured in [KR].

We shall omit the free field realization of the Ramond type sector of $N=4$ and other remaining superconformal algebras as being quite long. On the other hand, as in the simplest cases of $N=1$ and 2, they are straightforward applications of Theorem 4.1.

5.5. $N=3$ Ramond type sector

Recall that the $N=3$ vertex algebra is $W_k(\mathfrak{g},\theta/2)$, where $\mathfrak{g} = spo(2|3)$. (To get the "linear" $N=3$ superconformal algebra one needs to tensor the above vertex algebra with one free fermion, and the results of this section can easily be extended to the latter case as in [KW].) We shall keep the notation of [KW], Section 8.5. In particular, the simple roots are α_1 and α_2, where α_1 is odd and α_2 is even, the scalar products between them being $(\alpha_1|\alpha_1)=0$, $(\alpha_1|\alpha_2)=1/2$, $(\alpha_2|\alpha_2)=-1/2$.

Since $\theta = 2\alpha_1 + 2\alpha_2$, we have: $\alpha_2^\natural = -\alpha_1^\natural = \alpha_2$. Recall also that $S_0 = \{\pm\alpha_2\}$, $S_{1/2} = \{\alpha_1, \theta/2, \alpha_1 + 2\alpha_2\}$, $S_1 = \{\theta\}$. The dual Coxeter number $h^\vee = 1/2$.

Consider the Ramond type automorphisms $\sigma = \sigma_{a,b}$ of $\mathfrak{g}$ defined by $\sigma(e_{10}) = e^{2\pi i a}e_{10}$, $\sigma(e_{01}) = e^{2\pi i b}e_{01}$, $\sigma|_{\mathfrak{h}} = 1$, where $a, b \in \mathbb{R}$ are such that $a + b \in \frac{1}{2}\mathbb{Z}$. We consider the following three cases:

$$\text{I (resp. II): } a = -b, -1/2 < a \leqslant 0 \ \ (\text{resp. } a = -b\,, 0 < a \leqslant 1/2)\,,$$
$$\text{III: } a + b = 1/2, -1/2 < a \leqslant 1/2\,.$$

Note that $\epsilon(\sigma) = 0$ in cases I and II, and $\epsilon(\sigma) = 1$ in case III, when $\mathfrak{g}_{1/2}(\sigma)_0 = \mathbb{C}e_{11}$ in (2.5). We have the following possibilities for $\mathfrak{n}(\sigma)_\pm$:

I, II , $a \neq 0, 1/2 : \mathfrak{n}(\sigma)_- = \mathbb{C}e_{22}$, $\mathfrak{n}(\sigma)_+ = \mathbb{C}f_{22}$,
I, II, $a = 0 : \mathfrak{n}(\sigma)_- = \mathbb{C}e_{22} + \mathbb{C}f_{01}$, $\mathfrak{n}(\sigma)_+ = \mathbb{C}f_{22} + \mathbb{C}e_{01}$,
I, II, $a = 1/2 : \mathfrak{n}(\sigma)_- = \text{span}\,\{e_{22}, e_{10}, f_{12}\}$, $\mathfrak{n}(\sigma)_+ = \text{span}\,\{f_{22}, e_{12}, f_{10}\}$,
III, $a \neq 1/2 : \mathfrak{n}(\sigma)_- = \mathbb{C}e_{22} + \mathbb{C}e_{11}$, $\mathfrak{n}(\sigma)_+ = \mathbb{C}f_{22} + \mathbb{C}f_{11}$,
III, $a = 1/2 : \mathfrak{n}(\sigma)_- = \text{span}\,\{e_{22}, e_{11}, f_{12}, f_{01}, e_{10}\}$,
$\mathfrak{n}(\sigma)_+ = \text{span}\,\{f_{22}, f_{11}, e_{12}, e_{01}, f_{10}\}$.

In these cases the s_α are as follows (up to the relation (3.3)):

I. : $s_{\alpha_1} = a$, $s_{\alpha_2} = -a$, $s_{\theta/2} = 0$, $s_{\alpha_1+2\alpha_2} = -a$, $s_\theta = 0$;
II. : the same as in I, except for $s_{\alpha_2} = 1 - a$;
III. : $s_{\alpha_1} = a$, $s_{\alpha_2} = 1/2 - a$, $s_{\theta/2} = 1/2$, $s_{\alpha_1+2\alpha_2} = -a$, $s_\theta = 0$.

One finds that in these three cases:

I. : $\gamma'^\natural = (a - \frac{1}{2})\alpha_2$, $\gamma_{1/2}^\natural = a\alpha_2$, $s_{\mathfrak{g}} + s_{\text{gh}} = \frac{a(1-a)}{4k+2} + \frac{a}{2}$;
II. : $\gamma'^\natural = (a + \frac{1}{2})\alpha_2$, $\gamma_{1/2}^\natural = a\alpha_2$, $s_{\mathfrak{g}} + s_{\text{gh}} = -\frac{a(a+1)}{4k+2} + \frac{a}{2}$;
III. : $\gamma'^\natural = a\alpha_2$, $\gamma_{1/2}^\natural = a\alpha_2$, $s_{\mathfrak{g}} + s_{\text{gh}} = -\frac{a^2}{4k+2} - \frac{1}{16}$.

Consequently we have in these cases ($m \in \mathbb{Z}_+$):

I. : $\widehat{\Delta}^{\text{re}}_{++} = \{(m-a)\delta+\alpha_2,\ (m+1+a)\delta-\alpha_2,\ (m+a)\delta+\alpha_1,\ (m-a)\delta+\alpha_1+2\alpha_2,\ m\delta + \theta/2,\ m\delta + \theta\}$,
II. : $\widehat{\Delta}^{\text{re}}_{++} = \{(m+1-a)\delta+\alpha_2,\ (m+a)\delta-\alpha_2,\ (m+a)\delta+\alpha_1,\ (m-a)\delta+\alpha_1+2\alpha_2,\ m\delta + \theta/2,\ m\delta + \theta\}$,
III. : $\widehat{\Delta}^{\text{re}}_{++} = \{(m + 1/2 - a)\delta + \alpha_2,\ (m + 1/2 + a)\delta - \alpha_2,\ (m + a)\delta + \alpha_1,\ (m - a)\delta + \alpha_1 + 2\alpha_2,\ (m + 1/2)\delta + \theta/2,\ m\delta + \theta\}$.

We have: $\mathfrak{h}^\natural = \mathbb{C}\alpha_2$, hence $\mathfrak{h}_{W,\sigma} = \mathbb{C}\alpha_2 + \mathbb{C}L_0^{\text{tw}}$. Define $\alpha \in \mathfrak{h}^*_{W,\sigma}$ by $\alpha = \alpha_2|_{\mathfrak{h}^\natural}$, $\alpha(L_0^{\text{tw}}) = 0$. Then we have in the three cases ($m \in \mathbb{Z}_+$):

I. : $\Delta^+_{W,\sigma} = \{(m-a)\delta'+\alpha\,,\ (m+1+a)\delta'-\alpha,\ (m+1/2+a)\delta'-\alpha,\ (m+\frac{1}{2}-a)\delta'+\alpha,\ (m + 1/2)\delta',\ (m + 1)\delta'\}$;
II. : $\Delta^+_{W,\sigma} = \{(m+1-a)\delta'+\alpha,\ (m+a)\delta'-\alpha,\ (m+1/2+a)\delta'-\alpha,\ (m+1/2-a)\delta'+\alpha,\ (m + 1/2)\delta',\ (m + 1)\delta'\}$;
III. : $\Delta^+_{W,\sigma} = \{(m + 1/2 - a)\delta' + \alpha,\ (m + 1/2 + a)\delta' - \alpha,\ (m + 1)\delta'\}$.

The multiplicities of these positive roots of $W(\mathfrak{g}, \sigma, \theta/2)$ are 1, except for the following cases: mult $(m + 1)\delta' = 2$ in cases I and II, mult $(m + 1/2 \mp a)\delta' \pm \alpha = 2$

and mult $(m+1)\delta' = 3$ in case III ($m \in \mathbb{Z}_+$). Note, however, that in case III we have, in fact, one even root and one odd root equal $(m+1/2\mp a)\delta' \pm \alpha$, each having multiplicity 1, and an even (resp. odd) root $(m+1)\delta'$ of multiplicity 2 (resp. 1).

We have: $\alpha_2^\natural = -\alpha_1^\natural = \alpha$. Note that 0 is a (odd) root of $\Delta_{W,\sigma}^+$ only in case III.

Let $P_{N=3}^{a,b}(\eta)$ be the corresponding partition function. Let h and j be the respective eigenvalues of $L_0^{\rm tw}$ and $J_0^{\{-4\alpha_2\},{\rm tw}}$ on v_λ, so that $\lambda^\natural = \frac{j}{2}\alpha$.

Introduce the following notations for the factors of the determinant:

$$\begin{aligned}\varphi_{m,n} &= \varphi_{(m-1)\delta+\theta,n},\\ \varphi_{m,\alpha} &= \varphi_{m\delta+\alpha_1+2\alpha_2,1},\\ \varphi_{m,-\alpha} &= \varphi_{m\delta+\alpha_1,1},\\ \varphi_{m,n,\pm\alpha} &= \varphi_{m\delta\pm\alpha_2,n}.\end{aligned}$$

Formulas (4.8)–(4.10) give the following expressions in case I:

$$\begin{aligned}\varphi_{m,n}(k,h,j) &= h - \frac{1}{4k+2}\Big(\big(m(k+\frac{1}{2}) - \frac{n}{2}\big)^2 - \frac{(j+1)^2}{4}\Big) + \frac{1}{4}(k+\frac{3}{2}) + \frac{a}{2},\\ \varphi_{m,\pm\alpha}(k,h,j) &= h - (m+\frac{1}{2})^2(k+\frac{1}{2}) \pm \frac{1}{2}(m+\frac{1}{2})(j+1) + \frac{1}{4}(k+\frac{3}{2}) + \frac{a}{2},\\ \varphi_{m,n,\pm\alpha}(k,j) &= m(k+\frac{1}{2}) + \frac{n}{4} \mp \frac{j+1}{4}.\end{aligned}$$

By Theorem 4.2 and Remarks 4.2(a) and (b) we obtain the following formula for $\det_\eta(k,h,j)$ in case I (a special case of this formula was conjectured in [KMR] and partially proved in [M]):

$$\begin{aligned}&\prod_{m,n\in\mathbb{N}} (k+\frac{1}{2})^{P_{N=3}^a(\eta-mn\delta')} \prod_{\substack{m,n\in\mathbb{N}\\ m+n \text{ even}}} \varphi_{m,n}(k,h,j)^{P_{N=3}^a(\eta-\frac{1}{2}mn\delta')}\\ &\quad\times \prod_{m\in\mp a+\frac{1}{2}+\mathbb{Z}_+} \varphi_{m,\pm\alpha}(k,h,j)^{P_{N=3;m\delta'\pm\alpha}^a(\eta-(m\delta'\pm\alpha))}\\ &\times \prod_{\substack{m\in -a+\mathbb{Z}_+\\ n\in\mathbb{N}}} \varphi_{m,n,\alpha}(k,j)^{P_{N=3}^a(\eta-n(m\delta'+\alpha))} \prod_{\substack{m\in a+\mathbb{N}\\ n\in\mathbb{N}}} \varphi_{m,n,-\alpha}(k,j)^{P_{N=3}^a(\eta-n(m\delta'-\alpha))}.\end{aligned}$$

In case II the determinant formula is similar. It can be obtained from the above formula by replacing $j+1$ by $j-1$ and a by $-a$ in all factors and by changing the range of m in the last two factors by exchanging $\mathbb{Z}_+$ and $\mathbb{N}$.

In case III we have:

$$\begin{aligned}\varphi_{m,n}(k,h,j) &= h - \frac{1}{4k+2}\Big(\big(m(k+\frac{1}{2}) - \frac{n}{2}\big)^2 - \frac{j^2}{4}\Big) + \frac{k}{4} + \frac{3}{16},\\ \varphi_{m,\pm\alpha}(k,h,j) &= h - \big(m+\frac{1}{2}\big)^2\big(k+\frac{1}{2}\big) \pm \frac{j}{2}\big(m+\frac{1}{2}\big) + \frac{k}{4} + \frac{3}{16},\\ \varphi_{m,n,\pm\alpha}(k,j) &= m\big(k+\frac{1}{2}\big) + \frac{n}{4} \mp \frac{j}{4}.\end{aligned}$$

The extra factor is $\varphi_0 = h + \frac{1}{16}(4k+3+\frac{j^2}{k+1/2})$, which is computed, using formula (4.7) (in this case $h_0^\vee = -1/2$).

By Theorem 4.2 and Remarks 4.1 and 4.2(a) and (b) we obtain the following formula for $\det_\eta(k,h,j)$ in case III:

$$\Big(k+\frac{1}{2}\Big)^{\sum_{m,n\in\mathbb{N}} P^a_{N=3}(\eta-mn\delta')+\sum_{m\in\mathbb{N}} P^a_{N=3;m\delta'}(\eta-m\delta')}$$

$$\times\Big(h+\frac{1}{16}(4k+3+\frac{j^2}{k+1/2})\Big)^{P'^a_{N=3}(\eta)} \prod_{\substack{m,n\in\mathbb{N}\\ m+n\ \mathrm{odd}}} \varphi_{m,n}(k,h,j)^{P^a_{N=3}(\eta-\frac{1}{2}mn\delta')}$$

$$\times \prod_{m\in\frac{1}{2}\mp a+\mathbb{Z}_+} \varphi_{m,\pm\alpha}(k,h,j)^{P^a_{N=3;m\delta'\pm\alpha}(\eta-(m\delta'\pm\alpha))}$$

$$\times \prod_{\substack{m\in\frac{1}{2}\mp a+\mathbb{Z}_+\\ n\in\mathbb{N}}} \varphi_{m,n,\pm\alpha}(k,j)^{P^a_{N=3}(\eta-n(m\delta'\pm\alpha))}\,.$$

5.6. Big $N=4$ Ramond type sector

Recall that the big $N=4$ vertex algebra is $W_k(\mathfrak{g},\theta/2)$, where $\mathfrak{g}=D(2,1;a)$. (To get the "linear" N=4 superconformal algebra ([KL],[S],[STP]) one needs to tensor the above vertex algebra with four free fermions and one free boson [GS], and the results of this and the next section can easily be extended to the latter case as in [KW].) We shall keep the notation of [KW], Section 8.6. In particular, the simple roots are α_1, α_2, α_3, where α_1 and α_3 are even, and α_2 is odd, the non-zero scalar products between them being ($a\neq 0,-1$):

$$(\alpha_1|\alpha_2)=\frac{1}{a+1}\,,\qquad (\alpha_2|\alpha_3)=\frac{a}{a+1}\,,$$
$$(\alpha_1|\alpha_1)=-\frac{2}{a+1}\,,\qquad (\alpha_3|\alpha_3)=-\frac{2a}{a+1}\,.$$

We shall slightly simplify notation of [KW] by letting

$$e_1=e_{100},\qquad e_2=e_{010},\qquad e_3=e_{001},$$
$$f_1=f_{100},\qquad f_2=f_{010},\qquad f_3=f_{001},$$
$$e=e_{121},\qquad f=f_{121}.$$

In this subsection we consider the Ramond type automorphisms $\sigma=\sigma_{\mu,\nu}$ of $\mathfrak{g}$ defined by $\sigma(e_1)=e^{2\pi i\mu}e_1$, $\sigma(e_2)=e^{-\pi i(\mu+\nu)}e_2$, $\sigma(e_3)=e^{2\pi i\nu}e_3$, $\sigma|_{\mathfrak{h}}=1$, where $\mu,\nu\in\mathbb{R}$ are such that $-1\leqslant\mu\pm\nu<1$. We consider separately the following four cases: $(++):\mu,\nu\geqslant 0$; $(-+):\mu<0,\nu\geqslant 0$; $(+-):\mu\geqslant 0,\nu<0$; $(--):\mu,\nu<0$. In all cases, $\epsilon(\sigma)=0$ and $\mathfrak{h}^\sigma=\mathfrak{h}$. Since $\theta=\alpha_1+2\alpha_2+\alpha_3$, we have: $\mathfrak{h}^\natural=\mathbb{C}\alpha+\mathbb{C}\alpha'$, where $\alpha:=\alpha_1|_{\mathfrak{h}^\natural}=\alpha_1^\natural$, $\alpha':=\alpha_3|_{\mathfrak{h}^\natural}=\alpha_3^\natural$, and $\alpha_2^\natural=-(\alpha+\alpha')/2$. Recall also that $S_0=\{\pm\alpha_1,\pm\alpha_3\}$, $S_{1/2}=\{\alpha_2,\alpha_1+\alpha_2\,,\,\alpha_2+\alpha_3\,,\,\alpha_1+\alpha_2+\alpha_3\}$, $S_1=\{\theta\}$. The dual Coxeter number $h^\vee=0$.

We have the following possibilities for $\mathfrak{n}(\sigma)_\pm$:

(i) $\mu=-1$, $\nu=0$: $\mathfrak{n}(\sigma)_- = \text{span}\,\{e, e_2, e_{011}, f_1, f_3, f_{111}, f_{110}\}$, $\mathfrak{n}(\sigma)_+ = \text{span}\,\{f, f_2, f_{011}, e_1, e_3, e_{111}, e_{110}\}$;

(ii) $\mu+\nu=-1$, $\nu\neq 0$: $\mathfrak{n}(\sigma)_- = \text{span}\,\{e, e_2, f_{111}\}$, $\mathfrak{n}(\sigma)_+ = \text{span}\,\{f, e_{111}, f_2\}$;

(iii) $\mu-\nu=-1$, $\nu\neq 0$: $\mathfrak{n}(\sigma)_- = \text{span}\{e, e_{011}, f_{110}\}$, $\mathfrak{n}(\sigma)_+ = \text{span}\{f, e_{110}, f_{011}\}$;

(iv) $\mu=\nu=0$: $\mathfrak{n}(\sigma)_- = \text{span}\,\{e, f_1, f_3\}$, $\mathfrak{n}(\sigma)_+ = \text{span}\,\{f, e_1, e_3\}$;

(v) $\mu=0$, $\nu\neq 0$: $\mathfrak{n}(\sigma)_- = \mathbb{C}e+\mathbb{C}f_1$, $\mathfrak{n}(\sigma)_+ = \mathbb{C}f+\mathbb{C}e_1$;

(vi) $\mu\neq 0,-1$, $\nu=0$: $\mathfrak{n}(\sigma)_- = \mathbb{C}e+\mathbb{C}f_3$, $\mathfrak{n}(\sigma)_+ = \mathbb{C}f+\mathbb{C}e_3$;

(vii) in all other cases: $\mathfrak{n}(\sigma)_- = \mathbb{C}e$, $\mathfrak{n}(\sigma)_+ = \mathbb{C}f$.

The s_α are as follows:

$$s_\theta = 0, \qquad s_{\alpha_2} = -\frac{\mu+\nu}{2}, \qquad s_{\alpha_1+\alpha_2+\alpha_3} = \frac{\mu+\nu}{2},$$
$$s_{\alpha_1+\alpha_2} = \frac{\mu-\nu}{2}, \qquad s_{\alpha_2+\alpha_3} = -\frac{\mu-\nu}{2}$$

in all cases; the remaining s_α (up to the relation (3.3)) are: $s_{\alpha_1}=\mu$ in cases $(++)$ and $(+-)$, $s_{\alpha_1}=1+\mu$ in cases $(-+)$ and $(--)$; $s_{\alpha_3}=\nu$ in cases $(++)$ and $(-+)$, $s_{\alpha_3}=1+\nu$ in cases $(+-)$ and $(--)$.

Using these data one finds that $\gamma^\natural_{1/2} = -\frac{\mu\alpha_1+\nu\alpha_3}{2}$ in all cases and that in the four cases (ϵ,ϵ'), where each ϵ and ϵ' is $+$ or $-$ one has: $\gamma'^\natural = -\frac{\epsilon\alpha_1+\epsilon'\alpha_3}{2}$, $s_\mathfrak{g}+s_{\text{gh}} = -\frac{1}{4}\left((\mu-\epsilon 1)^2+(\nu-\epsilon' 1)^2\right)+\frac{1}{2}$.

Furthermore, let

$$\widehat{\Delta}^{(1/2)}_{++} = \{(m-\frac{\mu+\nu}{2})\delta+\alpha_2\,,\ (m+\frac{\mu+\nu}{2})\delta+\alpha_1+\alpha_2+\alpha_3\,,$$
$$(m+\frac{\mu-\nu}{2})\delta+\alpha_1+\alpha_2\,,\ (m-\frac{\mu-\nu}{2})\delta+\alpha_2+\alpha_3 | m\in\mathbb{Z}_+\}\,,$$

and define $\widehat{\Delta}^{(0)}_{++}$ in the four cases as follows ($m\in\mathbb{Z}_+$):

$$(++):\{(m+\mu)\delta+\alpha_1, (m+\nu)\delta+\alpha_3, (m+1-\mu)\delta-\alpha_1,$$
$$(m+1-\nu)\delta-\alpha_3\}\,,$$
$$(+-):\{(m+\mu)\delta+\alpha_1, (m+1+\nu)\delta+\alpha_3, (m+1-\mu)\delta-\alpha_1,$$
$$(m-\nu)\delta-\alpha_3\},$$
$$(-+):\{(m+1+\mu)\delta+\alpha_1, (m+\nu)\delta+\alpha_3, (m-\mu)\delta-\alpha_1,$$
$$(m+1-\nu)\delta-\alpha_3\},$$
$$(--):\{(m+1+\mu)\delta+\alpha_1, (m+1+\nu)\delta+\alpha_3, (m-\mu)\delta-\alpha_1,$$
$$(m-\nu)\delta-\alpha_3\}\,.$$

Then $\widehat{\Delta}^{\text{re}}_{++} = \widehat{\Delta}^{(0)}_{++} \cup \widehat{\Delta}^{(1/2)}_{++} \cup \{m\delta+\theta | \, m\in\mathbb{Z}_+\}$.

Next,

$$\mathfrak{h}_{W,\sigma} = \mathfrak{h}^\natural \oplus \mathbb{C}L_0^{\mathrm{tw}}, \quad \text{and} \quad \Delta_{W,\sigma}^{+,\mathrm{re}} = \Delta_{W,\sigma}^{+(1/2)} \cup \Delta_{W,\sigma}^{+(0)} \subset \mathfrak{h}_{W,\sigma}^*,$$

where

$$\Delta_{W,\sigma}^{+(1/2)} = \{(m+\frac{1}{2}-\frac{\mu+\nu}{2})\delta' - \frac{\alpha+\alpha'}{2}, (m+\frac{1}{2}+\frac{\mu+\nu}{2})\delta' + \frac{\alpha+\alpha'}{2},$$
$$(m+\frac{1}{2}+\frac{\mu-\nu}{2})\delta' + \frac{\alpha-\alpha'}{2}, (m+\frac{1}{2}-\frac{\mu-\nu}{2})\delta' - \frac{\alpha-\alpha'}{2} | m \in \mathbb{Z}_+\},$$

and $\Delta_{W,\sigma}^{+(0)}$ in the four cases is as follows ($m \in \mathbb{Z}_+$):

$$(++) : \{(m+\mu)\delta' + \alpha, (m+\nu)\delta' + \alpha', (m+1-\mu)\delta' - \alpha,$$
$$(m+1-\nu)\delta' - \alpha'\},$$
$$(+-) : \{(m+\mu)\delta' + \alpha, (m+1+\nu)\delta' + \alpha', (m+1-\mu)\delta' - \alpha,$$
$$(m-\nu)\delta' - \alpha'\},$$
$$(-+) : \{(m+1+\mu)\delta' + \alpha, (m+\nu)\delta' + \alpha', (m-\mu)\delta' - \alpha,$$
$$(m+1-\nu)\delta' - \alpha'\},$$
$$(--) : \{(m+1+\mu)\delta' + \alpha, (m+1+\nu)\delta' + \alpha', (m-\mu)\delta' - \alpha,$$
$$(m-\nu)\delta' - \alpha'\}.$$

The multiplicities of all these roots of $W(\mathfrak{g}, \sigma, \theta/2)$ are 1. There are, in addition, roots $m\delta'$ ($m \in \mathbb{N}$), all of multiplicity 3.

Let $P_{N=4}^{\mu,\nu}(\eta)$ be the corresponding partition function. Let h, j and j' be the respective eigenvalues of L_0^{tw}, J_0^{tw} and $J_0'^{\mathrm{tw}}$ on v_λ, so that $\lambda^\natural = \frac{1}{2}(j\alpha + j'\alpha')$.

Formulas (4.8)–(4.10) give the following expressions for the factors of the determinant in the $(++)$ case:

$$\varphi_{(m-1)\delta+\theta,n} = h - \frac{1}{4k}(n-mk)^2 + \frac{(j+1-\mu)^2}{4k(a+1)} + \frac{a(j'+1-\nu)^2}{4k(a+1)}$$
$$+\frac{k}{4} + \frac{(\mu-1)^2+(\nu-1)^2}{4};$$

$$\varphi_{m\delta+\beta,1} = h - \frac{1}{k}\Big((m+\frac{1}{2})k + \frac{j+1-\mu}{2}(\beta|\alpha_1) + \frac{j'+1-\nu}{2}(\beta|\alpha_3)\Big)^2$$
$$+\frac{(j+1-\mu)^2}{4k(a+1)} + \frac{a(j'+1-\nu)^2}{4k(a+1)} + \frac{k}{4} + \frac{(\mu-1)^2+(\nu-1)^2}{4}$$

if $\beta \in S_{1/2}$;

$$\varphi_{m\delta+\beta,n} = mk + \frac{j+1-\mu}{2}(\beta|\alpha_1) + \frac{j'+1-\nu}{2}(\beta|\alpha_3) - \frac{n(\beta|\beta)}{2}$$

if $\beta \in S_0$.

The factors in the remaining three cases are obtained from the above formulas by a shift of μ and ν as follows:

$$(-+) : \mu \to \mu + 2\,,\ \nu \to \nu\,;$$
$$(+-) : \mu \to \mu\,,\ \nu \to \nu + 2\,;$$
$$(--) : \mu \to \mu + 2\,,\ \nu \to \nu + 2\,.$$

By Theorem 4.2 and Remark 4.2(a) we obtain the following formula for $\det_\eta(k, h, j, j')$:

$$\prod_{m,n\in\mathbb{N}} (k^2\varphi_{(m-1)\delta+\theta,n}(h,k,j,j'))^{P^{\mu,\nu}_{N=4}(\eta-mn\delta')}$$
$$\times \prod_{m\delta+\beta\in\widehat{\Delta}^{(1/2)}_{++}} \varphi_{m\delta+\beta,1}(k,h,j,j')^{P^{\mu,\nu}_{N=4;(m+1/2)\delta'+\beta^\natural}(\eta-(m+1/2)\delta'-\beta^\natural)}$$
$$\times \prod_{\substack{m\delta+\beta\in\widehat{\Delta}^{(0)}_{++}\\ n\in\mathbb{N}}} \varphi_{m\delta+\beta,n}(k,h,j,j')^{P^{\mu,\nu}_{N=4}(\eta-n(m\delta'+\beta^\natural))}\,.$$

5.7. Big $N = 4$ twisted sector

In this subsection we consider the involutions $\sigma = \sigma_{\mathrm{tw},b}$ of $\mathfrak{g} = D(2,1;1) = osp(4,2)$ defined by:

$$\sigma(e_1) = e_3\,,\ \sigma(e_2) = e^{-\pi ib}e_2\,,\ \sigma(e_3) = e^{2\pi ib}e_1\,,$$
$$\sigma(f_1) = f_3\,,\ \sigma(f_2) = e^{\pi ib}f_2\,,\ \sigma(f_3) = e^{-2\pi ib}f_1\,,$$

where $b \in \mathbb{R}$, $-1 \leqslant b < 1$. Introduce the following elements of $\mathfrak{g}$: $e^{(1)} = \frac{1}{\sqrt{2}}(e_1 + e^{-\pi ib}e_3)$, $f^{(1)} = \frac{1}{\sqrt{2}}(f_1 + e^{\pi ib}f_3)$, $e^{(3)} = \frac{1}{\sqrt{2}}(e_1 - e^{-\pi ib}e_3)$, $f^{(3)} = \frac{1}{\sqrt{2}}(f_1 - e^{\pi ib}f_3)$, $e^{(110)} = \frac{1}{\sqrt{2}}(e_{110} + e^{-\pi ib}e_{011})$, $f^{(110)} = \frac{1}{\sqrt{2}}(f_{110} + e^{\pi ib}f_{011})$, $e^{(011)} = \frac{1}{\sqrt{2}}(e_{110} - e^{-\pi ib}e_{011})$, $f^{(011)} = \frac{1}{\sqrt{2}}(f_{110} - e^{\pi ib}f_{011})$. We have the following eigenspace decomposition of $\mathfrak{g}$ with respect to σ (here, as before, $\mathfrak{g}^\mu = \{a \in \mathfrak{g} | \sigma(a) = e^{2\pi i\mu}a\}$): $\mathfrak{g} = \mathfrak{g}^0 + \mathfrak{g}^{1/2} + \mathfrak{g}^{b/2} + \mathfrak{g}^{-b/2} + \mathfrak{g}^{(1+b)/2} + \mathfrak{g}^{(1-b)/2}$, where

$$\mathfrak{g}^0 = \mathrm{span}\{e^{(011)}, f^{(011)}, e, f, \alpha_2, \alpha_1 + \alpha_3\}, \mathfrak{g}^{b/2} = \mathrm{span}\{e^{(1)}, e_{111}, f_2\}\,,$$
$$\mathfrak{g}^{-b/2} = \mathrm{span}\{e_2, f^{(1)}, f_{111}\}, \mathfrak{g}^{(1+b)/2} = \mathbb{C}e^{(3)}, \mathfrak{g}^{-(1+b)/2} = \mathbb{C}f^{(3)}\,,$$
$$\mathfrak{g}^{1/2} = \mathrm{span}\{e^{(110)}, f^{(110)}, \alpha_1 - \alpha_3\}\,.$$

Then $\mathfrak{h}^\sigma = \mathbb{C}\theta + \mathbb{C}(\alpha_1 + \alpha_3)$ and the roots of $\widehat{\mathfrak{g}}^{\mathrm{tw}}$ are described in terms of $\tilde{\alpha}_i = \alpha_i|_{\mathfrak{h}^\sigma}$ $(i = 1, 2)$ and δ, the non-zero inner products between them being $(\tilde{\alpha}_1|\tilde{\alpha}_1) = -1/2$, $(\tilde{\alpha_1}|\tilde{\alpha}_2) = 1/2$. The union of the above bases of the eigenspaces of σ is a basis of $\mathfrak{g}$, compatible with the $\frac{1}{2}\mathbb{Z}$-gradation and the root space decomposition with respect to $\mathfrak{h}^\sigma$, which we denoted by S. Furthermore, $\mathfrak{h}^\natural = \mathbb{C}\alpha$, where $\alpha = \alpha_1^\natural = -\alpha_2^\natural$.

We have: $\epsilon(\sigma) = 1$, $\mathfrak{g}_{1/2}(\sigma)_0 = \mathbb{C}e^{(110)}$, and the following possibilities for $\mathfrak{n}(\sigma)_\pm$:

$$\begin{aligned}
&\text{(i) } b \in 2\mathbb{Z} : \mathfrak{n}(\sigma)_- = \text{span}\,\{e, e^{(110)}, f^{(1)}\}, \mathfrak{n}(\sigma)_+ = \text{span}\,\{f, f^{(110)}, e^{(1)}\};\\
&\text{(ii) } b \in 2\mathbb{Z}+1 : \mathfrak{n}(\sigma)_- = \text{span}\,\{e, e^{(110)}, e_2, f_{111}, f^{(3)}\},\\
&\qquad\qquad \mathfrak{n}(\sigma)_+ = \text{span}\,\{f, f^{(110)}, f_2, e_{111}, e^{(3)}\};\\
&\text{(iii) } b \notin \mathbb{Z} : \mathfrak{n}(\sigma)_- = \mathbb{C}e + \mathbb{C}e^{(110)}, \mathfrak{n}(\sigma)_+ = \mathbb{C}f + \mathbb{C}f^{(110)}.
\end{aligned}$$

We consider separately the following two cases: $(+)$: $0 \leqslant b < 1$; $(-)$: $-1 \leqslant b < 0$.

The s_i are as follows (in this case they depend not only on the root, but also on the root vector): $s_e = 0$, $s_{e_2} = -b/2$, $s_{e_{111}} = b/2$, $s_{e^{(110)}} = 1/2$, $s_{e^{(011)}} = 0$, $s_{e^{(3)}} = \frac{1}{2}(1+b)$, $s_{\alpha_1-\alpha_3} = 1/2$ in all cases; the remaining s_i (up to the relation (3.3)) are: $s_{e^{(1)}} = b/2$ in case $(+)$, $s_{e^{(1)}} = 1+b/2$ in case $(-)$. Using this, one finds that

$$\gamma^\natural_{1/2} = -\frac{b}{2}\tilde{\alpha}_1\,,\ \gamma'^\natural = \mp\frac{1}{2}\tilde{\alpha}_1 \text{ and } s_\mathfrak{g} + s_{\text{gh}} = -\frac{b^2}{8} \pm \frac{b}{4} \text{ in case } (\pm)\,.$$

Furthermore, let $(m \in \mathbb{Z}_+)$:

$$\widehat{\Delta}^{(1/2)}_{++} = \{(m-\frac{b}{2})\delta + \tilde{\alpha}_2\,,\ (m+\frac{b}{2})\delta + (2\tilde{\alpha}_1 + \tilde{\alpha}_2)\}\,,$$

and define $\widehat{\Delta}^{(0)}_{++}$ in cases $(\pm)$ as follows $(m \in \mathbb{Z}_+)$:

$$\widehat{\Delta}^{(0)}_{++} = \{\frac{m \pm b}{2}\delta \pm \tilde{\alpha}_1\,,\ \frac{m+1\mp b}{2}\delta \mp \tilde{\alpha}_1\}\,.$$

Then $\widehat{\Delta}^{\text{re}}_{++} = \widehat{\Delta}^{(0)}_{++} \cup \widehat{\Delta}^{(1/2)}_{++} \cup \{\frac{m\delta+\theta}{2}\,,\ m\delta+\theta\,|\,m \in \mathbb{Z}_+\}$.

Next, $\mathfrak{h}_{W,\sigma} = \mathfrak{h}^\natural \oplus \mathbb{C}L_0^{\text{tw}}$, and in cases $(\pm)$ we have $(m \in \mathbb{Z}_+)$:

$$\begin{aligned}
\Delta^+_{W,\sigma} = &\{\frac{m \pm b}{2}\delta' \pm \alpha\,,\ \frac{m+1\mp b}{2}\delta' \mp \alpha,\ (m+1)\delta'\,,\ (m+\frac{1}{2})\delta'\}\\
&\cup\{\frac{m+1}{2}\delta'\,,\ (m+\frac{1+b}{2})\delta' + \alpha\,,\ (m+\frac{1-b}{2})\delta' - \alpha\}\,,
\end{aligned}$$

the elements from the first (resp. second) set being even (resp. odd) roots, and the multiplicities of all roots being 1, except for $(m+1)\delta'$, whose multiplicity is 2.

Let $P^b_{N=4}(\eta)$ be the corresponding partition function. Let h and j be the respective eigenvalues of L_0^{tw} and $J_0^{\text{tw},\{-4\alpha_1\}}$, so that $\lambda^\natural = \frac{j}{2}\alpha$. Formulas (4.8)–(4.10) give the following expressions for the factors of the determinant in $(\pm)$ cases:

$$\varphi_{(m-1)\delta+\theta,n} = h - h_{n,m}(k,j),$$

where

$$h_{n,m}(k,j) = \frac{1}{4k}\Big(\big(\frac{n}{2} - mk\big)^2 - \frac{(j-b\pm 1)^2}{4} - k^2\Big) - \frac{(b\mp 1)^2+1}{8};$$

$$\varphi_{m\delta+\beta,1} = h - \frac{1}{k}\Big(\big(m+\frac{1}{2}\big)k + \frac{j-b\pm 1}{2}(\beta|\tilde{\alpha}_1)\Big)^2 - \frac{k^2}{4}$$
$$- \frac{(j-b\pm 1)^2}{16} \text{ if } \beta \in S_{1/2}\backslash\{\theta/2\},$$

$$\varphi_{m\delta+\tilde{\alpha}_1,n} = mk - \frac{j-b\pm 1+n}{4},$$

$$\varphi_{m\delta-\tilde{\alpha}_1,n} = mk + \frac{j-b\pm 1+n}{4}.$$

The extra factor is computed using formula (4.7) (in this case $h_0^\vee = -1$), which gives:

$$\varphi_{(\theta-\delta)/2,0} = h + \frac{(j-b\pm 1)^2}{16k} + \frac{(b\mp 1)^2}{8} + \frac{k}{4} + \frac{1}{8},$$

This again confirms our conjecture made in Remark 4.2(c).

By Theorem 4.2 and Remarks 4.2(a) and (b), we obtain the following formula for $\det_\eta(k,h,j)$:

$$\varphi_{(\theta-\delta)/2,0}(k,h,j)^{P'^b_{N=4}(\eta)} \prod_{m,n\in\mathbb{N}} k^{P^b_{N=4}(\eta - mn\delta')}$$
$$\times \prod_{\substack{m,n\in\mathbb{N} \\ m+n \text{ odd}}} (h - h_{n,m}(k,j))^{P^b_{N=4}(\eta-\frac{1}{2}mn\delta')}$$
$$\times \prod_{\substack{m\delta+\beta\in\widehat{\Delta}^{(1/2)}_{++} \\ \beta=\tilde{\alpha}_2,\tilde{\alpha}_1+\tilde{\alpha}_2,2\tilde{\alpha}_1+\tilde{\alpha}_2}} \varphi_{m\delta+\beta,1}(k,h,j)^{P^b_{N=4;(m+\frac{1}{2})\delta'+\beta^\natural}(\eta-(m+\frac{1}{2})\delta'-\beta^\natural)}$$
$$\times \prod_{\substack{m\delta\pm\tilde{\alpha}_1\in\widehat{\Delta}^{(0)}_{++} \\ n\in\mathbb{N}}} \varphi_{m\delta\pm\alpha_1,n}(k,j)^{P^b_{N=4}(\eta-n(m\delta'\pm\tilde{\alpha}_1^\natural))}.$$

References

[BK] B. Bakalov and V.G. Kac, Field algebras, IMRN 3 (2003) 123–159.

[BFK] W. Boucher, D. Friedan and A. Kent, Determinant formulae and unitarity for the $N = 2$ superconformal algebras in two dimensions or exact results on string compactification, Phys. Lett. 172B (1986) 316–322.

[BS] P. Bouwknegt and K. Schoutens, *W-symmetry*, Advanced Series Math. Phys., vol. 22, Singapore, World Sci., 1995.

[BT] J. de Boer and T. Tjin, The relation between quantum W-algebras and Lie algebras, Comm. Math. Phys. 160 (1994) 317–332 .

[EK] A.G. Elashvili and V.G. Kac, Classification of good gradings of simple Lie algebras, Amer. Math. Soc. Transl. (2) vol. 213, 2005. math-ph/0312030.

[FF] B.L. Feigin and E. Frenkel, Quantization of Drinfeld-Sokolov reduction, Phys. Lett., B 246 (1990) 75–81.

[FL] E.S. Fradkin and V.Ya. Linetsky, Classification of superconformal and quasisuperconformal algebras in two dimensions, Phys. Lett. B 291 (1992) 71–76.

[FB] E. Frenkel and D. Ben-Zvi, *Vertex algebras and algebraic curves*, AMS monographs 88, 2001.

[FKW] E. Frenkel, V. Kac and M. Wakimoto, Characters and fusion rules for W-algebras via quantized Drinfeld-Sokolov reduction, Comm. Math. Phys. 147 (1992) 295–328.

[GS] P. Goddard and A. Schwimmer, Factoring out free fermions and superconformal algebras, Phys. Lett. 214B (1988) 209–214.

[GMS] V. Gorbounov, F. Malikov and V. Schechtman, Gerbes of chiral differential operators II, preprint math. AG/0003170.

[KK] V.G. Kac and D.A. Kazhdan, Structure of representations with highest weight of infinite-dimensional Lie algebras, Adv. Math. 34 (1979) 97–108.

[K1] V.G. Kac, Lie superalgebras, Adv. Math. 26 (1977) 8–96.

[K2] V.G. Kac, Contravariant form for infinite-dimensional Lie algebras and superalgebras, Lecture Notes in Physics 94 (1979) 441–445.

[K3] V.G. Kac, *Infinite-dimensional Lie algebras*, 3rd edition, Cambridge University Press, 1990.

[K4] V.G. Kac, *Vertex algebras for beginners*, Providence: AMS, University Lecture Notes, Vol. 10, 1996, Second edition, 1998.

[KL] V.G. Kac and J.W. van de Leur, On classification of superconformal algebras, in *Strings 88*, ed. S.J. Gates *et al.*, World Scientific 1989, 77–106.

[KRW] V.G. Kac, S.-S. Roan and M. Wakimoto, Quantum reduction for affine superalgebras, Comm. Math. Phys. 241 (2003) 307–342.

[KW] V.G. Kac and M. Wakimoto, Quantum reduction and representation theory of superconformal algebras, Adv. Math. 185 (2004) 400–458. math-ph/0304011.

[KW1] V.G. Kac and M. Wakimoto, Unitarizable highest weight representations of the Virasoro, Neveu-Schwarz and Ramond algebras, in Lecture Notes in Phys. 261, 1986, pp 345–371.

[KW2] V.G. Kac and M. Wakimoto, Modular invariant representations of infinite-dimensional Lie algebras and superalgebras, Proc. Natl. Acad. Sci. USA 85 (1988) 4956–4960.

[KW3] V.G. Kac and M. Wakimoto, Classification of modular invariant representations of affine algebras, in Advanced ser. in *Infinite-dimensional Lie algebras and groups*, Math. Phys. vol. 7, World Scientific, 1989, 138–177.

[KW4] V.G. Kac and M. Wakimoto, Quantum reduction and characters of superconformal algebras, in preparation.

[KM] M. Kato and S. Matsuda, Null field construction and Kac formulae of $N = 2$ superconformal algebras in two dimensions, Phys. Lett. B184 (1987) 184–190.

[KR] A. Kent and H. Riggs, Determinant formulae for the $N = 4$ superconformal algebras, Phys. Lett. B198 (1987) 491–497.

[KMR] A. Kent, M. Mattis and H. Riggs, Highest weight representations of $N = 3$ superconformal algebras and their determinant formulae, Nucl. Phys. B301 (1988) 426–440.

[M] K. Miki, The representation theory of the $SO(3)$ invariant superconformal algebra, Int. J. Modern Phys. A, 5 (1990) 1293–1318.

[S] K. Schoutens, $O(N)$-extended superconformal field theory in superspace, Nucl. Phys. B295 (1988) 634–652.

[STP] A. Severin, W. Troost and A. Van Proyen, Superconformal algebras in two dimensions with $N = 4$, Phys. Lett. B208 (1988) 447–450.

[ST] A. Severin and W. Troost, Extensions of the Virasoro algebra and gauged WZW models, Phys. Lett B315 (1993), 304–310.

[R] P. Ramond, Dual theory for free fermions, Phys. Rev. D3 (1971) 2415–2418.

[T] C.B. Thorn, Computing Kac determinant using dual models techniques and more about the no-ghost theorem, Nucl. Phys. B248 (1984) 551–569.

Victor G. Kac
Department of Mathematics
M.I.T.
Cambridge
MA 02139, USA
e-mail: kac@math.mit.edu

Minoru Wakimoto
Graduate School of Mathematics
Kyushu University
Fukuoka 812-8581
Japan
e-mail: wakimoto@math.kyushu-u.ac.jp

Progress in Mathematics, Vol. 237, 133–156

Representation Theory and Quantum Integrability

A. Gerasimov, S. Kharchev and D. Lebedev

Abstract. We describe new constructions of the infinite-dimensional representations of $U(\mathfrak{g})$ and $U_q(\mathfrak{g})$ for $\mathfrak{g}$ being $\mathfrak{gl}(N)$ and $\mathfrak{sl}(N)$. The application of these constructions to the quantum integrable theories of Toda type is discussed. With the help of these infinite-dimensional representations we manage to establish a direct connection between the group theoretical approach and the Quantum Inverse Scattering Method based on the representation theory of the Yangian and its generalizations.

In the case of $U_q(\mathfrak{g})$ the considered representation is naturally supplied with the structure of a $U_q(\mathfrak{g}) \otimes U_{\tilde{q}}(\check{\mathfrak{g}})$-bimodule where $\check{\mathfrak{g}}$ is the Langlands dual to $\mathfrak{g}$ and $\log q/2\pi i = -(\log \tilde{q}/2\pi i)^{-1}$. This bimodule structure is a manifestation of the Morita equivalence of the algebra and its dual.

Mathematics Subject Classification (2000). 17B37, 33C80, 37J35, 81R50.

Keywords. Infinite-dimensional representations of $U_q(sl(N))$, Whittaker modules, Gelfand-Zetlin representation, Toda systems, Modular double.

1. Introduction

Since the early days of quantum mechanics, representation theory plays an important role as a succinct tool to describe and explicitly solve quantum theories. On the other hand, the problems in quantum theories serve as a source of new ideas in representation theory. The most notable recent example is the emergence of the notion of quantum groups [1], [2] from the algebraic formulation of the Quantum Inverse Scattering Method (QISM) [3], [4].

In this paper we describe several constructions in representation theory of classical and quantum groups inspired by our studies of the simple quantum integrable models. Our starting point was the desire to understand QISM in more standard representation theory terms. It was a surprise that one should instead invent new constructions in representation theory. It is well known to experts in integrable systems that there exist distinguished coordinates in which the descrip-

---

Extended talk by the third author at the satellite XIV ICMP workshop: "Infinite-Dimensional Algebras and Quantum Integrable systems", July 21–25, 2003, Faro, Portugal.

tion of the quantum system, although non-trivial, is drastically simplified. Multi-dimensional nonlinear spectral problems are reduced to the one-dimensional case and the solution of the quantum theory may be constructed explicitly. The recent interest in this approach initiated by Sklyanin [5] leads to an explosion of activity connected with a search of the explicit transforms to such coordinates. This Separation of Variables method (SoV) may be considered as an extension/generalization of QISM.

It was shown in [6], [7] that the separation of variables in the modern form has a clear group theoretic meaning and goes back to the natural parameterization of the regular group elements of the non-compact groups. This leads to the new construction of the representations of the universal enveloping of the Lie algebras in terms of the difference operators. A certain particular case of this construction leading to the finite-dimensional representations turns out to be the well-known Gelfand-Zetlin recursive construction of the representations of classical groups [8], [9], [10].

The explicit connection with the QISM approach appears through the closely related construction of the representations of the Yangian. Namely, let $\mathfrak{g}$ be the Lie algebra $\mathfrak{g}l(N)$ and let $Y(\mathfrak{g})$ be the Yangian [11]. There is a natural epimorphism $\pi_N : Y(\mathfrak{g}) \to \mathcal{U}(\mathfrak{g})$ compatible with the representation in terms of the difference operators. The constructed representation of the Yangian turns out to be a manifestation of the simultaneous existence of the RTT-type realization and Drinfeld's "new" realization [12]. The distinguished maximal commutative subalgebra of the Yangian is a key ingredient in the explicit connection between these two constructions and may be described by the set of the commuting operators $\mathcal{A}_n(\lambda)\,, n = 1, \ldots, N$. In the special class of representations $\mathcal{A}_n(\lambda)$ act as polynomials in λ. The zeros of these polynomials provide the variables appearing in SoV. On the other hand, these variables are exactly the variables that appear in our generalization of the Gelfand-Zetlin construction. Note that Drinfeld's "new" realization is known for the Yangian of an arbitrary simple Lie algebra and the generalized Gelfand-Zetlin construction of the representation of $\mathfrak{g}l(N)$ may be naturally extended to the case of the general simple Lie algebra.

There is a further generalization of the construction providing the explicit infinite-dimensional representation of the quantum group $U_q(\mathfrak{g})$ for $\mathfrak{g} = \mathfrak{gl}(N)$, $\mathfrak{sl}(N)$. However, a new essential phenomenon appears in this case. The infinite-dimensional representations of $U_q(\mathfrak{g})$ are naturally supplied with the structure of a $U_q(\mathfrak{g}) \otimes U_{\tilde{q}}(\check{\mathfrak{g}})$-bimodule where $\check{\mathfrak{g}}$ is the Langlands dual to $\mathfrak{g}$ and $\log q/2\pi i = -(\log \tilde{q}/2\pi i)^{-1}$ (for preliminary results in these direction see [13], [14], [15]). We give an explicit construction of this bimodule structure and show the natural appearance of the Langlands dual for $U_q(\mathfrak{g})$, when $\mathfrak{g} = \mathfrak{gl}(N), \mathfrak{sl}(N)$. Note that in this paper we essentially use various rational forms of the quantum groups (see [16], [17] and references therein). Thus, the Langlands dual to the minimal ("adjoint") rational form of $U_q(\mathfrak{sl}(N))$ turns out to be another form of $U_{\tilde{q}}(\mathfrak{sl}(N))$ in accordance with the fact that the Langlands dual to $PSL(N)$ is $SL(N)$ for the classical groups. Further study shows that the appearance of the Langlands dual in this construction is not accidental and is a manifestation of the Morita equivalence of certain two algebras naturally associated with $U_q(\mathfrak{g})$ and $U_{\tilde{q}}(\check{\mathfrak{g}})$ (compare with [18], [19]). This formulation appears to be very close to the initial problem of the classification of the irreducible representations in the decomposition of the regular representation of the group G in terms of the dual group $\check{G}$ [20]. We are planning to discuss this construction in full details elsewhere.

The plan of the paper is as follows. In Section 2 we review the part of [6] concerning the analytic continuation of the Gelfand-Zetlin construction to the case of Whittaker modules of $U(\mathfrak{gl}(N))$. An explicit description of a Whittaker module is straightforward as soon as we have an explicit expression for the cyclic Whittaker vector. In Section 3 we generalize this construction to the case of quantum groups. The structure of $U_q(\mathfrak{g}) \otimes U_{\tilde{q}}(\check{\mathfrak{g}})$-bimodule and its connection with Langlands duality is shortly discussed. In Section 4 we demonstrate the connection with SoV and QISM and describe the applications to the quantum integrable theories of Toda type.

The results of these paper are based on the work that was done for the past few years by the ITEP group in Moscow and was reported in the series of papers [21], [22], [23], [15], [6], [7].

2. Whittaker modules in the Gelfand-Zetlin representation

2.1. The representation of $U(\mathfrak{gl}(N))$

Let us remind the construction of the paper [6], where an analytical continuation of the Gelfand-Zetlin (GZ) theory to infinite-dimensional representations of the universal enveloping algebra $U(\mathfrak{gl}(N))$ was introduced.

Let $\hat{\mathbb{T}}_\hbar$ be an associative algebra generated by $\hat{\gamma}_{nj}, \hat{\beta}_{nj}^{\pm 1}$, $n = 1, \dots, N-1; 1 \leq j \leq n$, and $\hat{\gamma}_{Nj}$, $1 \leq j \leq N$, subject to the relations

$$[\hat{\gamma}_{nj}, \hat{\gamma}_{ml}] = [\hat{\beta}_{nj}, \hat{\beta}_{ml}] = 0 , \qquad [\hat{\beta}_{ml}, \hat{\gamma}_{nj}] = i\hbar\delta_{mn}\delta_{lj}\hat{\beta}_{ml} . \tag{2.1}$$

Theorem 2.1. *Let E_{nm}, $n, m = 1, \dots, N$ be the generators of $\mathfrak{gl}(N)$. The following explicit expressions defines the embedding π: $\mathfrak{gl}(N) \hookrightarrow \hat{\mathbb{T}}_\hbar$*

$$E_{nn} = \frac{1}{i\hbar}\Big(\sum_{j=1}^{n} \hat{\gamma}_{nj} - \sum_{j=1}^{n-1} \hat{\gamma}_{n-1,j}\Big), \qquad (n = 1, \dots, N), \tag{2.2a}$$

$$E_{n,n+1} = -\frac{1}{i\hbar}\sum_{j=1}^{n} \frac{\prod\limits_{r=1}^{n+1}(\hat{\gamma}_{nj} - \hat{\gamma}_{n+1,r} - \frac{i\hbar}{2})}{\prod\limits_{s\neq j}(\hat{\gamma}_{nj} - \hat{\gamma}_{ns})}\, \hat{\beta}_{nj}^{-1}, \qquad (n = 1, \dots, N-1), \tag{2.2b}$$

$$E_{n+1,n} = \frac{1}{i\hbar}\sum_{j=1}^{n} \frac{\prod\limits_{r=1}^{n-1}(\hat{\gamma}_{nj} - \hat{\gamma}_{n-1,r} + \frac{i\hbar}{2})}{\prod\limits_{s\neq j}(\hat{\gamma}_{nj} - \hat{\gamma}_{ns})}\, \hat{\beta}_{nj}, \qquad (n = 1, \dots, N-1). \tag{2.2c}$$

Let us consider the following natural representation of the quantum torus $\hat{\mathbb{T}}_\hbar$:

$$\begin{aligned} \hat{\gamma}_{nj} &= \gamma_{nj} \in \mathbb{C}\,, \quad n = 1, \dots, N\,, \qquad 1 \leq j \leq n\,, \\ \hat{\beta}_{nj} &= e^{i\hbar \frac{\partial}{\partial \gamma_{nj}}}\,, \quad n = 1, \dots, N-1\,, \quad 1 \leq j \leq n\,, \end{aligned} \tag{2.3}$$

where γ_{nj} and $e^{\pm i\hbar \frac{\partial}{\partial \gamma_{nj}}}$ are considered as operators acting on the space M of meromorphic functions of complex variables $\gamma_{nj}, n = 1, \dots, N-1, 1 \leq j \leq n$. The remaining $\gamma_{N1}, \dots, \gamma_{NN}$ are considered as constants. Thus the complex vector $\gamma_N = (\gamma_{N1}, \dots, \gamma_{NN})$ plays the role of a label which determines the above representation.

Let $\mathcal{Z}(\mathfrak{gl}(n))$ be the center of $U(\mathfrak{gl}(n))$. We say that a $U(\mathfrak{gl}(n))$-module V admits an infinitesimal character ξ if there is a homomorphism $\xi : \mathcal{Z}(\mathfrak{gl}(n)) \to \mathbb{C}$ such that $zv = \xi(z)v$ for all $z \in \mathcal{Z}(\mathfrak{gl}(n)), v \in V$. It is possible to show that the $U(\mathfrak{gl}(N))$-module M defined above admits an infinitesimal character and each central element of $U(\mathfrak{gl}(n))$ acts on M via multiplication by a symmetric polynomial in the variables γ_{nj} [6].

In the next subsection we calculate the explicit action of the central elements on M using the notion of Whittaker vectors.

2.2. Whittaker modules

Let us now give a construction for Whittaker modules using the representation of $U(\mathfrak{gl}(N))$ described above. We first recall some facts from [24].

Let n_+ and n_- be the subalgebras of $\mathfrak{gl}(N)$ generated, respectively, by positive and negative root generators. The homomorphisms (characters) $\chi_+ \colon n_+ \to \mathbb{C}$, $\chi_- \colon n_- \to \mathbb{C}$ are uniquely determined by their values on the simple root generators, and are called non-singular if the (complex) numbers $\chi_+(E_{n,n+1})$ and $\chi_-(E_{n+1,n})$ are non-zero for all $n = 1, \dots, N-1$.

Let V be any $U = U(\mathfrak{gl}(N))$-module. Denote the action of $u \in U$ on $v \in V$ by uv. A vector $w \in V$ is called a Whittaker vector with respect to the character χ_+ if

$$E_{n,n+1} w = \chi_+(E_{n,n+1}) w\,, \qquad (n = 1, \dots, N-1), \tag{2.4}$$

and an element $w' \in V'$ is called a Whittaker vector with respect to the character χ_- if

$$E_{n+1,n} w' = \chi_-(E_{n+1,n}) w'\,, \qquad (n = 1, \dots, N-1). \tag{2.5}$$

A Whittaker vector is cyclic for V if $Uw = V$, and a U-module is a Whittaker module if it contains a cyclic Whittaker vector. The U-modules V and V' are called dual if there exists a non-degenerate pairing $\langle .\,, . \rangle : V' \times V \to \mathbb{C}$ such that $\langle Xv', v \rangle = -\langle v', Xv \rangle$ for all $v \in V$, $v' \in V'$ and $X \in \mathfrak{gl}(N)$.

We proceed with explicit formulas for Whittaker vectors corresponding to the representation given by (2.2), (2.3).

Proposition 2.1. *The equations*

$$E_{n+1,n}w'_N = -i\hbar^{-1}w'_N, \tag{2.6}$$
$$E_{n,n+1}w_N = -i\hbar^{-1}w_N \tag{2.7}$$

for all $n = 1,\dots,N-1$, *admit the solutions*

$$w'_N = 1,$$
$$w_N = e^{-\frac{\pi}{\hbar}\sum\limits_{n=1}^{N-1}(n-1)\sum\limits_{j=1}^{n}\gamma_{nj}}\prod_{n=1}^{N-1}s_n(\gamma_n,\gamma_{n+1}), \tag{2.8}$$

where

$$s_n(\gamma_n,\gamma_{n+1}) = \prod_{k=1}^{n}\prod_{m=1}^{n+1}\hbar^{\frac{\gamma_{nk}-\gamma_{n+1,m}}{i\hbar}+\frac{1}{2}}\,\Gamma\Big(\frac{\gamma_{nk}-\gamma_{n+1,m}}{i\hbar}+\frac{1}{2}\Big). \tag{2.9}$$

(For the proof see [6].)

The solutions (2.8), (2.9) are not unique. Indeed, the set of Whittaker vectors is closed under the multiplication by an arbitrary $i\hbar$-periodic function in the variables γ_{nj}. Hence, there are infinitely many invariant subspaces in M corresponding to infinitely many Whittaker vectors.

To construct irreducible submodules, let us introduce the Whittaker modules W and W', generated cyclically by the Whittaker vectors w_N and w'_N, respectively. An explicit description of a Whittaker module is straightforward as soon as we have an explicit expression for the Whittaker vectors. Namely let $\boldsymbol{m}_n = (m_{n1},\dots,m_{nn})$ be the set of non-negative integers. The Whittaker module $W = Uw_N$ is spanned by the elements

$$w_{\boldsymbol{m}_1,\dots,\boldsymbol{m}_{N-1}} = \prod_{n=1}^{N-1}\prod_{k=1}^{n}\sigma_k^{m_{nk}}(\gamma_n)w_N, \tag{2.10}$$

where $\sigma_k(\gamma_n)$ is the elementary symmetric function of the variables $\gamma_{n1},\dots,\gamma_{nn}$ of order k:

$$\sigma_k(\gamma_n) = \sum_{j_1<\dots<j_k}\gamma_{nj_1}\cdots\gamma_{nj_k}. \tag{2.11}$$

Similarly, the Whittaker module $W' = Uw'_N$ is spanned by the polynomials

$$w'_{\boldsymbol{m}_1,\dots,\boldsymbol{m}_{N-1}} = \prod_{n=1}^{N-1}\prod_{k=1}^{n}\sigma_k^{m_{nk}}(\gamma_n). \tag{2.12}$$

The Whittaker modules W and W' are irreducible.

Let us note that for any subalgebra $U(\mathfrak{gl}(n)) \subset U(\mathfrak{gl}(N))$, $2 \le n < N$, the module over the ring of the polynomials in γ_n with the basis $\prod\limits_{l=1}^{n-1}\prod\limits_{k=1}^{l}\sigma_k^{m_{lk}}(\gamma_l)w_N$ is a $U(\mathfrak{gl}(n))$ Whittaker module. One can calculate the explicit form of the action of the central elements of $U(\mathfrak{gl}(n))$ on the space M. It is well known that the generating function $\mathcal{A}_n(\lambda)$ of the central elements of $U(\mathfrak{gl}(n))$ (the Casimir operators) can be represented as follows [25]:

$$\mathcal{A}_n(\lambda) = \sum_{s\in S_n} \operatorname{sign} s\Big[(\lambda - i\hbar\rho_1^{(n)})\delta_{s(1),1} - i\hbar E_{s(1),1}\Big]\cdots \\ \cdots\Big[(\lambda - i\hbar\rho_n^{(n)})\delta_{s(n),n} - i\hbar E_{s(n),n}\Big], \tag{2.13}$$

where $\rho_k^{(n)} = \frac{1}{2}(n-2k+1),\ k=1,\dots,n$ and the summation is over elements of the permutation group S_n. It can be proved [6] that the operators (2.13) have the following form on a space of meromorphic functions M:

$$\mathcal{A}_n(\lambda) = \prod_{j=1}^{n}(\lambda - \gamma_{nj}), \qquad (n=1,\dots,N). \tag{2.14}$$

It remains to construct a pairing between W and W', and to prove that the Whittaker modules W and W' are dual with respect to this pairing. Let $\phi \in W'$ and $\psi \in W$. Define the pairing $\langle .\,,.\rangle$: $W'\otimes W \to \mathbb{C}$ by

$$\langle \phi, \psi\rangle = \int\limits_{\mathbb{R}^{\frac{N(N-1)}{2}}} \mu_0(\gamma)\overline{\phi}(\gamma)\,\psi(\gamma) \prod_{\substack{n=1\\ j\leq n}}^{N-1} d\gamma_{nj}\,, \tag{2.15}$$

where

$$\mu_0(\gamma) = \prod_{n=2}^{N-1}\prod_{p<r}(\gamma_{np}-\gamma_{nr})(e^{\frac{2\pi\gamma_{np}}{\hbar}} - e^{\frac{2\pi\gamma_{nr}}{\hbar}}). \tag{2.16}$$

The integral (2.15) converges absolutely. (For the proof see [6].)

To construct the pair of the dual Whittaker modules, we should restrict the label of representation to the real values: $\boldsymbol{\gamma}_N \in \mathbb{R}^N$. Then the Whittaker modules W and W' will be dual with respect to the pairing defined by (2.15). That is for any $\phi \in W'$ and $\psi \in W$, the generators $X \in \mathfrak{gl}(N,\mathbb{R})$ possess the property

$$\langle \phi, X\psi\rangle = -\langle X\phi, \psi\rangle. \tag{2.17}$$

This property will be important in the derivation of the wave function in Section 4.

3. Construction of the representation of $U_q(\mathfrak{g})$

In this section we outline an extension of our approach to the quantum groups. We start with the construction of the embedding of $U_q(\mathfrak{g})$ in the product of a commutative and non-commutative tori. The skew field of fractions constructed from $U_q(\mathfrak{g})$ coincides with the functions on the product of the tori invariant under the product of the symmetric groups. Then we describe explicitly the structure of the $U_q(\mathfrak{g}) \otimes U_{\tilde{q}}(\check{\mathfrak{g}})$ bimodule.

3.1. The rational forms of $U_q(\mathfrak{g})$

We start with the definition of the quantum groups following [16], [17]. Let q be an indeterminate, and let $\mathbb{C}(q)$ be the field of rational functions of q with coefficients in $\mathbb{C}$. The associative $\mathbb{C}(q)$–algebra $U_q(\mathfrak{gl}(N))$ is generated by the elements $K_{nn}^{\pm 1}\,, E_{nm}\,; n \neq m\,; n,m = 1,\ldots,N$ subjected to the relations[1]:

$$\begin{aligned} K_{nn}K_{nn}^{-1} &= K_{nn}^{-1}K_{nn} = 1\,, \quad K_{nn}K_{mm} = K_{mm}K_{nn}\,, \\ K_{nn}E_{m,m+1}K_{nn}^{-1} &= q^{\delta_{nm}-\delta_{n,m+1}}E_{m,m+1}\,, \\ K_{nn}E_{m+1,m}K_{nn}^{-1} &= q^{\delta_{n,m+1}-\delta_{nm}}E_{m+1,m}\,, \\ E_{n,n+1}E_{m+1,m} - E_{m+1,m}E_{n,n+1} &= \delta_{nm}\frac{K_{nn}K_{n+1,n+1}^{-1} - K_{nn}^{-1}K_{n+1,n+1}}{q-q^{-1}} \end{aligned} \tag{3.1}$$

together with quantum analogues of the Serre relations

$$E_{n,n+1}E_{m,m+1} - E_{m,m+1}E_{n,n+1} = 0\,, \qquad m \neq n \pm 1\,, \tag{3.2a}$$

$$E_{n,n+1}^2E_{n+1,n+2} - (q+q^{-1})E_{n,n+1}E_{n+1,n+2}E_{n,n+1} + E_{n+1,n+2}E_{n,n+1}^2 = 0\,,$$

$$E_{n+1,n+2}^2E_{n,n+1} - (q+q^{-1})E_{n+1,n+2}E_{n,n+1}E_{n+1,n+2} + E_{n,n+1}E_{n+1,n+2}^2 = 0\,,$$

$$E_{n+1,n}E_{m+1,m} - E_{m+1,m}E_{n+1,n} = 0\,, \qquad m \neq n \pm 1\,, \tag{3.2b}$$

$$E_{n+1,n}^2E_{n+2,n+1} - (q+q^{-1})E_{n+1,n}E_{n+2,n+1}E_{n+1,n} + E_{n+2,n+1}E_{n+1,n}^2 = 0\,,$$

$$E_{n+2,n+1}^2E_{n+1,n} - (q+q^{-1})E_{n+2,n+1}E_{n+1,n}E_{n+2,n+1} + E_{n+1,n}E_{n+2,n+1}^2 = 0\,.$$

We will also need the explicit description of the quantum group $U_q(\mathfrak{sl}(N))$. Note that there exist various rational forms of quantum groups [16], [17]. In the case of $U_q(\mathfrak{sl}(N))$ we have the following description. Let $a_{nm} = 2\delta_{nm} - \delta_{n,m+1} - \delta_{n+1,n}$ be the Cartan matrix of $\mathfrak{sl}(N)$. Let Q and P denote the root lattice and the lattice of weights, respectively. The smallest rational form, the adjoint rational form $U_q^Q(\mathfrak{sl}(N))$, is the associative $\mathbb{C}(q)$-algebra with generators E_n, F_n and $K_n^{\pm 1}\,, n = 1,\ldots,N-1$, and relations

$$K_nK_n^{-1} = K_n^{-1}K_n = 1\,, \qquad K_nK_m = K_mK_n\,, \tag{3.3a}$$

$$K_nE_mK_n^{-1} = q^{a_{nm}}E_m\,, \qquad K_nF_mK_n^{-1} = q^{-a_{nm}}F_m\,, \tag{3.3b}$$

$$E_nF_m - F_mE_n = \delta_{nm}\frac{K_n - K_n^{-1}}{q-q^{-1}}\,, \tag{3.4}$$

$$\begin{aligned} \sum_{r=0}^{1-a_{nm}} (-1)^r \begin{bmatrix} 1-a_{nm} \\ r \end{bmatrix}_q E_n^{1-a_{nm}-r}E_mE_n^r = 0, \quad \text{if } n \neq m, \\ \sum_{r=0}^{1-a_{nm}} (-1)^r \begin{bmatrix} 1-a_{nm} \\ r \end{bmatrix}_q F_n^{1-a_{nm}-r}F_mF_n^r = 0, \quad \text{if } n \neq m. \end{aligned} \tag{3.5}$$

---

[1] We will not use the Hopf structure of the quantum groups and, therefore, we omit the comultiplication formulas in what follows.

Here we have used the standard notations

$$\left[\begin{array}{c} m \\ n \end{array}\right]_q = \frac{[m]_q!}{[n]_q![m-n]_q!}\,, \qquad [m]_q! = \prod_{1\leq j\leq m} \frac{q^j - q^{-j}}{q-q^{-1}}\,. \tag{3.6}$$

The largest $\mathbb{C}(q)$–algebra, the simply-connected rational form $U_q^P(\mathfrak{sl}(N))$, is obtained by adjoining to $U_q^Q(\mathfrak{sl}(N))$ the invertible elements $L_n\,, n = 1\ldots, N-1$, such that $K_n = \prod_m L_m^{a_{mn}}$. The relations (3.3b) are now replaced by

$$L_n E_m L_n^{-1} = q^{\delta_{nm}} E_m\,, \qquad L_n F_m L_n^{-1} = q^{-\delta_{nm}} F_m\,. \tag{3.7}$$

Denote by M any lattice such that $Q \subseteq M \subseteq P$. A general rational form $U_q^M(\mathfrak{g})$ is obtained by adjoining to $U_q^Q(\mathfrak{g})$ the elements $K_{\beta_i} = \prod_j L_j^{m_{ij}}$ for every $\beta_i = \sum_j m_{ij}\lambda_j \in M$, where λ_j are the fundamental weights. Note that various rational forms U_q^M have interesting applications to the construction of quantum integrable systems of Toda type.

3.2. $U_q(\mathfrak{g})$ in terms of the quantum tori

We start with the definition of the quantum torus algebra $\hat{\mathbb{T}}_q$. Let $\hat{\mathbb{T}}_q$ be the associative $\mathbb{C}(q)$– algebra of the rational functions of invertible elements $\mathbf{v}_{nj}$, $n = 1,\ldots,N$; $j = 1,\ldots,n$ and $\mathbf{u}_{nj}$, $n = 1,\ldots,N-1$; $j = 1,\ldots,n$ which are subjected to the relations

$$\begin{gathered} \mathbf{v}_{nj}\mathbf{v}_{mk} = \mathbf{v}_{mk}\mathbf{v}_{nj}\,, \qquad \mathbf{u}_{nj}\mathbf{u}_{mk} = \mathbf{u}_{mk}\mathbf{u}_{nj}\,, \\ \mathbf{u}_{nj}\mathbf{v}_{mk} = q^{\delta_{nm}\delta_{jk}}\mathbf{v}_{mk}\mathbf{u}_{nj}\,. \end{gathered} \tag{3.8}$$

Theorem 3.1.

(i) *Let $K_{nn}^{\pm 1}$, E_{nm} be the generators of $U_q(\mathfrak{gl}(N))$. The following explicit expressions define the embedding $\pi\colon U_q(\mathfrak{gl}(N)) \hookrightarrow \hat{\mathbb{T}}_q$*

$$\pi(K_{nn}) = \prod_{j=1}^{n} \mathbf{v}_{nj} \prod_{j=1}^{n-1} \mathbf{v}_{n-1,j}^{-1}\,, \tag{3.9}$$

$$\pi(E_{n,n+1}) = -\frac{q^{-1}}{q-q^{-1}} \prod_{j=1}^{n+1} \mathbf{v}_{n+1,j}^{-1} \prod_{j=1}^{n} \mathbf{v}_{n,j} \sum_{j=1}^{n} \mathbf{v}_{nj}^{-3} \frac{\prod\limits_{r=1}^{n+1}(\mathbf{v}_{nj}^2 - q\mathbf{v}_{n+1,r}^2)}{\prod\limits_{s\neq j}(\mathbf{v}_{nj}^2 - \mathbf{v}_{ns}^2)}\,\mathbf{u}_{nj}^{-1}\,,$$

$$\pi(E_{n+1,n}) = \frac{1}{q-q^{-1}} \prod_{j=1}^{n} \mathbf{v}_{nj} \prod_{j=1}^{n-1} \mathbf{v}_{n-1,j}^{-1} \sum_{j=1}^{n} \mathbf{v}_{nj}^{-1} \frac{\prod\limits_{r=1}^{n-1}(\mathbf{v}_{nj}^2 - q^{-1}\mathbf{v}_{n-1,r}^2)}{\prod\limits_{s\neq j}(\mathbf{v}_{nj}^2 - \mathbf{v}_{ns}^2)}\,\mathbf{u}_{nj}\,.$$

(ii) *Let $L_n^{\pm 1}, K_n^{\pm 1}, E_n, F_n$ be the generators of $U_q(\mathfrak{sl}(N))$. The following explicit expressions define the embedding $\pi\colon U_q(\mathfrak{sl}(N)) \hookrightarrow \hat{\mathbb{T}}_q$:*

$$\begin{aligned} \pi(L_n) &= \prod_{j=1}^{n} \mathbf{v}_{nj}\,, \qquad \prod_{j=1}^{N} \mathbf{v}_{Nj} = 1\,, \\ \pi(K_n) &= \prod_{m}(\prod_{j=1}^{m} \mathbf{v}_{mj})^{a_{mn}}\,, \end{aligned} \tag{3.10}$$

$$\pi(E_n) = -\frac{q^{-1}}{q-q^{-1}} \prod_{j=1}^{n+1} \mathbf{v}_{n+1,j}^{-1} \prod_{j=1}^{n} \mathbf{v}_{nj} \sum_{j=1}^{n} \mathbf{v}_{nj}^{-3} \frac{\prod\limits_{r=1}^{n+1} (\mathbf{v}_{nj}^2 - q\mathbf{v}_{n+1,r}^2)}{\prod\limits_{s\neq j} (\mathbf{v}_{nj}^2 - \mathbf{v}_{ns}^2)}\, \mathbf{u}_{nj}^{-1}\,, \tag{3.11}$$

$$\pi(F_n) = \frac{1}{q-q^{-1}} \prod_{j=1}^{n} \mathbf{v}_{nj} \prod_{j=1}^{n-1} \mathbf{v}_{n-1,j}^{-1} \sum_{j=1}^{n} \mathbf{v}_{nj}^{-1} \frac{\prod\limits_{r=1}^{n-1} (\mathbf{v}_{nj}^2 - q^{-1}\mathbf{v}_{n-1,r}^2)}{\prod\limits_{s\neq j} (\mathbf{v}_{nj}^2 - \mathbf{v}_{ns}^2)}\, \mathbf{u}_{nj}\,. \tag{3.12}$$

The algebra $\hat{\mathbb{T}}_q$ is very close to the minimal skew field of fractions of $U_q(\mathfrak{gl}(N))$ for which the inclusion of the universal enveloping algebra is possible. Consider the skew field of fractions $\mathcal{D}(U_q(\mathfrak{g}))$ of $U_q(\mathfrak{g})$ which consists of the elements of the form $u \cdot v^{-1}$ or $x^{-1} \cdot y$, where $u, v, x, y \in U_q(\mathfrak{g})$ (for more details see, e.g., [26]). It appears that the skew field of fractions $\mathcal{D}(U_q(\mathfrak{gl}(N)))$ is obtained by partial symmetrization of the algebra $\hat{\mathbb{T}}_q$

$$\mathcal{D}(U_q(\mathfrak{gl}(N))) = (\hat{\mathbb{T}}_q)^{\otimes_{n=1}^{N} S_n} \tag{3.13}$$

where, for $n < N$, the group S_n acts as:

$$\sigma\colon \mathbf{v}_{nj} \to \mathbf{v}_{n,\sigma(j)}, \tag{3.14a}$$

$$\sigma\colon \mathbf{u}_{nj} \to \mathbf{u}_{n,\sigma(j)}, \tag{3.14b}$$

and the group S_N acts as:

$$\sigma\colon \mathbf{v}_{Nj} \to \mathbf{v}_{N,\sigma(j)}. \tag{3.15}$$

This proposition can be considered as a generalization of the Gelfand-Kirillov theorem for $U(\mathfrak{g})$ [26].

3.3. $U_q(\mathfrak{g}) \otimes U_{\tilde{q}}(\breve{\mathfrak{g}})$-bimodule structure

It turns out that the consideration of the infinite-dimensional representation of various rational forms (3.3)-(3.7) of quantum groups reveals new phenomena. Namely some representations of the quantum group $U_q(\mathfrak{g})$ poses a natural structure of a $U_q(\mathfrak{g}) \otimes U_{\tilde{q}}(\breve{\mathfrak{g}})$-bimodule where $\breve{\mathfrak{g}}$ is Langlands dual to $\mathfrak{g}$ and $\log q/2\pi i = -(\log \tilde{q}/2\pi i)^{-1}$. The key point is the interpretation of the appropriately defined centralizer of the image of $U_q(\mathfrak{g})$ in these representations in terms of the representation of $U_{\tilde{q}}(\breve{\mathfrak{g}})$. The presence of the Langlands dual will be verified below in the case of $U_q(\mathfrak{sl}(N))$. It turns out that the centralizer construction leads to the connection between various forms of $U_q(\mathfrak{sl}(N))$. In particular, when we start from the minimal (adjoint) form, we come to what may be called the maximal form over the quadratic extension, which is in agreement with the classical picture of the duality between $SL(N)$ and $PSL(N)$.

Let us begin with the construction of a representation of $U_q(\mathfrak{gl}(N))$ generalizing the representation of $U(\mathfrak{gl}(N))$ described in Section 2. Let $\mathcal{V}$ be the space of functions of γ_{nj} with $n=1,\ldots,N-1$; $1\leq j\leq n$ and $\mathcal{V}^s$ be the space of functions of γ_{nj} invariant under the action of $\otimes_{n=1}^{N-1}S_n$ according to (3.14a). Let $\mathcal{R}$ be the algebra of rational functions of the exponents of the linear functions of γ_{nj} with $n=1,\ldots N$; $1\leq j\leq n$ and $\partial_{\gamma_{nj}}:=\frac{\partial}{\partial\gamma_{nj}}$ with $n=1,\ldots N-1$; $1\leq j\leq n$ and $\mathcal{R}^s$ be the algebra of rational functions of the exponents of the linear functions of γ_{nj} and $\partial_{\gamma_{nj}}$ invariant under the action of $\otimes_{n=1}^{N}S_n$ according to (3.14), (3.15). One has the following embedding $\varrho:\hat{\mathbb{T}}_q\hookrightarrow\mathcal{R}$

$$\varrho(\mathbf{u}_{nj})=e^{i\omega_1\partial_{\gamma_{nj}}}\,,\qquad \varrho(\mathbf{v}_{nj})=e^{\frac{2\pi\gamma_{nj}}{\omega_2}}\,,\qquad \varrho(q)=e^{\frac{2\pi i\omega_1}{\omega_2}}\,. \tag{3.16}$$

Using Theorem 3.1 we obtain the representation of $U_q(\mathfrak{gl}(N))$ in terms of the difference operators. By an abuse of notation we shall also denote by ϱ the composition $\varrho\circ\pi$.

Proposition 3.1. *The following expressions define a representation of* $U_q(\mathfrak{gl}(N))$ *with* $q=e^{\frac{2\pi i\omega_1}{\omega_2}}$ *:*

$$\varrho(K_{nn})=e^{\frac{2\pi}{\omega_2}\left(\sum\limits_{j=1}^{n}\gamma_{nj}-\sum\limits_{j=1}^{n-1}\gamma_{n-1,j}\right)}\,, \tag{3.17}$$

$$\varrho(E_{n,n+1})=\frac{2ie^{\frac{\pi i\omega_1}{\omega_2}(n-1)}}{\sin\frac{2\pi\omega_1}{\omega_2}}\sum_{j=1}^{n}\frac{\prod\limits_{r=1}^{n+1}\sinh\frac{2\pi}{\omega_2}(\gamma_{nj}-\gamma_{n+1,r}-\frac{i\omega_1}{2})}{\prod\limits_{s\neq j}\sinh\frac{2\pi}{\omega_2}(\gamma_{nj}-\gamma_{ns})}\,e^{-i\omega_1\partial_{\gamma_{nj}}}\,,$$

$$\varrho(E_{n+1,n})=-\frac{ie^{-\frac{\pi i\omega_1}{\omega_2}(n-1)}}{2\sin\frac{2\pi\omega_1}{\omega_2}}\sum_{j=1}^{n}\frac{\prod\limits_{r=1}^{n-1}\sinh\frac{2\pi}{\omega_2}(\gamma_{nj}-\gamma_{n-1,r}+\frac{i\omega_1}{2})}{\prod\limits_{s\neq j}\sinh\frac{2\pi}{\omega_2}(\gamma_{nj}-\gamma_{ns})}\,e^{i\omega_1\partial_{\gamma_{nj}}}\,.$$

Consider the dual quantum torus $\hat{\mathbb{T}}_{\tilde{q}}$ using the invertible elements $\tilde{\mathbf{v}}_{nj}$, $n=1,\ldots,N$; $j=1,\ldots,n$ and $\tilde{\mathbf{u}}_{nj}$, $n=1,\ldots,N-1$; $j=1,\ldots,n$ subjected to the relations

$$\begin{gathered}\tilde{\mathbf{v}}_{nj}\tilde{\mathbf{v}}_{mk}=\tilde{\mathbf{v}}_{mk}\tilde{\mathbf{v}}_{nj}\,,\qquad \tilde{\mathbf{u}}_{nj}\tilde{\mathbf{u}}_{mk}=\tilde{\mathbf{u}}_{mk}\tilde{\mathbf{u}}_{nj}\,,\\ \tilde{\mathbf{u}}_{nj}\tilde{\mathbf{v}}_{mk}=\tilde{q}^{\,\delta_{nm}\delta_{jk}}\tilde{\mathbf{v}}_{mk}\tilde{\mathbf{u}}_{nj}\,.\end{gathered} \tag{3.18}$$

One has the dual embedding $\tilde{\varrho}:\hat{\mathbb{T}}_{\tilde{q}}\hookrightarrow\mathcal{R}$

$$\tilde{\varrho}(\tilde{\mathbf{u}}_{nj})=e^{i\omega_2\partial_{\gamma_{nj}}}\,,\qquad \tilde{\varrho}(\tilde{\mathbf{v}}_{nj})=e^{-\frac{2\pi\gamma_{nj}}{\omega_1}}\,,\qquad \tilde{\varrho}(\tilde{q})=e^{-\frac{2\pi i\omega_2}{\omega_1}}\,. \tag{3.19}$$

and the two actions of the torus and its dual on the same space of functions of γ_{ni} mutually commute. Hence, we have the following

Proposition 3.2. *The operators*

$$\tilde{\varrho}(K_{nn}) = e^{-\frac{2\pi}{\omega_1}\left(\sum\limits_{j=1}^{n}\gamma_{nj}-\sum\limits_{j=1}^{n-1}\gamma_{n-1,j}\right)}, \tag{3.20}$$

$$\tilde{\varrho}(E_{n,n+1}) = -\frac{2i\, e^{-\frac{\pi i\omega_2}{\omega_1}(n-1)}}{\sin\dfrac{2\pi\omega_2}{\omega_1}} \sum_{j=1}^{n} \frac{\prod\limits_{r=1}^{n+1}\sinh\frac{2\pi}{\omega_1}(\gamma_{nj}-\gamma_{n+1,r}-\frac{i\omega_2}{2})}{\prod\limits_{s\neq j}\sinh\frac{2\pi}{\omega_1}(\gamma_{nj}-\gamma_{ns})}\, e^{-i\omega_2\partial_{\gamma_{nj}}},$$

$$\tilde{\varrho}(E_{n+1,n}) = \frac{i\, e^{\frac{\pi i\omega_2}{\omega_1}(n-1)}}{2\sin\dfrac{2\pi\omega_2}{\omega_1}} \sum_{j=1}^{n} \frac{\prod\limits_{r=1}^{n-1}\sinh\frac{2\pi}{\omega_1}(\gamma_{nj}-\gamma_{n-1,r}+\frac{i\omega_2}{2})}{\prod\limits_{s\neq j}\sinh\frac{2\pi}{\omega_1}(\gamma_{nj}-\gamma_{ns})}\, e^{i\omega_2\partial_{\gamma_{nj}}},$$

generate a representation of the dual quantum group $U_{\tilde{q}}(\mathfrak{gl}(N))$, where $\tilde{q} = e^{-\frac{2\pi i\omega_2}{\omega_1}}$. The operators $\varrho(U_q(\mathfrak{gl}(N)))$ and $\tilde{\varrho}(U_{\tilde{q}}(\mathfrak{gl}(N)))$ commute by construction.

Thus we have the structure of a $U_q(\mathfrak{gl}(N)) \otimes U_{\tilde{q}}(\mathfrak{gl}(N))$-bimodule. More precisely, this bimodule may be characterized by the condition

$$\varrho(K_{nn}) = \tilde{\varrho}(K_{nn})^{-\tau}, \qquad n = 1, \ldots, N, \tag{3.21}$$

where $\tau = \omega_1/\omega_2$. We will discuss a better way to formulate this condition later in this section.

Let us remark that we actually constructed the embeddings $\varrho : U_q(\mathfrak{gl}(N)) \hookrightarrow \mathcal{R}^s$ and $\tilde{\varrho} : U_{\tilde{q}}(\mathfrak{gl}(N)) \hookrightarrow \mathcal{R}^s$. Consider the centralizer $[\mathcal{D}(\varrho(U_q\mathfrak{gl}(N)))]'$ of the algebras $\mathcal{D}(\varrho(U_q\mathfrak{gl}(N)))$ as a subalgebra of $\mathcal{R}^s$. Then we have

Theorem 3.2.

$$[\mathcal{D}(\varrho(U_q\mathfrak{gl}(N)))]' = \mathcal{D}(\tilde{\varrho}(U_{\tilde{q}}\mathfrak{gl}(N))). \tag{3.22}$$

This Theorem is proved by direct calculation. Consider now the more interesting case of $U_q(\mathfrak{sl}(N))$. It appears that the centralizer of the minimal (adjoint) rational form of the quantum group $U_q(\mathfrak{sl}(N))$ is described in terms of the different forms of the same quantum group.

This may be considered as an indication of the fact that the Langlands dual of $PSL(N)$ is $SL(N)$. One may conjecture that the Langlands dual quantum group $U_{\tilde{q}}(\check{\mathfrak{g}})$ may be generally obtained as a result of explicit calculations of the centralizer in the appropriate generalization of the above construction to arbitrary quantum groups.

Below we give the explicit formulas for $U_q(\mathfrak{sl}(2))$. The adjoint rational form of the quantum group $U_q^Q(\mathfrak{sl}(2))$ is generated by elements K, K^{-1}, E, F subjected to the relations

$$\begin{gathered} KEK^{-1} = q^2 E, \quad KFK^{-1} = q^{-2}F, \\ EF - FE = \frac{K-K^{-1}}{q-q^{-1}}. \end{gathered} \tag{3.23}$$

The representation is given by

$$\begin{aligned}
&\varrho(K)=e^{\frac{4\pi\gamma_{11}}{\omega_2}}\,,\\
&\varrho(E)=\frac{2i}{\sin\frac{2\pi\omega_1}{\omega_2}}\sinh\frac{2\pi}{\omega_2}(\gamma_{11}-\nu-\frac{i\omega_1}{2})\sinh\frac{2\pi}{\omega_2}(\gamma_{11}+\nu-\frac{i\omega_1}{2})\,e^{-i\omega_1\partial\gamma_{11}}\,,\\
&\varrho(F)=-\frac{i}{\sin\frac{2\pi\omega_1}{\omega_2}}\,e^{i\omega_1\partial\gamma_{11}}\,,\qquad \varrho(q)=e^{\frac{2\pi i\omega_1}{\omega_2}}\,.
\end{aligned}\tag{3.24}$$

This representation is obtained from those of $U_q(\mathfrak{gl}(2))$ by the restriction $\gamma_{21}=-\gamma_{22}:=\nu$. The centralizer in $\mathcal{R}$ is generated by the algebra of functions of the dual torus $\mathbb{T}_{\tilde{q}^{1/2}}$:

$$\tilde{u}\tilde{v}=\tilde{q}^{1/2}\tilde{v}\tilde{u}\,,\tag{3.25}$$

where

$$\tilde{\varrho}(\tilde{u})=e^{\frac{i\omega_2}{2}\partial\gamma_{11}},\qquad \tilde{\varrho}(\tilde{v})=e^{-\frac{2\pi\gamma_{11}}{\omega_1}},\qquad \tilde{\varrho}(\tilde{q})=e^{-\frac{2\pi i\omega_2}{\omega_1}}.\tag{3.26}$$

The centralizer in $\mathcal{R}^s$ may be described as an image of the skew field of fractions of the following algebra generated by $\tilde{L},\tilde{E},\tilde{F},\tilde{K}=\tilde{L}^4$ over the quadratic extension $\mathbb{C}(\tilde{q}^{1/2})$:

$$\begin{aligned}
\tilde{L}\tilde{E}\tilde{L}^{-1}=\tilde{q}^{1/2}\tilde{E}\,,\qquad \tilde{L}\tilde{F}\tilde{L}^{-1}=\tilde{q}^{-1/2}\tilde{F},\\
\tilde{E}\tilde{F}-\tilde{F}\tilde{E}=\frac{\tilde{K}-\tilde{K}^{-1}}{\tilde{q}-\tilde{q}^{-1}}\,.
\end{aligned}\tag{3.27}$$

under the representation

$$\begin{aligned}
&\tilde{\varrho}(L)=e^{-\frac{2\pi\gamma_{11}}{\omega_1}}\,,\\
&\tilde{\varrho}(E)=-\frac{2i}{\sin\frac{2\pi\omega_2}{\omega_1}}\sinh\frac{2\pi}{\omega_1}(2\gamma_{11}-\nu-\frac{i\omega_2}{2})\sinh\frac{2\pi}{\omega_1}(2\gamma_{11}+\nu-\frac{i\omega_2}{2})\,e^{-\frac{i\omega_2}{2}\partial\gamma_{11}}\,,\\
&\tilde{\varrho}(F)=\frac{i}{2\sin\frac{2\pi\omega_2}{\omega_1}}\,e^{\frac{i\omega_2}{2}\partial\gamma_{11}}.
\end{aligned}\tag{3.28}$$

This algebra may be considered as a maximal form of $U_{\tilde{q}}(\mathfrak{sl}(2))$ over the quadratic extension $\mathbb{C}(\tilde{q}^{1/2})$. Let us stress that the reconstruction of the algebra from its skew field of fractions is not unique. Thus, in this example the same centralizer may be interpreted as the image of the skew field of fractions of the simply-connected form $U^P_{\tilde{q}^{1/2}}(\mathfrak{sl}(2))$:

$$\begin{aligned}
&\tilde{\varrho}(L)=e^{-\frac{2\pi\gamma_{11}}{\omega_1}}\,,\\
&\tilde{\varrho}(E)=-\frac{2i}{\sin\frac{\pi\omega_2}{\omega_1}}\sinh\frac{2\pi}{\omega_1}(\gamma_{11}-\nu-\frac{i\omega_2}{4})\sinh\frac{2\pi}{\omega_1}(\gamma_{11}+\nu-\frac{i\omega_2}{4})\,e^{-\frac{i\omega_2}{2}\partial\gamma_{11}}\,,\\
&\tilde{\varrho}(F)=\frac{i}{2\sin\frac{\pi\omega_2}{\omega_1}}\,e^{\frac{i\omega_2}{2}\partial\gamma_{11}}.
\end{aligned}\tag{3.29}$$

where the operators satisfy the relations:

$$\tilde{L}\tilde{E}\tilde{L}^{-1} = \tilde{q}^{1/2}\tilde{E}\,, \qquad \tilde{L}\tilde{F}\tilde{L}^{-1} = \tilde{q}^{-1/2}\tilde{F},$$
$$\tilde{E}\tilde{F} - \tilde{F}\tilde{E} = \frac{\tilde{K} - \tilde{K}^{-1}}{\tilde{q}^{1/2} - \tilde{q}^{-1/2}}\,, \tag{3.30}$$

and $\tilde{K} = \tilde{L}^2$. Moreover, there is an isomorphism of the algebras $\mathcal{D}(U^P_{\tilde{q}\,1/2}(\mathfrak{sl}(2)))$ and $\mathcal{D}(U^Q_{\tilde{q}\,1/4}(\mathfrak{sl}(2)))$. This allows to reformulate the duality in a more symmetric form. Taking $p = e^{\pi i\tau}$ and $\tilde{p} = e^{-\pi i/\tau}$ with $\tau = \frac{2\omega_1}{\omega_2}$ one gets the duality between algebra $U^Q_p(\mathfrak{sl}(2))$ and $U^Q_{\tilde{p}}(\mathfrak{sl}(2))$ which leads to the modular double considered in [13], [14]. Let us stress however that the dual quantum deformation parameters enter here in a non-standard way.

Finally, consider the algebra $U^P_q(\mathfrak{sl}(2))$ such that $\mathcal{D}(U^P_q(\mathfrak{sl}(2))) = \mathbb{T}_q := \{v = e^{\frac{2\pi\gamma_{11}}{\omega_2}},\, u = e^{i\omega_1\partial_{\gamma_{11}}}\}$ with $q = e^{\frac{2\pi i\omega_1}{\omega_2}}$. Obviously, $[\mathcal{D}(U^P_q(\mathfrak{sl}(2)))]' = \mathbb{T}_{\tilde{q}} := \{\tilde{v} = e^{-\frac{2\pi\gamma_{11}}{\omega_1}},\, \tilde{u} = e^{i\omega_2\partial_{\gamma_{11}}}\}$ with $\tilde{q} = e^{-\frac{2\pi i\omega_2}{\omega_1}}$. In other words, the algebras $\mathcal{D}(U^P_q(\mathfrak{sl}(2)))$ and $\mathcal{D}(U^P_{\tilde{q}}(\mathfrak{sl}(2)))$ with $q = e^{2\pi i\tau}$ and $\tilde{q} = e^{-2\pi i/\tau}$, $\tau = \frac{\omega_1}{\omega_2}$ centralize each other.

Clearly to use the centralizers as a way to describe the Langlands dual pairs deserves additional structures on the representation space comparing to what was discussed above. We are going to consider these matters elsewhere. Let us also remark that the use of the continuous powers of the Cartan generators in (3.21) is not quite appropriate in our setting. A possible solution is to identify instead the actions of the centers of both algebras. It can be shown that the center is described in terms of the symmetric polynomials of $\mathbf{v}_{Ni}$ and thus the bimodule structure may be described equivalently as

$$\varrho(\mathbf{v}_{Nj}) = \tilde{\varrho}(\tilde{\mathbf{v}}_{Nj})^{-\tau}\,. \tag{3.31}$$

If we consider $\log \mathbf{v}_{Nj}$ as legitimate operators then the relations (3.31) make sense. However we believe that a proper description of the structure of this bimodule which solves this problem should be given in a different way. Let us notice that the description of the universal enveloping algebra in terms of the quantum tori has obvious asymmetry. The variables $\mathbf{v}_{Ni}$ are coordinate functions on the commutative sub-torus and thus there is no natural definition of the dual torus through the centralizer. Thus it is natural to guess that some generalization of the universal enveloping algebra provides the proper setting for the discussion of the Langlands duality for quantum groups through the centralizers. The most natural candidate is the quantum group analogs $\mathcal{A}(G_q)$ of the differential operators on the group $Diff(G)$ and on the basic affine space $Diff(G/N)$. This leads to the interpretation of the resulting $\mathcal{A}(G_q) \otimes \mathcal{A}(\check{G}_{\tilde{q}})$-bimodule as an explicit realization of the Morita equivalence of the algebras $\mathcal{A}(G_q)$ and $\mathcal{A}(\check{G}_{\tilde{q}})$. Preliminary results based on the generalized Gelfand-Zetlin representation (see [27]) support this conjecture. We are going to discuss this approach in the future.

3.4. Whittaker modules

Let us define the special class of the representations of $U_q(\mathfrak{gl}(N))\otimes U_{\tilde{q}}(\mathfrak{gl}(N))$, generalizing the Whittaker module for the classical algebras described in the previous section. The Whittaker vectors for $U_q(\mathfrak{gl}(N))\otimes U_{\tilde{q}}(\mathfrak{gl}(N))$-bimodule are defined using the generalization of the definition in [15]. [2]

Definition 3.1. *The Whittaker vectors w_N and w'_N are defined by equations*

$$\begin{aligned} E_{n,n+1}w_N &= \frac{\chi_n}{q-q^{-1}}\,K_{nn}\prod_{m=1}^{N}K_{mm}^{c_{nm}-c_{n+1,m}}w_N\,,\\ \tilde{E}_{n,n+1}w_N &= \frac{\chi_n}{\tilde{q}-\tilde{q}^{-1}}\,\tilde{K}_{nn}\prod_{m=1}^{N}\tilde{K}_{mm}^{c_{nm}-c_{n+1,m}}w_N\,, \end{aligned}\qquad(3.32)$$

$$\begin{aligned} E_{n+1,n}w'_N &= \frac{\chi'_n}{q-q^{-1}}\,K_{nn}^{-1}\prod_{m=1}^{N}K_{mm}^{c'_{nm}-c'_{n+1,m}}w'_N\,,\\ \tilde{E}_{n+1,n}w'_N &= \frac{\chi'_n}{\tilde{q}-\tilde{q}^{-1}}\,\tilde{K}_{nn}^{-1}\prod_{m=1}^{N}\tilde{K}_{mm}^{c'_{nm}-c'_{n+1,m}}w'_N\,, \end{aligned}\qquad(3.33)$$

where $||c_{nm}||$, $||c'_{nm}||$ are the $N\times N$ symmetric matrices such that $c_{nm}-c_{n+1,m}$, $c'_{nm}-c'_{n+1,m}$ are integers, and χ_n, χ'_n are arbitrary parameters.

The direct calculations show that the following statement is true:

Proposition 3.3. *The defining equations* (3.32), (3.33) *are consistent with the full set of the Serre relations* (3.2) *and their dual analogues.*

The results of Section 2 can be naturally extended to the quantum group case. The structure of the Whittaker vectors and Whittaker modules remains essentially the same. In particular, the Whittaker vectors can be written in a form similar to equations (2.8), (2.9). For example, there is a solution to (3.33) which is unique up to multiplication by an arbitrary double-periodic function:

$$w'_N=e^{-\frac{\pi i}{\omega_1\omega_2}\sum\limits_{n,m=1}^{N}c'_{nm}h_nh_m}\prod_{n=1}^{N-1}e^{\frac{\pi i}{\omega_1\omega_2}\sum\limits_{p=1}^{n}\gamma_{np}^2+\frac{\pi(\omega_1+\omega_2)d_n}{\omega_1\omega_2}\sum\limits_{p=1}^{n}\gamma_{np}}\,,\qquad(3.34)$$

where

$$h_n=\sum_{j=1}^{n}\gamma_{nj}-\sum_{j=1}^{n-1}\gamma_{n-1,j}\,,\qquad(3.35)$$

and $d_n=2n-c'_{nn}+2c'_{n,n+1}-c'_{n+1,n+1}-1$. In the present example $\chi'_n=(-1)^{d_n}$.

Let us denote by W'_N the Whittaker module with the cyclic vector (3.34). It can be proved that it is spanned by the symmetric polynomials of the variables

[2] For the case of $U_q(\mathfrak{g})$, where $\mathfrak{g}$ is an arbitrary simple Lie algebra, the construction of the non-degenerate characters of nilpotent subalgebras was done by Sevostyanov [28]. However, the appearance of the bimodule structure reveals larger symmetry in the representation theory of quantum groups.

$e^{\pm\frac{2\pi\gamma_{nj}}{\omega_1}}, e^{\pm\frac{2\pi\gamma_{nj}}{\omega_2}}$ (compare with Section 2.2). The matrix elements of the particular group elements between the Whittaker vectors leads to the explicit expressions for the wave functions of the generalized quantum Toda theories and will be considered elsewhere.

4. The QISM's eigenfunction via representation theory

4.1. An infinite-dimensional representations of the Yangian

The aim of this section is to introduce a special type of infinite-dimensional representations of the $Y(\mathfrak{gl}(N))$. This allows to connect the QISM methods of the solution of the integrable system based on the representation theory of the Yangian and the solution based on the representation discussed in Section 2.

We start with some well-known facts of the Yangian theory [11], [12] (see also the recent review [29]). The Yangian $Y(\mathfrak{gl}(N))$ is an associative Hopf algebra generated by the elements $T_{ij}^{(r)}$, where $i, j = 1, \dots, N$ and $r = 0, 1, 2, \dots$, subject to the following relations. Consider the $N \times N$ matrix $T(\lambda) = ||T_{ij}(\lambda)||_{i,j=1}^N$ with operator-valued entries

$$T_{ij}(\lambda) = \lambda\delta_{ij} + \sum_{r=0}^{\infty} T_{ij}^{(r)}\lambda^{-r}. \tag{4.1}$$

Let

$$R(\lambda) = I \otimes I + i\hbar P/\lambda\,, \qquad P_{ik,jl} = \delta_{il}\delta_{kj}\,, \tag{4.2}$$

be an $N^2 \times N^2$ numerical matrix (the Yang R-matrix). Then the relations between the generators $T_{ij}^{(r)}$ can be written in the standard form

$$R(\lambda-\mu)(T(\lambda)\otimes I)(I\otimes T(\mu)) = (I\otimes T(\mu))(T(\lambda)\otimes I)R(\lambda-\mu)\,. \tag{4.3}$$

The center of the Yangian is generated by the coefficients of the formal Laurent series (the quantum determinant of $T(\lambda)$ in the sense of [4]):

$$\begin{aligned}\det{}_q T(\lambda) = \sum_{s\in S_N} \operatorname{sign} s\; T_{s(1),1}(\lambda - i\hbar\rho_1^{(N)}) \dots T_{s(k),k}(\lambda - i\hbar\rho_k^{(N)}) \dots \\ \dots T_{s(N),N}(\lambda - i\hbar\rho_N^{(N)})\,,\end{aligned} \tag{4.4}$$

where $\rho_n^{(N)} = \frac{1}{2}(N-2n+1)$, $n = 1, \dots, N$ and the summation is over the elements of the permutation group S_N. Let $X(\lambda) = ||X_{ij}(\lambda)||_{i,j=1}^n$ be an $n \times n$ submatrix of the matrix $||T_{ij}(\lambda)||_{i,j=1}^N$. It is obvious from the explicit form of $R_N(\lambda)$ that this submatrix satisfies an analogue of the relations (4.3). The quantum determinant $\det_q X(\lambda)$ is defined similarly to (4.4) (with the evident change $N \to n$).

The following way to describe the Yangian $Y(\mathfrak{gl}(N))$ was introduced in [11]. Let $\mathbf{A}_n(\lambda)$, $n = 1, \dots, N$, be the quantum determinants of the submatrices, determined by the first n rows and columns, and let the operators $\mathbf{B}_n(\lambda), \mathbf{C}_n(\lambda)$, $n = 1, \dots, N-1$, be the quantum determinants of the submatrices with elements $T_{ij}(\lambda)$, where $i = 1, \dots, n$; $j = 1, \dots, n-1, n+1$ and $i = 1, \dots, n-1, n+1$;

$j = 1,\dots,n$, respectively. The expansion coefficients of $\mathbf{A}_n(\lambda), \mathbf{B}_n(\lambda), \mathbf{C}_n(\lambda)$, $n = 1,\dots,N-1$, with respect to λ, together with those of $\mathbf{A}_N(\lambda)$, generate the algebra $Y(\mathfrak{gl}(N))$. The parts of the relations, which we use below, are as follows:

$$\begin{aligned}
&[\mathbf{A}_n(\lambda), \mathbf{A}_m(\mu)] = 0\,; \qquad (n,m = 1,\dots,N),\\
&[\mathbf{B}_n(\lambda), \mathbf{B}_m(\mu)] = 0; \quad [\mathbf{C}_n(\lambda), \mathbf{C}_m(\mu)] = 0\,; \qquad (m \neq n \pm 1),\\
&(\lambda - \mu + i\hbar)\mathbf{A}_n(\lambda)\mathbf{B}_n(\mu) = (\lambda-\mu)\mathbf{B}_n(\mu)\mathbf{A}_n(\lambda) + i\hbar\mathbf{A}_n(\mu)\mathbf{B}_n(\lambda),\\
&(\lambda - \mu + i\hbar)\mathbf{A}_n(\mu)\mathbf{C}_n(\lambda) = (\lambda-\mu)\mathbf{C}_n(\lambda)\mathbf{A}_n(\mu) + i\hbar\mathbf{A}_n(\lambda)\mathbf{C}_n(\mu).
\end{aligned} \tag{4.5}$$

Let $A(\mathfrak{gl}(N))$ be the commutative subalgebra of $Y(\mathfrak{gl}(N))$ generated by $\mathbf{A}_n(\lambda)$, $n = 1,\dots,N$. It was proved in [31] that $A(\mathfrak{gl}(N))$ is the maximal commutative subalgebra of $Y(\mathfrak{gl}(N))$.

There is another realization of the $Y(\mathfrak{gl}(N))$ introduced by Drinfeld [12]. The algebra $Y(\mathfrak{gl}(N))$ is generated by the coefficients of the formal series

$$k_n(\lambda) = \lambda + \sum_{a=0}^{\infty} k_n^{(a)}\lambda^{-a}, \quad e_n(\lambda) = \sum_{a=0}^{\infty} e_n^{(a)}\lambda^{-a-1}, \quad f_n(\lambda) = \sum_{a=0}^{\infty} f_n^{(a)}\lambda^{-a-1}, \tag{4.6}$$

subjected to the commutation relations

$$\begin{aligned}
&[k_n(\lambda), k_m(\mu)] = 0\,,\\
&[k_n(\lambda), e_m(\mu)] = i\hbar(\delta_{nm} - \delta_{n,m+1})\, k_n(\lambda)\frac{e_m(\lambda) - e_m(\mu)}{\lambda - \mu}\,,\\
&[k_n(\lambda), f_m(\mu)] = -i\hbar(\delta_{nm} - \delta_{n,m+1})\frac{f_m(\lambda) - f_m(\mu)}{\lambda - \mu}\, k_n(\lambda)\,,\\
&[e_n(\lambda), f_m(\mu)] = i\hbar\frac{k_n^{-1}(\mu)k_{n+1}(\mu) - k_n^{-1}(\lambda)k_{n+1}(\lambda)}{\lambda - \mu}\,\delta_{nm}\,,\\
&[e_n^{(a+1)}, e_m^{(b)}] - [e_n^{(a)}, e_m^{(b+1)}] = \tfrac{i\hbar}{2}\, a_{nm}(e_n^{(a)}e_m^{(b)} + e_m^{(b)}e_n^{(a)})\,,\\
&[f_n^{(a+1)}, f_m^{(b)}] - [f_n^{(a)}, f_m^{(b+1)}] = -\tfrac{i\hbar}{2}\, a_{nm}(f_n^{(a)}f_m^{(b)} + f_m^{(b)}f_n^{(a)})\,,\\
&[e_n^{(a)}, [e_n^{(b)}, e_{n\pm1}^{(c)}]] + [e_n^{(b)}, [e_n^{(b)}, e_{n\pm1}^{(c)}]] = 0\,,\\
&[f_n^{(a)}, [f_n^{(b)}, f_{n\pm1}^{(c)}]] + [f_n^{(b)}, [f_n^{(b)}, f_{n\pm1}^{(c)}]] = 0\,,
\end{aligned} \tag{4.7}$$

where $a_{nm} = 2\delta_{nm} - \delta_{n,m+1} - \delta_{n+1,n}$. The relation between two realization is given by

$$\begin{aligned}
&k_n(\lambda) = \frac{\mathbf{A}_n(\lambda - \frac{i(n-1)\hbar}{2})}{\mathbf{A}_{n-1}(\lambda - \frac{in\hbar}{2})}\,,\\
&e_n(\lambda) = \mathbf{A}_n^{-1}(\lambda - \tfrac{i(n-1)\hbar}{2})\mathbf{B}_n(\lambda - \tfrac{i(n-1)\hbar}{2})\,,\\
&f_n(\lambda) = \mathbf{C}_n(\lambda - \tfrac{i(n-1)\hbar}{2})\mathbf{A}_n^{-1}(\lambda - \tfrac{i(n-1)\hbar}{2})\,.
\end{aligned} \tag{4.8}$$

Let us stress that Drinfeld's realization is known for the $Y(\mathfrak{g})$, where $\mathfrak{g}$ is any simple Lie algebra [11]–[12].

There is a natural epimorphism $\pi_N : Y(\mathfrak{gl}(N)) \to U(\mathfrak{gl}(N))$

$$\pi_N(T_{jk}(\lambda)) = \lambda\delta_{jk} - i\hbar E_{jk}\,, \qquad (j\,,\,k = 1,\dots,N). \tag{4.9}$$

Denote the images under π_N of the generators $\mathbf{A}_n(\lambda)$ and $\mathbf{B}_n(\lambda)$, $\mathbf{C}_n(\lambda)$ by $\mathcal{A}_n(\lambda)$ and $\mathcal{B}_n(\lambda)$, $\mathcal{C}_n(\lambda)$, respectively. Note that the images are polynomials in λ of orders n and $n-1$, respectively.

To obtain the representation of the Yangian $Y(\mathfrak{gl}(N))$ we start with the construction of a natural representation of the Cartan subalgebra generated by $k_n(\lambda)$. It can be represented by the rational functions as follows:

$$\prod_{s=1}^{n} k_s(\lambda - i\hbar\rho_s^{(n)}) = \prod_{j=1}^{n}(\lambda - \gamma_{nj})\,, \tag{4.10}$$

where

$$\rho_s^{(n)} = \frac{1}{2}(n - 2s + 1) \quad \text{and} \quad n = 1,\dots,N.$$

Then the operators $A(\mathfrak{gl}(N))$ act by the polynomials

$$\mathcal{A}_n(\lambda) = \prod_{j=1}^{n}(\lambda - \gamma_{nj}), \quad n = 1,\dots,N.$$

We resolve the rest of the Yangian relations and find the explicit expressions for the generators $e_n(\lambda)$, $f_n(\lambda)$ (and $\mathcal{B}_n(\lambda)$ and $\mathcal{C}_n(\lambda)$) in terms of the operators acting on the space of functions depending on the variables $\gamma_{nj}\,, j = 1,\dots,n;\; n = 1,\dots,N$.

Theorem 4.1. *The operators*

$$k_n(\lambda) = \frac{\prod\limits_{j=1}^{n}(\lambda - \gamma_{nj} - \frac{i(n-1)\hbar}{2})}{\prod\limits_{j=1}^{n-1}(\lambda - \gamma_{n-1,j} - \frac{in\hbar}{2})}\,,$$

$$e_n(\lambda) = \sum_{j=1}^{n} \frac{1}{\lambda - \gamma_{nj} - \frac{i(n-1)\hbar}{2}} \frac{\prod\limits_{r=1}^{n+1}(\gamma_{nj} - \gamma_{n+1,r} - \frac{i\hbar}{2})}{\prod\limits_{s\neq j}(\gamma_{nj} - \gamma_{ns})} e^{-i\hbar\partial_{\gamma_{nj}}}\,, \tag{4.11}$$

$$f_n(\lambda) = \sum_{j=1}^{n} \frac{1}{\lambda - \gamma_{nj} - i\hbar - \frac{i(n-1)\hbar}{2}} \frac{\prod\limits_{r=1}^{n-1}(\gamma_{nj} - \gamma_{n-1,r} + \frac{i\hbar}{2})}{\prod\limits_{s\neq j}(\gamma_{nj} - \gamma_{ns})} e^{i\hbar\partial_{\gamma_{nj}}}\,,$$

satisfy the complete set of relations (4.7) *and, therefore, define a representation of the Yangian* $Y(\mathfrak{gl}(N))$.

As a consequence we have

$$\mathcal{A}_n(\lambda) = \prod_{j=1}^{n}(\lambda - \gamma_{nj}),$$
$$\mathcal{B}_n(\lambda) = \sum_{j=1}^{n}\prod_{s\neq j}\frac{\lambda - \gamma_{ns}}{\gamma_{nj} - \gamma_{ns}}\prod_{r=1}^{n+1}(\gamma_{nj} - \gamma_{n+1,r} - \frac{i\hbar}{2})\, e^{-i\hbar\partial_{\gamma_{nj}}}, \tag{4.12}$$
$$\mathcal{C}_n(\lambda) = -\sum_{j=1}^{n}\prod_{s\neq j}\frac{\lambda - \gamma_{ns}}{\gamma_{nj} - \gamma_{ns}}\prod_{r=1}^{n-1}(\gamma_{nj} - \gamma_{n-1,r} + \frac{i\hbar}{2})\, e^{i\hbar\partial_{\gamma_{nj}}}.$$

Note that we also obtain the representation discussed in Section 2 through the following integral formulas which express the generators E_{ij} of the Lie algebra $\mathfrak{gl}(N)$ in terms of the Yangian generators

$$\begin{aligned} E_{n,n+1} &= \frac{1}{2\pi\hbar}\oint e_n(\lambda)d\lambda \ , (n = 1, \ldots, N-1), \\ E_{n+1,n} &= \frac{1}{2\pi\hbar}\oint f_n(\lambda)d\lambda \ , (n = 1, \ldots, N-1), \\ E_{nn} &= \frac{1}{2\pi\hbar}\oint k_n(\lambda)\frac{d\lambda}{\lambda} - \frac{1}{2}(n-1) \ , (n = 1, \ldots, N). \end{aligned} \tag{4.13}$$

The non-simple root generators are defined recursively as $E_{jk} = [E_{jm}, E_{mk}]$ for $j < m < k$ and $j > m > k$. Here, the integrands are understood as Laurent series and the contours of integrations are taken around ∞.

Let us finally remark, that there is a direct generalization of the Yangian to the case of quantum groups [30] and it is possible to extend the results of this section to the quantum group case.

4.2. The Toda chains and R-matrix formalism

The quantum Toda chain is one of the popular examples of an integrable system. It is described by the Hamiltonian

$$H = \sum_{n=1}^{N}\Big(\frac{p_n^2}{2} + e^{x_n - x_{n+1}}\Big), \tag{4.14}$$

where $[x_n, p_m] = i\hbar\delta_{nm}$. There are two different ways to fix the boundary conditions: The choice of $x_{N+1} = \infty$ corresponds to the open $(GL(N))$ Toda chain; while the choice of $x_{N+1} = x_1$ corresponds to the periodic (affine) Toda chain.

The open and periodic Toda chains can be described, uniformly, by using the R-matrix formalism [32]. Introduce the Lax operators

$$L_n(\lambda) = \begin{pmatrix} \lambda - p_n & e^{-x_n} \\ -e^{x_n} & 0 \end{pmatrix}, \qquad n = 1, \ldots, N, \tag{4.15}$$

satisfying the following commutation relations

$$\begin{aligned} &R(\lambda-\mu)(L_n(\lambda))\otimes I)(I\otimes L_m(\mu)) \\ &\qquad = (I\otimes L_m(\mu))(L_n(\lambda)\otimes I)R(\lambda-\mu)\delta_{nm}\,, \end{aligned} \tag{4.16}$$

with the rational 4×4 R-matrix

$$R(\lambda) = I\otimes I + \frac{i\hbar}{\lambda}P\,. \tag{4.17}$$

The monodromy matrix

$$T_{_N}(\lambda) = L_N(\lambda)\dots L_1(\lambda) := \begin{pmatrix} A_N(\lambda) & B_N(\lambda) \\ C_N(\lambda) & D_N(\lambda) \end{pmatrix} \tag{4.18}$$

satisfies the equation

$$R(\lambda-\mu)(T(\lambda)\otimes I)(I\otimes T(\mu)) = (I\otimes T(\mu))(T(\lambda)\otimes I)R(\lambda-\mu)\,. \tag{4.19}$$

In particular, the following commutation relations hold:

$$\begin{aligned} [A_N(\lambda), A_N(\mu)] = [C_N(\lambda), C_N(\mu)] = 0\,, \\ (\lambda-\mu+i\hbar)A_N(\mu)C_N(\lambda) = (\lambda-\mu)C_N(\lambda)A_N(\mu) + i\hbar A_N(\lambda)C_N(\mu)\,. \end{aligned} \tag{4.20}$$

From (4.19) it can be easily shown that the trace of the monodromy matrix

$$\widehat{t}_N(\lambda) = A_N(\lambda) + D_N(\lambda) \tag{4.21}$$

satisfies $\left[\widehat{t}(\lambda), \widehat{t}(\mu)\right] = 0$ and is a generating function for the Hamiltonians of the periodic Toda chain:

$$\widehat{t}_N(\lambda) = \sum_{k=0}^{N}(-1)^k\lambda^{N-k}H_k\,. \tag{4.22}$$

We formulate the spectral problems for the periodic Toda chain as follows:

$$\widehat{t}_N(\lambda)\Psi_{\boldsymbol{E}} = t_N(\lambda;\boldsymbol{E})\Psi_{\boldsymbol{E}}\,, \tag{4.23}$$

where

$$t_N(\lambda;\boldsymbol{E}) = \sum_{k=0}^{N}(-1)^k\lambda^{N-k}E_k\,. \tag{4.24}$$

On the other hand, the operator

$$A_N(\lambda) := \sum_{k=0}^{N}(-1)^k\lambda^{N-k}h_k \tag{4.25}$$

is a generating function of the Hamiltonians h_k of the N particles open Toda chain.

The generating functions for the open Toda chains are connected by the recursive relations:

$$\begin{aligned} A_N(\lambda) &= (\lambda - p_N)A_{N-1}(\lambda) + e^{-x_N}C_{N-1}(\lambda)\,, \\ C_N(\lambda) &= -e^{x_N}A_{N-1}(\lambda)\,. \end{aligned} \tag{4.26}$$

4.3. The spectral problem for the open Toda chain

The main goal of this subsection is to apply the results from the Section 2 to the solution of the spectral problem of the open Toda chain.

Denote $\boldsymbol{x} = (x_1, \dots, x_N)$. To solve the spectral problem we should find the common eigenfunction of the system of differential-difference equations:

$$\begin{aligned} A_N(\lambda)\psi_{\boldsymbol{\gamma}_N}(\boldsymbol{x}) &= \prod_{m=1}^{N} (\lambda - \gamma_{Nm})\, \psi_{\boldsymbol{\gamma}_N}(\boldsymbol{x})\,, \\ A_{N-1}(\gamma_{Nj})\psi_{\boldsymbol{\gamma}_N}(\boldsymbol{x}) &= i^{1-N} e^{-x_N} e^{-i\hbar\partial_{\gamma_{Nj}}} \psi_{\boldsymbol{\gamma}_N}(\boldsymbol{x})\,, \end{aligned} \tag{4.27}$$

$j = 1, \dots, N$. It is worth mentioning that the system (4.27) is the quantum counterpart of the Flashka and McLaughlin [33] Darboux transform to separated variables $(p, x) \to (\gamma, \theta)$. For the first time the system (4.27) was introduced and solved in the framework of QISM ([22]). Below we describe the representation theory solution of the system (4.27).

Let W' and W be the dual irreducible Whittaker modules and $w'_N \in W'$, $w_N \in W$ be the corresponding cyclic Whittaker vectors. The representation of the Cartan subalgebra is integrated to the action of the Cartan torus, so the following function is well defined

$$\psi_{\gamma_{N1},\dots,\gamma_{NN}} = e^{-\boldsymbol{x}\cdot\boldsymbol{\rho}^{(N)}} \left\langle w'_N, e^{-\sum\limits_{k=1}^{N} x_k E_{kk}} w_N \right\rangle, \tag{4.28}$$

where $\boldsymbol{x}\cdot\boldsymbol{\rho}^{(N)}$ is the standard product in $\mathbb{R}^N$.

Definition 4.1. *The radial projections $A_n(\lambda)$ of the generation functions $\mathcal{A}_n(\lambda)$* (2.13) *of the central elements of $\mathcal{U}(\mathfrak{gl}(N))$ are defined by*

$$A_n(\lambda)\psi_{\gamma_{N1},\dots,\gamma_{NN}} = e^{-\boldsymbol{x}\cdot\boldsymbol{\rho}^{(N)}} \left\langle w'_N, e^{-\sum\limits_{k=1}^{N} x_k E_{kk}} \mathcal{A}_n(\lambda - \tfrac{i(N-n)\hbar}{2}) w_N \right\rangle. \tag{4.29}$$

There is the relation between the operators $A_n(\lambda)$ of different levels:

$$A_n(\lambda) = (\lambda - p_n)A_{n-1}(\lambda) - e^{x_{n-1}-x_n}A_{n-2}(\lambda)\,, \tag{4.30}$$

where $n = 1, \dots, N$ and $A_{-1} = 0$, $A_0 = 1$. Therefore the $A_n(\lambda)$ is the generating function of the Hamiltonians of the n-particles open Toda chain.

The following theorem identifies our construction of the matrix element (4.28) with the integral formula for the eigenfunction of the open Toda chain in terms of the Mellin-Barnes integrals [22].

Theorem 4.2. *The matrix element* (4.28) *satisfies the set of equations* (4.27).

It remains to express the matrix element (4.28) in the integral form. Substituting the expressions (2.8), (2.9), and (2.2a) into (4.28), we obtain

$$\psi_{\gamma_N}(\boldsymbol{x}) = e^{-\boldsymbol{x}\cdot\boldsymbol{\rho}^{(N)}} \times \int\limits_{\mathbb{R}^{\frac{N(N-1)}{2}}} \prod_{n=1}^{N-1} \frac{\prod\limits_{k=1}^{n}\prod\limits_{m=1}^{n+1} \hbar^{\frac{\gamma_{nk}-\gamma_{n+1,m}}{i\hbar}+\frac{1}{2}}\, \Gamma\left(\frac{\gamma_{nk}-\gamma_{n+1,m}}{i\hbar}+\frac{1}{2}\right)}{\prod\limits_{s<p}\left|\Gamma\left(\frac{\gamma_{ns}-\gamma_{np}}{i\hbar}\right)\right|^2}$$
$$\times\, e^{\frac{i}{\hbar}\sum\limits_{n,j=1}^{N}(\gamma_{nj}-\gamma_{n-1,j})x_n} \prod_{\substack{n=1\\ j\leq n}}^{N-1} d\gamma_{nj}\,, \tag{4.31}$$

where by definition $\gamma_{nj} = 0$ for $j > n$.

In the study of the analytic properties of this solution with respect to $\boldsymbol{\gamma}_N$, it is useful to transform (4.31). Let us change the variables of integration in (4.31):

$$\gamma_{nj} \;\to\; \gamma_{nj} - \frac{i\hbar}{n}\sum_{s=1}^{n}\rho_s^{(N)}, \qquad (n = 1, \ldots, N-1)\,. \tag{4.32}$$

After the change of variables (4.32) we shift the domain of integration $\mathbb{R}^{\frac{N(N-1)}{2}}$ to the complex plane in such a way that the domain of integration over the variables $\gamma_{n-1,j}$ lies above the domain of integration over the variables γ_{nj}. Thus, we arrive at the analytic continuation equal to:

$$\psi_{\gamma_N}(x_1, \ldots, x_N) \tag{4.33}$$
$$= \int\limits_{\mathcal{C}} \prod_{n=1}^{N-1} \frac{\prod\limits_{k=1}^{n}\prod\limits_{m=1}^{n+1} \hbar^{\frac{\gamma_{nk}-\gamma_{n+1,m}}{i\hbar}}\, \Gamma\left(\frac{\gamma_{nk}-\gamma_{n+1,m}}{i\hbar}\right)}{\prod\limits_{s\neq p}\Gamma\left(\frac{\gamma_{ns}-\gamma_{np}}{i\hbar}\right)}\; e^{\frac{i}{\hbar}\sum\limits_{n,j=1}^{N}(\gamma_{nj}-\gamma_{n-1,j})x_n} \prod_{\substack{n=1\\ j\leq n}}^{N-1} d\gamma_{nj}\,,$$

where the domain of integration $\mathcal{C}$ is defined by the conditions

$$\min_j\{\operatorname{Im}\gamma_{kj}\} > \max_m\{\operatorname{Im}\gamma_{k+1,m}\}$$

for all $k = 1, \ldots, N-1$. The integral (4.33) converges absolutely. Thus, we obtain the integral representation [22] for the eigenfunction of open Toda chain by purely representation theory methods.

Finally, let us note that the Gelfand-Zetlin type representation may be generalized to the case of $Y(\mathfrak{g})$, with $\mathfrak{g}$ being an arbitrary simple Lie algebra. This provides the uniform approach to the solution of the various integrable systems based on various (quantum) Lie groups. We are planning to discuss these results elsewhere.

Acknowledgments

The research was partly supported by grants CRDF RM1-2545-MO-03; INTAS 03-513350; grant 1999.2003.2 for support of scientific schools, and by grants RFBR 03-02-17554 (A. Gerasimov, D. Lebedev), RFBR 03-02-17373 (S. Kharchev). We are grateful to M. Kontsevich, and M. Semenov-Tian-Shansky for their interest in this work and we are grateful to A. Rosly for useful discussion. D.L. is also grateful to IHES for warm hospitality and to organizers and participants of the ICMP workshop in Faro for the creation of the stimulating atmosphere. D.L. thanks to 21 COE RIMS Research Project 2004: Quantum Integrable Systems and Infinite-Dimensional Algebras, where the paper was finished, for support and warm hospitality.

References

[1] V.G. Drinfeld, *Hopf algebras and the quantum Yang-Baxter equation*, Soviet Math. Dokl. **32**, 254–258, (1985).

[2] M. Jimbo, *A q-difference analogue of* $U(\mathfrak{g})$ *and the Yang-Baxter equation*, Lett. Math. Phys. **10**, (1985), 63–69.

[3] L.D. Faddeev, *Quantum completely integrable models in field theory*, Sov. Sci. Rev., Sect. C (Math. Phys. Rev.) **1**, (1980), 107–155.

[4] P.P. Kulish, E.K. Sklyanin, *Quantum spectral transform method. Recent developments*, Lecture Notes in Phys. **151**, pp. 61–119, Springer, Berlin-New York, 1982.

[5] E.K. Sklyanin, *Separation of variables - new trends*, Quantum field theory, integrable models and beyond, (Kyoto, 1994). Progr. Theor. Phys. Suppl. **118**, (1995), 35–60.

[6] A. Gerasimov, S. Kharchev, D. Lebedev, *Representation Theory and Quantum Inverse Scattering Method: The Open Toda Chain and the Hyperbolic Sutherland Model,* Int. Math. Res. Notices **2004**, no.17, (2004), 823–854.

[7] A. Gerasimov, S. Kharchev, D. Lebedev, *On a class of integrable systems connected with* $GL(N,\mathbb{R})$, arXiv:math.QA/0301025.

[8] I.M. Gelfand, M.L. Tsetlin, *Finite-dimensional representations of the group of unimodular matrices*, Dokl. Akad. Nauk SSSR **71**, (1950), 825–828.

[9] I.M. Gelfand, M.I. Graev, *Finite-dimensional irreducible representations of the unitary and the full linear groups, and related special functions*, Izv. Akad. Nauk SSSR, Ser. Mat. **29**, (1965), 1329–1356; [Transl., II Ser., Am. Math. Soc. **64**, (1965), 116–146].

[10] F. Lemire, J. Patera, *Formal analytic continuation of Gelfand's finite-dimensional representations of gl(n,C)*, J. Math. Phys. **20**, (1979), 820–829.

[11] V.G. Drinfeld, *Quantum groups,* Proc. Int. Congr. Math. Berkeley, California, (1986), Providence (1987), 718–820.

[12] V.G. Drinfeld, *A new realization of Yangians and of quantum affine algebras*, (Russian) Dokl. Akad. Nauk SSSR **296**, (1987), no. 1, 13–17; translation in Soviet Math. Dokl. **36**, (1988), 212–216.

[13] L.D. Faddeev, *Discrete Heisenberg-Weyl group and modular group*, Lett. Math. Phys. **34**, (1995), 249–254.

[14] L.D. Faddeev, *Modular double of a quantum group*, in: Conférence Moshé Flato 1999, Quantization, Deformations, and Symmetries. Vol. I, 149–156, Kluwer Acad. Publ., Dordrecht, 2000.

[15] S. Kharchev, D. Lebedev, M. Semenov-Tian-Shansky, *Unitary representations of $U_q(\mathfrak{sl}(2,\mathbb{R}))$, the modular double and the multiparticle q-deformed Toda chains,* Comm. Math. Phys. **225**, (2002), 573–609.

[16] G. Lusztig, *Introduction to quantum groups*, Progress in Mathematics, 110, Birkhäuser Boston, Inc., Boston, MA, 1993

[17] V. Chari, A. Pressley, *A guide to quantum groups*, Cambridge Univ. Press, Cambridge, 1994.

[18] A. Connes, *Non commutative geometry*, Academic Press, 1994.

[19] M. Rieffel, *C^*-algebras associated with irrational rotations*, Pacific J. Math. **93**, (1981), 415–430.

[20] J. Adams, D. Barbasch and D. Vogan Jr. *The Langlands Classification and Irreducible Characters for Real Reductive Groups*, Progress in Mathematics **104**, Birkhäuser 1992.

[21] S. Kharchev, D. Lebedev, *Integral representation for the eigenfunctions of a quantum periodic Toda chain*, Lett. Math. Phys. **50**, (1999), 53–77.

[22] S. Kharchev, D. Lebedev, *Eigenfunctions of $GL(N,\mathbb{R})$ Toda chain: The Mellin-Barnes representation*, JETP Lett. **71**, (2000), 235–238.

[23] S. Kharchev, D. Lebedev, *Integral representations for the eigenfunctions of quantum open and periodic Toda chains from QISM formalism*, J.Phys. **A34**, (2001), 2247–2258.

[24] B. Kostant, *On Whittaker vectors and representation theory*, Inventiones Math. **48**, (1978), 101–184.

[25] R. Howe, T. Umeda, *The Capelli identity, the double commutant theorem, and multiplicity-free actions*, Math. Ann. **290**, (1991), 569–619.

[26] I. Gelfand, A. Kirillov, *Sur les corps liés aux algèbres enveloppantes des algèbres de Lie*, Publ. Math. de l'IHÉS **31**, (1966), 509–523.

[27] A. Gerasimov, S. Kharchev, D. Lebedev, *in preparation.*

[28] A. Sevostyanov, *Regular nilpotent elements and quantum groups*, Comm. Math. Phys. **204**, (1999), 1–16.

[29] A. Molev, M. Nazarov, G. Olshanski, *Yangians and classical Lie algebras*, Russian Math. Surveys **51**, (1996), 205–282.

[30] M. Nazarov, V. Tarasov, *Yangians and Gelfand-Zetlin bases,* Publ. Res. Inst. Math. Sci. **30**, (1994), 459–478.

[31] I.V. Cherednik, *A new interpretation of Gelfand-Tzetlin bases,* Duke Math. J. **54**, (1987), 563–577.

[32] E.K. Sklyanin, *The quantum Toda chain*, Lect. Notes in Phys. **226**, (1985), 196–233.

[33] H. Flaschka, D. McLaughlin, *Canonically conjugate variables for the Korteweg-de Vries equation and the Toda lattice with periodic boundary conditions*, Progr. Theor. Phys. **55**, (1976), 438–456.

A. Gerasimov
Institute for Theoretical and Experimental Physics
Moscow, Russia

and

Hamilton Mathematical Institute at Trinity College
Dublin, Ireland

S. Kharchev and D. Lebedev
Institute for Theoretical and Experimental Physics
Moscow, Russia

Progress in Mathematics, Vol. 237, 157–173

Connecting Lattice and Relativistic Models via Conformal Field Theory

H.E. Boos[1], V.E. Korepin and F.A. Smirnov[2]

Abstract. We consider the quantum group invariant XXZ-model. In the infrared limit it describes a Conformal Field Theory (CFT) with modified energy-momentum tensor. The correlation functions are related to solutions of level -4 of the qKZ equations. We describe these solutions relating them to level 0 solutions. We further consider general matrix elements (form factors) containing local operators and asymptotic states. We explain that the formulae for solutions of the qKZ equations suggest a decomposition of these matrix elements with respect to states of the corresponding CFT.

Mathematics Subject Classification (2000). 82B20, 82B23, 20G10, 14H70, 32G34, 17B37.

Keywords. qKZ, XXZ model, CFT, correlation functions, quantum group.

1. Quantum group invariant XXZ-model

Let us recall some well-known facts concerning the XXZ-model and its continuous limit. Usually the XXZ-model is considered as thermodynamic limit of a finite spin chain. Consider the space $\left(\mathbb{C}^2\right)^{\otimes N}$. The finite spin chain in question is described by the Hamiltonian:

$$H_{XXZ} = \sum_{k=1}^{N} (\sigma_k^1 \sigma_{k+1}^1 + \sigma_k^2 \sigma_{k+1}^2 + \Delta \sigma_k^3 \sigma_{k+1}^3) \tag{1}$$

where the periodic boundary conditions are implied: $\sigma_{N+1} = \sigma_1$. We consider the critical case $|\Delta| < 1$ and parametrize it as follows:

$$\Delta = \cos \pi\nu.$$

[1] On leave of absence from the Institute for High Energy Physics, Protvino, 142284, Russia.
[2] Member of CNRS.

It is well known that in the infrared limit the model describes a Conformal Field Theory (CFT) with $c = 1$ and coupling constant equal to ν. The correlation functions in the thermodynamic limit were found by Jimbo and Miwa [2].

It is equally matter of common knowledge that the model is closely related to the R-matrix:

$$R(\beta,\nu) = \begin{pmatrix} a(\beta) & 0 & 0 & 0 \\ 0 & b(\beta) & c(\beta) & 0 \\ 0 & c(\beta) & b(\beta) & 0 \\ 0 & 0 & 0 & a(\beta) \end{pmatrix} \tag{2}$$

where

$$a(\beta) = R_0(\beta), \quad b(\beta) = R_0(\beta)\frac{\sinh\nu\beta}{\sinh\nu(\pi i - \beta)}$$

$$c(\beta) = R_0(\beta)\frac{\sinh\nu\pi i}{\sinh\nu(\pi i - \beta)}$$

$$R_0(\beta) = \exp\left\{ i\int_0^\infty dk \frac{\sin(\beta k)\sinh\frac{\pi k(\nu-1)}{2\nu}}{k\sinh\frac{\pi k}{2\nu}\cosh\frac{\pi k}{2}} \right\}.$$

The coupling constant ν will be often omitted from $R(\beta,\nu)$. The relation between R-matrix and XXZ-model is explained later.

From the point of view of mathematics the R-matrix (2) is the R-matrix for two-dimensional evaluation representations of the quantum affine algebra $U_q(\widehat{sl}_2)$. The latter algebra contains two sub-algebras $U_q(sl_2)$. Let us perform a gauge transformation with the R-matrix in order to make the invariance with respect to one of them transparent:

$$\begin{aligned} \mathcal{R}(\beta_1,\beta_2,\nu) &= e^{\frac{\nu}{2}\beta_1\sigma^3} \otimes e^{\frac{\nu}{2}\beta_2\sigma^3}\, R(\beta_1-\beta_2,\nu)\, e^{-\frac{\nu}{2}\beta_1\sigma^3} \otimes e^{-\frac{\nu}{2}\beta_2\sigma^3} \\ &= \frac{R_0(\beta_1-\beta_2)}{2\sinh\nu(\pi i - \beta_1+\beta_2)} \left(e^{\nu(\beta_1-\beta_2)} R_{21}^{-1}(q) - e^{\nu(\beta_2-\beta_1)} R_{12}(q) \right) \end{aligned} \tag{3}$$

where

$$q = e^{2i\pi(\nu+1)}.$$

Adding 1 to ν is important since we will use fractional powers of q. Here $R(q)$ is the usual R-matrix for $U_q(sl_2)$:

$$R_{12}(q) = \begin{pmatrix} q^{\frac{1}{2}} & 0 & 0 & 0 \\ 0 & 1 & q^{\frac{1}{2}} - q^{-\frac{1}{2}} & 0 \\ 0 & 0 & 1 & 0 \\ 0 & 0 & 0 & q^{\frac{1}{2}} \end{pmatrix}.$$

We want to use this quantum group symmetry. Unfortunately, the Hamiltonian (1) is not invariant with respect to the action of the quantum group which is represented in the space $\left(\mathbb{C}^2\right)^{\otimes N}$ by

$$S^3 = \sum_{k=1}^{N} \sigma_k^3$$

$$S^{\pm} = \sum_{k=1}^{N} q^{-\frac{\sigma_1^3}{4}} \cdots q^{-\frac{\sigma_{k-1}^3}{4}} \sigma_k^{\pm} q^{\frac{\sigma_{k+1}^3}{4}} \cdots q^{\frac{\sigma_N^3}{4}}.$$

A solution of this problem of quantum group invariance was found by Pasquier and Saleur [3]. They proposed to consider another integrable model on the finite lattice with Hamiltonian corresponding to open boundary conditions:

$$H_{RXXZ} = \sum_{k=1}^{N-1} (\sigma_k^1 \sigma_{k+1}^1 + \sigma_k^2 \sigma_{k+1}^2 + \Delta \sigma_k^3 \sigma_{k+1}^3) + i\sqrt{1-\Delta}\,(\sigma_1^3 - \sigma_N^3). \qquad (4)$$

This Hamiltonian is manifestly invariant under the action of a quantum group on the finite lattice. After the thermodynamic limit one obtains a model with the same spectrum as the original XXZ, but different scattering (this point will be described later). The infrared limit corresponds to a CFT with modified energy-momentum tensor of central charge

$$c = 1 - \frac{6\nu^2}{1-\nu}$$

especially interesting when ν is rational and additional restriction takes place.

In the present paper we shall consider the RXXZ-model. We shall propose formulae for correlators for this model showing their similarity with correlators for the XXX-model. The latter can be expressed in terms of values of the Riemann zeta-function at odd natural arguments. We shall obtain an analogue of this statement for the RXXZ-model.

Let us say a few words about the hypothetic relation between the XXZ and RXXZ models in the thermodynamic limit. The argument that this limit should not depend on the boundary conditions must be dismissed in our situation since we consider a critical model with long-range correlations. Still we would expect that the following relation between the two models in infinite volume exists. The quantum group $U_q(sl_2)$ acts on the infinite XXZ-model and commutes with the Hamiltonian. Consider a projector $\mathcal{P}$ on the invariant subspace. We had the XXZ-vacuum $|\text{vac}\rangle_{XXZ}$. We suppose that the RXXZ-model is obtained by projection, in particular:

$$|\text{vac}\rangle_{RXXZ} = \mathcal{P}|\text{vac}\rangle_{XXZ}.$$

The correlators in the RXXZ-model are

$$_{RXXZ}\langle\text{vac}|\mathcal{O}|\text{vac}\rangle_{RXXZ} = \ _{XXZ}\langle\text{vac}|\mathcal{P}\mathcal{O}\mathcal{P}|\text{vac}\rangle_{XXZ}$$

which can be interpreted in two ways: either as correlator in the RXXZ-model or as correlator of the $U_q(sl_2)$-invariant operator $\mathcal{P}\mathcal{O}\mathcal{P}$ in XXZ-model. This assumption explains the notation RXXZ standing for Restricted XXZ-model. So, we assume that in the lattice case a phenomenon close to the one taking place in massive models occurs [12].

Let us explain in some more details the set of operators in XXZ model for which we are able to calculate the correlators in simple form provided the above reasoning holds. Under $\mathcal{O}$ we understand some local operator of XXZ-chain, i.e., a product of several local spins σ_k^a, $a = 1,2,3$. Under the above action of the quantum group these spins transform with respect to the three-dimensional adjoint representation. The projection $\mathcal{POP}$ extracts all the invariant operators, i.e., projects over the subspace of singlets in the tensor product of the three-dimensional representations.

Let us explain more explicitly the relation between the R-matrix and the XXZ, and RXXZ Hamiltonians. Both of them can be constructed from the transfer-matrix with different boundary conditions constructed via the monodromy matrix:

$$R_{01}(\lambda)R_{02}(\lambda)\cdots R_{0,N-1}(\lambda)R_{0,N}(\lambda).$$

In some cases it is very convenient to consider the inhomogeneous model for which the monodromy matrix contains a fragment:

$$R_{0k}(\lambda-\lambda_k)\cdots R_{0,k+n}(\lambda-\lambda_{k+n}).$$

As we shall see many formulae become far more transparent for the inhomogeneous case.

2. QKZ on level −4 and correlators

The main result of the Kyoto group [1, 2] is that the correlators in XXZ-model are related to solutions of the qKZ-equations [6, 8] on level −4. We formulate the equations first and then explain the relation. The equations for the function $g(\beta_1,\dots,\beta_{2n})\in\mathbb{C}^{\otimes 2n}$ are

$$\begin{aligned} R(\beta_j-\beta_{j+1})g(\beta_1,\dots,\beta_{j+1},\beta_j,\dots,\beta_{2n})& \\ = g(\beta_1,\dots,\beta_j,\beta_{j+1},\dots,\beta_{2n})& \end{aligned} \tag{5}$$

$$g(\beta_1,\dots,\beta_{2n-1},\beta_{2n}+2\pi i) = g(\beta_{2n},\beta_1,\dots,\beta_{2n-1}). \tag{6}$$

For application to correlators a particular solution is needed which satisfies additional requirement:

$$\begin{aligned} g(\beta_1,\dots,\beta_j,\beta_{j+1},\dots,\beta_{2n})|_{\beta_{j+1}=\beta_j-\pi i}& \\ = s_{j,j+1}\otimes g(\beta_1,\dots,\beta_{j-1},\beta_{j+2},\dots,\beta_{2n})& \end{aligned} \tag{7}$$

where $s_{j,j+1}$ is the vector $(\uparrow\downarrow)+(\downarrow\uparrow)$ in the tensor product of jth and $(j+1)$th spaces.

The relation of these equations to correlators is conjectured by Jimbo and Miwa [2]. It cannot be proved for the critical model under consideration as it was done for the XXZ-model with $|q|<1$ in [1]. However, later arguments based on Bethe Anzatz technique were proposed by Maillet and collaborators [4, 5] which can be considered as a proof of Jimbo and Miwa conjecture.

Jimbo and Miwa find the solution needed [2] in the form:

$$g(\beta_1,\dots,\beta_{2n})=\frac{1}{\sum e^{\beta_j}}\prod_{i<j}\zeta^{-1}(\beta_i-\beta_j)\int\limits_{-\infty}^{\infty}d\alpha_1\cdots\int\limits_{-\infty}^{\infty}d\alpha_{n-1}\prod_{i,j}\varphi(\alpha_i,\beta_j,\nu)$$

$$\times\prod_{i<j}\frac{A_i^2-A_j^2}{a_i-a_jq}\ D(a_1,\dots,a_{n-1}|b_1,\dots,b_{2n})$$

where

$$\varphi(\alpha,\beta,\nu)=\exp\left\{-(1+\nu)\frac{\alpha+\beta}{2}-2\int\limits_0^{\infty}dk\frac{\sin^2(\frac{\alpha-\beta}{2}k)\sinh\frac{\pi k(\nu+1)}{2\nu}}{k\sinh\frac{\pi k}{2\nu}\sinh\pi k}\right\}.$$

$\zeta(\beta)$ is some complicated function, we shall not need it. We use the notations:

$$a_j=e^{2\nu\alpha_j},\quad b_j=e^{2\nu\beta_j},\quad A_j=e^{\alpha_j}\quad B_j=e^{\beta_j}.$$

$D(a_1,\dots,a_{n-1}|b_1,\dots,b_{2n})$ is a Laurent polynomial of all its variables taking values in $\mathbb{C}^{2n}$. We shall not use explicit formula for this polynomial in the present paper.

For application to correlators in the homogeneous XXZ-model one has to specify:

$$\beta_1=\beta_2=\cdots=\beta_n=-\frac{\pi i}{2}$$

$$\beta_{n+1}=\beta_{n+2}=\cdots=\beta_{2n}=\frac{\pi i}{2}.$$

Then

$$g\left(-\frac{\pi i}{2},\dots,-\frac{\pi i}{2},\frac{\pi i}{2},\dots,\frac{\pi i}{2}\right)$$

$$=\int\limits_{-\infty}^{\infty}d\alpha_1\cdots\int\limits_{-\infty}^{\infty}d\alpha_{n-1}\prod_{i<j}\frac{A_i^2-A_j^2}{a_i-a_jq}\prod_i\frac{1}{A_i+A_i^{-1}}\ \widetilde{D}(a_1,\dots,a_{n-1}) \qquad (8)$$

with some Laurent polynomial $\widetilde{D}(a_1,\dots,a_{n-1})$. The trouble with this integral is that it is an essentially multi-fold one. In our previous papers we have shown that the integrals can be simplified and essentially reduced to products of one-fold ones in the XXX case. For the moment we cannot state the same for XXZ-model, but we shall explain that the simplification can be done in the RXXZ case. Let us consider this in some more details.

According to the understanding of the relation between XXZ and RXXZ models explained in the Introduction we expect that the correlators for the RXXZ model are related to certain solution of the same equations (6,7), invariant under the quantum group. In order to make the quantum group symmetry transparent we make the transformation:

$$\widehat{g}(\beta_1,\dots,\beta_{2n})=\exp\left(\frac{\nu}{2}\sum\beta_j\sigma_j^3\right)g(\beta_1,\dots,\beta_{2n}).$$

With this notation the equations (6,7) take the form:

$$\begin{aligned}&\mathcal{R}(\beta_j,\beta_{j+1})\widehat{g}(\beta_1,\dots,\beta_{j+1},\beta_j,\dots,\beta_{2n})\\&\quad=\widehat{g}(\beta_1,\dots,\beta_j,\beta_{j+1},\dots,\beta_{2n})\end{aligned} \tag{9}$$

$$\widehat{g}(\beta_1,\dots,\beta_{2n-1},\beta_{2n}+2\pi i)=-q^{-\frac{1}{2}\sigma^3_{2n}}\widehat{g}(\beta_{2n},\beta_1,\dots,\beta_{2n-1}) \tag{10}$$

and

$$\begin{aligned}&\widehat{g}(\beta_1,\dots,\beta_j,\beta_{j+1},\dots,\beta_{2n})|_{\beta_{j+1}=\beta_j-\pi i}\\&\quad=i\ \widehat{s}_{j,j+1}\otimes\widehat{g}(\beta_1,\dots,\beta_{j-1},\beta_{j+2},\dots,\beta_{2n})\end{aligned} \tag{11}$$

where $\widehat{s}_{j,j+1}$ is the quantum group singlet in the tensor product of the corresponding spaces:

$$q^{\frac{1}{4}}(\uparrow\downarrow)-q^{-\frac{1}{4}}(\downarrow\uparrow).$$

These equations respect the invariance under the quantum group. This fact is obvious for the first and the third equations. To see this in the second equation one has to keep in mind that $q^{\frac{1}{2}\sigma^3}$ gives in the two-dimensional representation the element which realizes the square of the antipode as inner automorphism.

From the Jimbo-Miwa solution (8) one can obtain a solution to (10, 11) by projection on the $U_q(sl_2)$-invariant subspace which will suffer of the same problems related to the denominators. The main goal of this paper is to show that at least in this case corresponding to the RXXZ-model another form of solution is possible.

3. QKZ on level 0

Consider the qKZ equations on level 0 which are the same as two out of three basic equations (axioms) for the form factors. We write these equations in $U_q(sl_2)$-invariant form which corresponds to form factors of RSG-model [12]. Consider a co-vector $\widehat{f}(\beta_1,\dots,\beta_{2n})\in\left(\mathbb{C}^{\otimes 2n}\right)^*$. The equations are

$$\begin{aligned}&\widehat{f}(\beta_1,\dots,\beta_{j+1},\beta_j,\dots,\beta_{2n})\\&\quad=\widehat{f}(\beta_1,\dots,\beta_j,\beta_{j+1},\dots,\beta_{2n})\mathcal{R}(\beta_j-\beta_{j+1})\\&\widehat{f}(\beta_1,\dots,\beta_{2n-1},\beta_{2n}+2\pi i)=-q^{-\frac{1}{2}\sigma^3_{2n}}\widehat{f}(\beta_{2n},\beta_1,\dots,\beta_{2n-1}).\end{aligned}$$

We need a solution belonging to the singlet with respect to the action of the $U_q(sl_2)$ subspace as has been explained in the level -4 case. The application to form factors imposes the additional requirement which connects sectors with different number of particles:

$$\begin{aligned}&2\pi i\,\mathrm{res}_{\beta_{2n}=\beta_{2n-1}+\pi i}\widehat{f}(\beta_1,\dots,\beta_{2n-2},\beta_{2n-1},\beta_{2n})\\&\quad=\widehat{s}^{\,*}_{2n-1,2n}\otimes\widehat{f}(\beta_1,\dots,\beta_{2n-2})\left(1-\mathcal{R}(\beta_{2n-1}-\beta_1)\cdots\right.\\&\qquad\left.\cdots\mathcal{R}(\beta_{2n-1}-\beta_{2n-2})\right).\end{aligned} \tag{12}$$

The difference with the level -4 case seems to be minor, but the formulae for solutions are much nicer. Many solutions can be written which are counted sets of integers: $\{k_1, \ldots, k_{n-1}\}$ such that $0 \le k_1 < \cdots < k_{n-1} \le 2n-2$:

$$f^{\{k_1,\ldots,k_{n-1}\}}(\beta_1, \ldots, \beta_{2n}) = \prod_{i<j} \zeta(\beta_i - \beta_j) \int_{-\infty}^{\infty} d\alpha_1 \cdots \int_{-\infty}^{\infty} d\alpha_{n-1} \prod_{i,j} \varphi(\alpha_i, \beta_j)$$
$$\times \det\|A_i^{k_j}\|_{1\le i,j\le n-1} \ \ h(a_1, \ldots, a_{n-1}|b_1, \ldots, b_{2n}) \prod_j a_j A_j$$

where h is a polynomial, skew-symmetric w.r. to the α's. Notice that there are no denominators mixing the integration variables in the integrand, so, effectively the integral is reduced to one-fold integrals of the form:

$$\langle P \mid p \rangle = \int_{-\infty}^{\infty} \prod_j \varphi(\alpha, \beta_j) \ P(A) \ p(a) a A d\alpha \tag{13}$$

where $p(\alpha)$ and $P(A)$ are polynomials. This is what we would like to have for the correlators!

Again we do not describe explicitly the functions h which take values in $\left(\mathbb{C}^{\otimes 2n}\right)^*$. As has been said they are skew-symmetric polynomials of $a's$. They are also rational functions of $b's$ with simple poles at $b_i = qb_j$ only. But there is one important property of h which we need to mention.

First, the integral (13) is such that the degree of any polynomial $s(a)$ can be reduced to $2n-2$ or less. For the polynomials of degree $\le 2n-2$ there is a basis (choice is not unique)

$$s_j(\alpha), \ j = -(n-1), \ldots, (n-1), \ \deg(s_j) = j + (n-1)$$

with special properties described later. We shall not write down explicit formulae. Then

$$h(a_1, \ldots, a_{n-1}) = \sum_{j_1 \ne 0, \ldots, j_{n-1} \ne 0} h_{j_1,\ldots,j_{n-1}} \det\|s_{j_p}(a_q)\|_{1\le p,q\le n-1}$$

and the skew-symmetric tensor h belongs to the subspace of maximal irreducible representation of the symplectic group $Sp(2n-2)$ of dimension

$$\dim(\mathcal{H}_{\text{irreducible}}) = \binom{2n-2}{n-1} - \binom{2n-2}{n-3}.$$

Let

$$J = 1, \ldots, \binom{2n-2}{n-1} - \binom{2n-2}{n-3}.$$

Consider the basis e^J in $\mathcal{H}_{\text{irreducible}}$ with components $e^J_{j_1,\ldots,j_{n-1}}$. Then we define h_J by

$$h_{j_1,\ldots,j_{n-1}} = \sum_J h_J \ e^J_{j_1,\ldots,j_{n-1}}.$$

Recall that $h(a_1,\dots,a_{n-1})$ takes values in the singlet subspace, so, it has components $h^I(a_1,\dots,a_{n-1})$ where I counts the basis of this subspace:

$$I=1,\dots,\binom{2n}{n}-\binom{2n}{n-1}.$$

Notice that

$$\binom{2n}{n}-\binom{2n}{n-1}=\binom{2n-2}{n-1}-\binom{2n-2}{n-3}$$

which means that there is a square matrix h^I_J defined by

$$h^I(a_1,\dots,a_{n-1})=\sum_J h^I_J s^J(a_1,\dots,a_{n-1})$$

where $s_J(a_1,\dots,a_{n-1})$ are the following anti-symmetric polynomials:

$$s^J(a_1,\dots,a_{n-1})=\sum_{j_1,\dots,j_{n-1}} e^J_{j_1,\dots,j_{n-1}}\det\|s_{j_p}(a_q)\|_{1\le p,q\le n-1}.$$

If we do not consider $s_0(a)$ the degrees of the polynomials $P(A)$ can be reduced to $2n-3$ or less. We consider a special basis

$$S_j(A),\quad |j|=1,\dots,(n-1),$$
$$\deg(S_{-k})=2k-1,\ k=1,\dots,n-1,\qquad \deg(S_k)=2k-2,\ k=1,\dots,n-1$$

which we do not describe explicitly, again.

The most important property of the integrals $\langle S_i\mid s_j\rangle$ is the deformed Riemann bilinear relation:

$$\sum_{k=1}^{n-1}\left(\langle S_k\mid s_i\rangle\langle S_{-k}\mid s_j\rangle-\langle S_k\mid s_j\rangle\langle S_{-k}\mid s_i\rangle\right)=\delta_{i,-j}$$
$$\sum_{k=1}^{n-1}\left(\langle S_i\mid s_k\rangle\langle S_j\mid s_{-k}\rangle-\langle S_j\mid s_k\rangle\langle S_i\mid s_{-k}\rangle\right)=\delta_{i,-j}.$$

These relations and the properties of $h(\alpha_1,\ \dots,\ \alpha_{n-1})$ imply that among $f^{\{k_1,\dots,k_{n-1}\}}$ only $\dim(\mathcal{H}_{\text{irreducible}})$ are linearly independent which are spanned by action of $Sp(2n-2)$ on $\{1,3,\dots,2n-3\}$. The basis in this space is denoted by

$$S_J(A_1,\dots,A_{n-1})=\sum_{j_1,\dots,j_{n-1}} e_J^{j_1,\dots,j_{n-1}}\det\|S_{j_p}(A_q)\|_{p,q=1,\dots,n-1}.$$

The result is that the solutions are combined into a square matrix (there is the same number of solutions as the dimension of space):

$$F_I^J=P_I^K H_K^J$$

where H_{KJ} is a polynomial function of β_j, the transcendental dependence on β_j is hidden in the period matrix P_I^J which is defined as

$$P_I^J=\langle S^J\mid s_I\rangle$$

where the notation has the obvious meaning:

$$\langle P_1 \wedge \cdots \wedge P_{n-1} \mid p_1 \wedge \cdots \wedge p_{n-1} \rangle = \det \| \langle P_i \mid p_j \rangle \|_{1 \le i,j \le n-1}.$$

4. New formula for level -4 from level 0

Recall that solutions to qKZ on level 0 are co-vectors while solutions on level -4 are vectors. Consider the scalar product for two solutions:

$$f(\beta_1, \ldots, \beta_{2n}) g(\beta_1, \ldots, \beta_{2n})$$

it is a quasi-constant (symmetric function of e^{β_j}).
So, we can construct singlet solutions of qKZ on level -4 from those on level 0. Indeed we have square matrix F:

$$G = F^{-1} = H^{-1} P^{-1}.$$

The matrix H^{-1} is complicated but rational function of β_j.
Due to the deformed Riemann relation it is easy to invert P! Indeed

$$\left(P^{-1}\right)^I_J = \langle S^I \mid s^{\dagger}_J \rangle$$

where $s^{\dagger}_J$ is obtained from s_J by replacing all

$$s_j \longrightarrow \mathrm{sgn}(j) s_{-j}.$$

So, the transcendental part almost does not change, and we prove that the new formula for solutions on level -4 is possible:

$$g^{\{k_1,\ldots,k_{n-1}\}}(\beta_1, \ldots, \beta_{2n}) = \prod_{i<j} \zeta(\beta_i - \beta_j) \int_{-\infty}^{\infty} d\alpha_1 \cdots \int_{-\infty}^{\infty} d\alpha_{n-1} \prod_{i,j} \varphi(\alpha_i, \beta_j)$$
$$\times \det \| A_i^{k_j} \|_{1 \le i,j \le n-1} \;\; \tilde{h}(a_1, \ldots, a_{n-1} | b_1, \ldots, b_{2n}) \prod_j a_j A_j$$

where $\tilde{h}$ are skew-symmetric w.r. to the a_i polynomials. Actually, they are polynomials in b_j as well. The proof is based on the following calculation:

$$H^{-1} = H^* \left(HH^*\right)^{-1}.$$

The operator HH^* is nicer than H itself because it acts from $\mathcal{H}_{\text{irreducible}}$ to itself. We were able to calculate its determinant:

$$\det(HH^*) = Const \left(\prod_{i,j} (b_i - q b_j) \right)^{-\left(\binom{2n-4}{n-2} - \binom{2n-4}{n-4}\right)}.$$

Also the rank of the residue of H at $b_j = q\beta_i$ equals the dimension of the singlet subspace in $\mathbb{C}^{\otimes(2n-2)}$.

5. Cohomological meaning of new formula

"Classical" limit: $\nu \to 0$ and β_j are rescaled in such a way that b_j are finite. In this limit

$$\langle P \mid p\rangle = \int_{-\infty}^{\infty} \prod_j \varphi(\alpha, \beta_j)\ P(A)\ p(a) d\alpha$$

$$\to \int_\gamma \frac{p(a)}{c} da$$

where the hyper-elliptic surface X is defined by

$$c^2 = \prod (a - b_j).$$

The genus equals $n-1$. The contour γ is defined by P. In particular,

$$S_{-k} \leftrightarrow \mathrm{b}_k, \quad S_k \leftrightarrow \mathrm{a}_k.$$

Consider

$$\mathrm{Symm}(X^{n-1})$$

the points on this variety are divisors:

$$\{P_1, \dots, P_{n-1}\} \qquad P_j = \{a_j, c_j\} \in X.$$

Consider the non-compact variety

$$\mathrm{Symm}(X^{n-1}) - D$$

where

$$D = \{\{P_1, \dots, P_{n-1}\} | P_j = \infty^\pm, P_i = \sigma(P_j)\}.$$

This is an affine variety isomorphic to the affine Jacobian.

The integrand of the classical limit of the invariant part of the Jimbo-Miwa solution (8) gives a $(n-1)$-differential form (maximal dimension) on

$$\mathrm{Symm}(X^{n-1}) - D$$

of the kind:

$$\Omega = \frac{F(a_1, c_1, \dots, a_{n-1}, c_{n-1})}{\prod_{i<j}(a_i - a_j)} \frac{da_1}{c_1} \wedge \dots \wedge \frac{da_{n-1}}{c_{n-1}}$$

where the polynomial $F(a_1, c_1, \dots, a_{n-1}, c_{n-1})$ vanishes when $a_i = a_j$ and $c_i = c_j$. The question arises concerning cohomologies.

Theorem (A. Nakayashiki) *The elements of* $H^{(n-1)}$ *can be realized as*

$$\Omega_{k_1, \dots, k_{n-1}} = \det \|a_p^{k_q}\|_{p,q=1,\dots,n-1} \quad \frac{da_1}{c_1} \wedge \dots \wedge \frac{da_{n-1}}{c_{n-1}}$$

where $k_q = 0, \dots, 2n-2$.

Remark. Actually some of these forms are linearly dependent (mod exact forms), we do not describe all details.

6. Back to correlators

Comparing with the Jimbo-Miwa solution one makes sure that the solution needed for the correlators is

$$g^{\{0,2,4,\ldots,2n-4\}}$$

so, it corresponds to "a-cycles". We need to put

$$\beta_k = \lambda_k - \frac{\pi i}{2} + i\delta_k, \quad \beta_{2n-k+1} = \lambda_k + \frac{\pi i}{2} - i\delta_k$$

and to take the limit $\delta_k \to 0$. The calculation of integrals is similar to the XXX case, the result can be expressed in terms of the function:

$$\chi(\alpha) = \frac{d}{d\alpha}\left(\log\frac{\varphi(\alpha - \frac{\pi i}{2})}{\varphi(\alpha + \frac{\pi i}{2})}\right) = i\int_0^\infty \frac{\cos(\alpha k)\sinh\frac{\pi k(\nu-1)}{2\nu}}{\sinh\frac{\pi k}{2\nu}\cosh\frac{\pi k}{2}}dk$$

$$= i\sum_{m=0}^\infty \alpha^{2m}\frac{(-1)^m}{(2m)!}\int_0^\infty \frac{k^{2m}\sinh\frac{\pi k(\nu-1)}{2\nu}}{\sinh\frac{\pi k}{2\nu}\cosh\frac{\pi k}{2}}dk.$$

Finally for the correlator in the inhomogeneous case:

$$g(\lambda_1 - \frac{\pi i}{2}\cdots\lambda_n - \frac{\pi i}{2}, \lambda_n + \frac{\pi i}{2}\cdots\lambda_1 + \frac{\pi i}{2})^{\epsilon_1\cdots\epsilon_n\epsilon_{n+1}\cdots\epsilon_{2n}}$$

$$= \sum_{m=0}^{[\frac{n}{2}]}\sum_{k_1,\ldots,k_{2m}} Q^{\epsilon_1\cdots\epsilon_n\epsilon_{n+1}\cdots\epsilon_{2n}}_{k_1k_2\cdots k_{2m-1}k_{2m}}(\lambda_1,\ldots,\lambda_n)\,\chi(\lambda_{k_1}-\lambda_{k_2})\cdots\chi(\lambda_{k_{2m-1}}-\lambda_{k_{2m}}).$$

7. General matrix elements

When we pass to the description of the XXZ-model in terms of particles a common phenomenon known nowadays as "modular double" [18] occurs. The essence of this phenomenon is that another quantum group with dual q enters the game. In a sense the RXXZ model is invariant with respect to the "modular double" which is a quite non-trivial, and not completely understood, combination of two quantum groups. The particle description of the model is as follows.

For coupling constants not very far from 0 the spectrum of the model contains one particle (magnon). This particle is parametrized by the rapidity θ carrying momentum and energy:

$$p(\theta) = \log\tanh\frac{1}{2}\left(\theta - \frac{\pi i}{2}\right), \quad e(\theta) = \frac{dp(\theta)}{d\theta}.$$

The particle has internal degrees living in isotopic space $\mathbb{C}^2$. The S-matrix is given by

$$S(\theta_1 - \theta_2) = R(\theta_1 - \theta_2, \tfrac{\nu}{1-\nu}).$$

This is where the second quantum group appears. The RXXZ model is invariant under the action of the two quantum groups:

$$U_q(sl_2)\ ,\ U_{\widetilde{q}}(sl_2),\quad \text{with}\quad q=e^{2\pi i(\nu+1)}\ ,\ \widetilde{q}=e^{\frac{2\pi i}{1-\nu}}.$$

For the asymptotic states it means that they must be taken as invariant under the action of the second quantum group. All that is familiar from the consideration of massive models and its restrictions [12].

Consider the matrix elements

$$_{RXXZ}\langle \text{vac} \mid \mathcal{O} \mid \theta_1,\dots,\theta_n \rangle_{RXXZ}$$

where $\mathcal{O}$ is some operator of the type

$$E_{\epsilon_1}^{\epsilon_1'}\cdots E_{\epsilon_n}^{\epsilon_n'}$$

It can be obtained from the "Kyoto generalization" which is the function

$$\widehat{f}(\beta_1,\dots,\beta_{2n},\theta_1,\dots,\theta_{2m})\in \mathbb{C}^{\otimes 2n}\otimes(\mathbb{C}^*)^{\otimes 2m}$$

which satisfies level -4 qKZ with R-matrix $\mathcal{R}(\cdot,\nu)$ (denoted by $\mathcal{R}(\cdot)$) with respect to β's and level 0 qKZ with gauge transformed S-matrix $\mathcal{R}(\cdot,\ \frac{\nu}{1-\nu})$ (denoted by $\mathcal{S}(\cdot)$) with respect to θ's. Actually, both equations are slightly modified. In addition it must satisfy the following normalization conditions. All together we have:

$$\begin{aligned}&\mathcal{R}(\beta_{j+1}-\beta_j)\widehat{f}(\beta_1,\dots,\beta_{j+1},\beta_j,\dots,\beta_{2n},\theta_1,\dots,\theta_{2m})\\ &=\ \widehat{f}(\beta_1,\dots,\beta_j,\beta_{j+1},\dots,\beta_{2n},\theta_1,\dots,\theta_{2m})\end{aligned}\tag{14}$$

$$\begin{aligned}&\widehat{f}(\beta_1,\dots,\beta_{2n-1},\beta_{2n}+2\pi i,\theta_1,\dots,\theta_{2m})\\ &=-\prod_{j=1}^{2m}\tanh\frac{1}{2}\left(\beta_{2n}-\theta_j+\frac{\pi i}{2}\right)q^{\frac{1}{2}\sigma^3_{2n}}\widehat{f}(\beta_{2n},\beta_1,\dots,\beta_{2n-1},\theta_1,\dots,\theta_{2m})\end{aligned}\tag{15}$$

$$\begin{aligned}&\widehat{f}(\beta_1,\dots,\beta_{2n-2},\beta_{2n-1},\beta_{2n},\theta_1,\dots,\theta_{2m})|_{\beta_{2n}=\beta_{2n-1}+\pi i}\\ &=\widehat{s}_{2n-1,2n}\otimes\widehat{f}(\beta_1,\dots,\beta_{2n-2},\theta_1,\dots,\theta_{2m})\end{aligned}\tag{16}$$

$$\begin{aligned}&\widehat{f}(\beta_1,\dots,\beta_{2n},\theta_1,\dots,\theta_{j+1},\theta_j,\dots,\theta_{2m})\\ &=\widehat{f}(\beta_1,\dots,\beta_{2n},\theta_1,\dots,\theta_j,\theta_{j+1},\dots,\theta_{2m})\mathcal{S}(\theta_j-\theta_{j+1})\end{aligned}\tag{17}$$

$$\begin{aligned}&\widehat{f}(\beta_1,\dots,\beta_{2n},\theta_1,\dots,\theta_{2m-1},\theta_{2m}+2\pi i)\\ &=-\prod_{j=1}^{2n}\tanh\frac{1}{2}\left(\theta_{2m}-\beta_j+\frac{\pi i}{2}\right)\widehat{f}(\beta_1,\dots,\beta_{2n},\theta_{2m},\theta_1,\dots,\theta_{2m-1})q^{-\frac{1}{2}\sigma^3_{2m}}\end{aligned}\tag{18}$$

$$2\pi i \mathrm{res}_{\theta_{2m}=\theta_{2m-1}+\pi i}\widehat{f}(\beta_1,\dots,\beta_{2n},\theta_1,\dots,\theta_{2m-2},\theta_{2m-1},\theta_{2m})$$
$$= \widehat{s}\,^{*}_{2m-1,2m}\otimes \widehat{f}(\beta_1,\dots,\beta_{2n},\theta_1,\dots,\theta_{2m-2}) \qquad (19)$$
$$\times\left(1-\prod_{j=1}^{2n}\tanh\frac{1}{2}\left(\theta_{2m-1}-\beta_j+\frac{\pi i}{2}\right)\mathcal{S}(\theta_{2m-1}-\theta_1)\cdots\mathcal{S}(\theta_{2m-1}-\theta_{2m-2})\right).$$

The equations (14, 15, 17, 18) are slightly different from respectively level -4 and level 0 qKZ equations because of multipliers containing tanh's. This difference, however, is easily taken care of by the multiplier

$$\prod_{i=1}^{2n}\prod_{j=1}^{2m}\psi(\beta_i,\theta_j)$$

where the function

$$\psi(\beta,\theta)=2^{-\frac{3}{4}}\exp\left(-\frac{\beta+\theta}{4}-\int_0^\infty\frac{\sin^2\frac{1}{2}(\beta-\theta+\pi i)k+\sinh^2\frac{\pi k}{2}}{k\sinh\pi k\cosh\frac{\pi k}{2}}dk\right)$$

satisfies the equations:

$$\psi(\beta,\theta+2\pi i)=\tanh\frac{1}{2}(\theta-\beta+\frac{\pi i}{2})\psi(\beta,\theta)$$
$$\psi(\beta,\theta)\psi(\beta,\theta+\pi i)=\frac{1}{e^{\beta}-ie^{\theta}}.$$

For the function $f(\beta_1,\dots,\beta_{2n},\theta_1,\dots,\theta_{2m})$ in the XXZ-model Jimbo-Miwa give a formula of the following kind:

$$f(\beta_1,\dots,\beta_{2n},\theta_1,\dots,\theta_{2m})=\prod_{i<j}\zeta(\theta_i-\theta_j,\tfrac{\nu}{1-\nu})\prod_{i<j}\zeta^{-1}(\beta_i-\beta_j,\nu)\prod_{i,j}\psi(\beta_i,\tau_j)$$
$$\times\int_{-\infty}^{\infty}d\alpha_1\cdots\int_{-\infty}^{\infty}d\alpha_n\int_{-\infty}^{\infty}d\sigma_1\cdots\int_{-\infty}^{\infty}d\sigma_m\prod\varphi(\alpha_i-\beta_j,\nu)\prod\varphi(\sigma_i-\theta_j,\tfrac{\nu}{1-\nu})$$
$$\times\prod_{i<j}\frac{A_i^2-A_j^2}{a_i-qa_j}\prod_{i<j}\frac{S_i^2-S_j^2}{\mathrm{s}_i-\widetilde{q}\mathrm{s}_j}\prod\frac{1}{A_i^2-S_j^2}$$
$$\times D(a_1,\dots,a_n|b_1,\dots,b_{2n})F(\mathrm{s}_1,\dots,\mathrm{s}_m|\mathrm{t}_1,\dots,\mathrm{t}_{2m})$$

where we use the notations:

$$a_j=e^{2\nu\alpha_j},\qquad b_j=e^{2\nu\beta_j},\qquad A_j=e^{\alpha_j},\qquad B_j=e^{\beta_j}$$
$$\mathrm{s}_j=e^{\frac{2\nu}{1-\nu}\sigma_j},\qquad \mathrm{t}_j=e^{\frac{2\nu}{1-\nu}\theta_j},\qquad S_j=e^{\sigma_j},\qquad T_j=e^{\theta_j}.$$

The functions D, F are polynomials of their variables. For us the main problem with this formula is in the denominators. Here we are concerned not only about the denominators a_i-qa_j and $\mathrm{s}_i-\widetilde{q}\mathrm{s}_j$ which are unpleasant for technical reasons as explained above. Our main trouble is in the denominators $A_i^2-S_j^2$ because due to certain physical intuition we would expect another kind of formula. Let us explain the point.

At this point it would be more clear to talk about the lattice SOS-model instead of the RXXZ-model. These two models are equivalent due to the usual Onzager relation between 2D classical statistical physics and 1D quantum mechanics. The advantage of the lattice model is due to the fact that it allows intuitively a clear relation to Euclidian Quantum Field Theory. Our physical intuition about the general matrix element is based on the following picture:

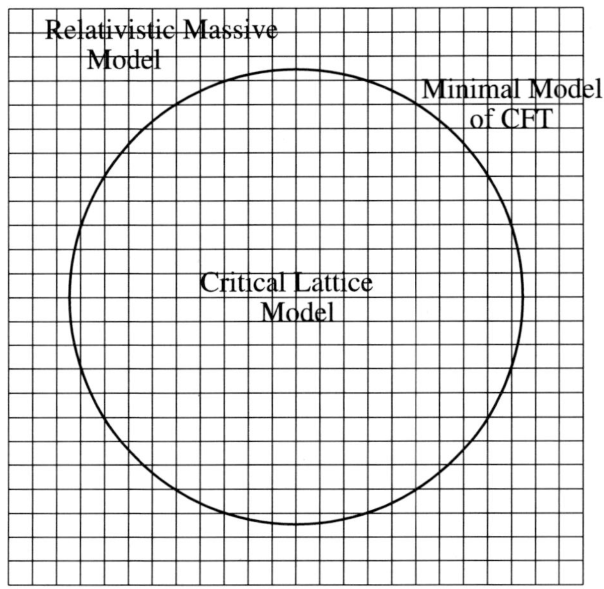

Let us give some explanations. Suppose we consider instead of the critical model the SOS-model out of criticality corresponding to the elliptic R-matrix. Suppose further that we are very close to the critical temperature. Then microscopically we have already critical lattice SOS-model. On the scales much bigger than the lattice size but much less than the correlation length we have massless relativistic field theory which is nothing but CFT with the central charge

$$c = 1 - \frac{6\nu^2}{\nu - 1}.$$

Finally on scales of the order of the correlation length we have the massive relativistic field theory which is the RSG-model with coupling constant $\frac{1-\nu}{\nu}$. The role of the CFT is clear: it describes the infrared limit of the lattice model and the ultraviolet limit of the massive model. The local operators of the massive model are counted by the states of the CFT. These local operators are described by form factors in the asymptotic states description. On the other hand one should be able to consider the lattice critical model with boundary conditions corresponding to different states of the CFT. That is why we expect the following kind of formula for the general matrix element:

$$_{RXXZ}\langle \text{vac}|\mathcal{O}|\theta_1,\dots,\theta_{2m}\rangle = \sum_{\Psi} {}_{RXXZ}\langle \text{vac}|\mathcal{O}|\Psi\rangle\langle\Psi\ |\theta_1,\dots,\theta_{2m}\rangle \tag{20}$$

where Ψ are states of the CFT, $\langle\Psi\ |\theta_1,\dots,\theta_{2m}\rangle$ are the form factors of the local operator corresponding to Ψ in the RSG-model, $_{RXXZ}\langle \text{vac}|\mathcal{O}|\Psi\rangle$ are correlators of a local operator $\mathcal{O}$ in the lattice model (of the usual kind $E_{\epsilon_1}^{\epsilon_1'}\cdots E_{\epsilon_n}^{\epsilon_n'}$). The latter object requires a more careful definition, we hope to return to it in future.

Notice that the formula (20) is in nice correspondence with the system of equations (14, 15, 16, 17, 18, 19) because passing from the Kyoto generalized correlator to the usual one we put $\beta_j = \beta_{2n-j+1} + \pi i$, so, the θ in equations with respect to θ_j cancel, and we get the usual Form Factor Axioms. In our case of RSG-model a complete set of solutions to these axioms is known [15, 20], so, a formula of the kind (20) must hold.

So, there must be a formula of the type:

$$\begin{aligned} f(\beta_1,\dots,\beta_{2n},\theta_1,\dots,\theta_{2m}) &= \prod_{i<j}\zeta(\theta_i-\theta_j,\tfrac{\nu}{1-\nu})\prod_{i<j}\zeta^{-1}(\beta_i-\beta_j,\nu)\prod_{i,j}\psi(\beta_i,\tau_j)\\ &\times\int_{-\infty}^{\infty}d\alpha_1\cdots\int_{-\infty}^{\infty}d\alpha_n\int_{-\infty}^{\infty}d\sigma_1\cdots\int_{-\infty}^{\infty}d\sigma_m\prod\varphi(\alpha_i-\beta_j,\nu)\prod\varphi(\sigma_i-\theta_j,\tfrac{\nu}{1-\nu})\\ &\times M(A_1,\dots,A_{n-1}|T_1,\dots,T_{m-1})\\ &\times\widetilde{h}(a_1,\dots,a_n|b_1,\dots,b_{2n})h(\mathrm{s}_1,\dots,\mathrm{s}_m|\mathrm{t}_1,\dots,\mathrm{t}_{2m}), \end{aligned} \tag{21}$$

where $M(A_1,\dots,A_{n-1}|S_1,\dots,S_{m-1})$ is a polynomial, skew-symmetric with respect to A_1, …, A_{n-1} and T_1, …, T_{m-1}, which depends on B_j, S_j as parameters. This polynomial must satisfy certain equations in order that the relations (16, 19) hold. We do not write down explicitly these bulky equations, but fortunately they coincide with the equations for similar polynomials for quite a different problem which is the calculation of form factors for massless flows [19]. The solution to these equations is not unique, but there is a "minimal" one which has minimal possible degree with respect to variables A_j and S_j. Our conjecture is that this is the solution we need. It satisfies all simple checks that we were able to carry out. Denote the sets $S=\{1,\dots,2n\}$, $S'=\{1,\dots,2m\}$. The polynomial is:

$$\begin{aligned} &M(A_1,\dots,A_{n-1}|S_1,\dots,S_{m-1})\\ &=\prod_{i<j}(A_i-A_j)\prod_{i<j}(S_i-S_j)\prod_{j=1}^{2n}B_j\prod_{j=1}^{n-1}A_j\\ &\times\sum_{\substack{T\subset S\\ \#T=n-1}}\ \sum_{\substack{T'\subset S'\\ \#T'=m-1}}\prod_{j\in T}B_j\prod_{i=1}^{n-1}\prod_{j\in T}(A_i+iB_j)\prod_{i=1}^{m-1}\prod_{j\in T'}(S_i+iT_j) \end{aligned}$$

$$\times \prod_{\substack{i,j\in S\setminus T\\ i<j}} (B_i+B_j) \prod_{\substack{i,j\in S'\setminus T'\\ i<j}} (T_i+T_j) \prod_{\substack{i\in T\\ j\in S\setminus T}} \frac{1}{B_i-B_j} \prod_{\substack{i\in T'\\ j\in S'\setminus T'}} \frac{1}{T_i-T_j}$$

$$\times \prod_{\substack{i\in T\\ j\in S'\setminus T'}} (B_i+iT_j) \prod_{\substack{i\in T'\\ j\in S\setminus T}} (T_i+iB_j) X_{T,T'}(B_1,\dots,B_{2n}|T_1,\dots,B_{2m})$$

where

$$X_{T,T'}(B_1,\dots,B_{2n}|T_1,\dots,B_{2m}) = \sum_{i_1,i_2\in S\setminus T} \prod_{p=1}^{2} \left(\frac{\prod_{j\in T}(B_{i_p}+B_j)\prod_{j\in T'}(B_{i_p}+iT_j)}{\prod_{j\in S'\setminus T'\setminus\{i_1,i_2\}}(B_{i_p}-B_j)\prod_{j\in S'\setminus T'}(B_{i_p}-iT_j)} \right).$$

Obviously, the formula (21) is in agreement with the intuitive formula (20). After specialization $\beta_k = \lambda_k + \frac{\pi i}{2}$, $\beta_{2n-k+1} = \lambda_k - \frac{\pi i}{2}$ (21) will turn into a sum of form factors of the RSG-model with coefficients constructed via the functions $\chi(\lambda_i - \lambda_j)$ which correspond to correlators of the RXXZ-model with boundary conditions. The identification of RSG-form factors with operators counted by the CFT is known at least to some extent [20, 21, 22]. So, it should be possible to make the correspondence between (21) and (20) more explicit, but this problem goes beyond the scope of the present paper.

Acknowledgments

HEB would like to thank Masahiro Shiroishi, Pavel Pyatov and Minoru Takahashi for useful discussions. This research has been supported by the following grants: the Russian Foundation of Basic Research under grant # 01–01–00201, by INTAS under grants #00-00055 and # 00-00561 and by EC network "EUCLID", contract number HPRN-CT-2002-00325. HEB would also like to thank the administration of the ISSP of Tokyo University for hospitality and perfect work conditions. The research of VEK was supported by NSF Grant PHY- 0354683. This paper is based on the talk given by FAS at "Infinite-Dimensional Algebras and Quantum Integrable Systems" (Faro, Portugal, July 21–25, 2003), FAS is grateful to organisers for their kind hospitality.

References

[1] M. Jimbo, K. Miki, T. Miwa, A. Nakayashiki, *Phys. Lett.* **A168** (1992) 256–263.

[2] M. Jimbo, T. Miwa, J. Phys. **A29** (1996) 2923–2958.

[3] V. Pasquier, H. Saleur, *Nucl. Phys.* **B330** (1990) 523.

[4] J.M. Maillet, V. Terras, *Nucl. Phys.* **B575** (2000) 627–647.

[5] N. Kitanine, J.M. Maillet, V. Terras, *Nucl. Phys.* **B567** (2000), no. 3, 554–582.

[6] F.A. Smirnov, Int. J. Mod. Phys. **A7**, (1992) S813–858.

[7] F.A. Smirnov, J. Phys. **A19** (1986) L575–L578.

[8] I. Frenkel, N. Reshetikhin, Comm. Math. Phys. **146** (1992) 1–60.

[9] H.E. Boos, V.E. Korepin, *J. Phys.* **A34** (2001) 5311–5316.

[10] H.E. Boos, V.E. Korepin, *Evaluation of integrals representing correlators in XXX Heisenberg spin chain.* in. MathPhys Odyssey 2001, Birkhäuser (2001) 65–108.

[11] H.E. Boos, V.E. Korepin, F.A. Smirnov, *Nucl. Phys.* **B** 658/3 (2003) 417–439.

[12] N.Yu. Reshetikhin, F.A. Smirnov, Comm. Math. Phys., 1990, v. 132, p. 415.

[13] F.A. Smirnov, *Form Factors in Completely Integrable Models of Quantum Field Theory.* Adv. Series in Math. Phys. 14, World Scientific, Singapore (1992).

[14] F.A. Smirnov, Lett. Math. Phys. **36** (1996) 267.

[15] F.A. Smirnov, Nucl. Phys. B **453[FS]** (1995) 807.

[16] A. Nakayashiki, F.A. Smirnov, Comm. Math. Phys., **217** (2001), 623.

[17] A. Nakayashiki, *On the cohomology of theta divisor of hyperelliptic Jacobian.* Contemporary mathematics, **309**, in Integrable systems, topology and physics, M.Guest et al. ed., AMS (2002).

[18] L.D. Faddeev, *Modular double of quantum group.* math.qa/9912078 , 11 pp.

[19] P. Mejean, F.A. Smirnov IJMPA 12 ,1997, no. 19, p. 3383–3395.

[20] O. Babelon, D. Bernard, F.A. Smirnov Commun. Math. Phys. 186 ,1997, 601–648.

[21] A. Nakayashiki, *The Chiral Space of Local Operators in SU(2)-Invariant Thirring Model.* math.QA/0303192.

[22] M. Jimbo, T. Miwa, Y. Takeyama, *Counting minimal form factors of the restricted sine-Gordon model* math-ph/03030.

H.E. Boos
Institute for Solid State Physics
University of Tokyo
Kashiwa, Chiba 277-8581, Japan

V.E. Korepin
C.N. Yang Institute for Theoretical Physics
State University of New York at Stony Brook
Stony Brook, NY 11794-3840, USA

F.A. Smirnov
LPTHE, Tour 16, 1-er étage
4, pl. Jussieu
F-75252, Paris Cedex 05, France

Progress in Mathematics, Vol. 237, 175–203

Elliptic Spectral Parameter and Infinite-Dimensional Grassmann Variety

Kanehisa Takasaki

Abstract. Recent results on the Grassmannian perspective of soliton equations with an elliptic spectral parameter are presented along with a detailed review of the classical case with a rational spectral parameter. The nonlinear Schrödinger hierarchy is picked out for illustration of the classical case. This system is formulated as a dynamical system on a Lie group of Laurent series with factorization structure. The factorization structure induces a mapping to an infinite-dimensional Grassmann variety. The dynamical system on the Lie group is thereby mapped to a simple dynamical system on a subset of the Grassmann variety. Upon suitable modification, almost the same procedure turns out to work for soliton equations with an elliptic spectral parameter. A clue is the geometry of holomorphic vector bundles over the elliptic curve hidden (or manifest) in the zero-curvature representation.

Mathematics Subject Classification (2000). 35Q58, 37K10, 58F07.

Keywords. soliton equation, elliptic curve, holomorphic bundle, Grassmann variety.

1. Introduction

Since the first proposal two decades ago by Sato [20], Segal and Wilson [21], the Grassmannian perspective of soliton equations has been successful for a variety of cases, even including higher-dimensional analogues such as the Bogomolny equation and the self-dual Yang-Mills equations [23]. The fundamental observation of this perspective is that a soliton equation can be translated to a simple (essentially linear) dynamical system on a subset of an infinite-dimensional "universal" Grassmann variety. Almost all of the cases thus examined, however, are equations with a *rational* zero-curvature representation, namely, equations whose zero-curvature equation is made of matrices depending rationally on a spectral parameter. The status of soliton equations related to an elliptic or higher genus algebraic curve has still remained rather obscure, though a few notable studies [4, 2] were done on the Landau-Lifshitz equation (a typical soliton equation with a zero-curvature representation made of elliptic functions [22, 3]).

Recent advances [1, 12, 14] have revealed the existence of a wide class of new integrable PDE's with a zero-curvature representation constructed on an algebraic curve of arbitrary genus. These equations, too, may be called "soliton equations" in a loose sense, namely, without implying the existence of soliton or soliton-like solutions. The works of Ben-Zvi and Frenkel [1] and Levin, Olshanetsky and Zotov [14] both stem from the notion of the Hitchin systems [7], and aim to obtain an integrable PDE as a $1+1$-dimensional analogue of the Hitchin systems. On the other hand, Krichever [12] uses the so called Tyurin parameters to construct Lax or zero-curvature equations on an algebraic curve. The notion of Tyurin parameters originates in algebraic geometry of holomorphic vector bundles over algebraic curves [26], and was applied by Krichever and Novikov in 1970's to the study of commutative rings of differential operators [9, 10, 11]. Krichever and Levin et al. illustrate their general scheme with several examples related to an elliptic curve. These examples can be used as valuable material for case studies.

One will naturally ask whether these new "soliton equations" can be understood in the Grassmannian perspective. An affirmative answer to this question has been obtained in the simplest case [24, 25], namely, a few examples that have a zero-curvature prepresentation with 2×2 matrices defined on an elliptic curve. Although this is indeed a case study, the upshot clearly shows that a similar result holds in a general and universal form. What distinguishes between the new and conventional soliton equations is the structure of a holomorphic bundle on the relevant algebraic curve. The aforementioned new equations are accompanied by a nontrivial bundle, which plays a central role in both the zero-curvature representation and the Grassmannian perspective. This article presents an outline of these results.

This article is organized as follows. The first half (Sections 2, 3 and 4) of this article is a review on conventional soliton equations with a rational zero-curvature representation. The nonlinear Schrödinger hierarchy is picked out for illustration. This system consists of an infinite number of evolution equations including the nonlinear Schrödinger equation itself in the lowest $1+1$-dimensional sector. One can reformulate this system as a dynamical system on a Lie group of Laurent series with factorization structure. The factorization induces a mapping to an infinite-dimensional Grassmann variety. The dynamical system on the Lie group is thereby mapped to a simple dynamical system on a subset of the Grassmann variety. This example shows a typical way the usual soliton equations are treated in the Grassmannian perspective. The second half (Section 5, 6 and 7) of this article presents the results on elliptic analogues [24, 25]. Two different types of elliptic analogues are considered here. The first case is an elliptic analogue of the nonlinear Schrödinger hierarchy. This system is constructed along the line of Krichever's scheme based on Tyurin parameters. The second case is concerned with the Landau-Lifshitz equation and an associated hierarchy of evolution equations. In both cases, a variant of the factorization is formulated as a Riemann-Hilbert problem with respect to the holomorphic bundle structure, and used to define a mapping to an infinite-dimensional Grassmann variety.

2. Nonlinear Schrödinger hierarchy

The construction of the nonlinear Schrödinger hierarchy starts from the first order matrix differential operator $\partial_x - A(\lambda)$, where $A(\lambda)$ a 2×2 matrix of the form

$$A(\lambda) = \begin{pmatrix} \lambda & u \\ v & -\lambda \end{pmatrix}. \tag{1}$$

u and v are fields on the one-dimensional space, $u = u(x)$, $v = v(x)$, and λ is a rational spectral parameter. From the point of view of affine Lie algebras, it is also natural to express $A(\lambda)$ as

$$A(\lambda) = J\lambda + A^{(1)}, \tag{2}$$

where

$$J = \begin{pmatrix} 1 & 0 \\ 0 & -1 \end{pmatrix}, \quad A^{(1)} = \begin{pmatrix} 0 & u \\ v & 0 \end{pmatrix}.$$

Generalities and backgrounds of this kind of soliton equations can be found in Frenkel's lectures [5].

2.1. Generating functions

The first step of the formulation of the hierarchy is to construct a 2×2 matrix of generating functions

$$U(\lambda) = \sum_{n=0}^{\infty} U_n \lambda^{-n}, \quad U_0 = J,$$

that satisfies the differential equation

$$[\partial_x - A(\lambda),\ U(\lambda)] = 0. \tag{3}$$

This reduces to the differential equations

$$\partial_x U_{n-1} = [J, U_n] + [A^{(1)}, U_{n-1}] \tag{4}$$

for U_n's. One can, in principle, solve these equations, by a subtle procedure decomposing the equations into the diagonal and off-diagonal part; a similar procedure is used below to construct another generating function $\phi(\lambda)$. This, however, leaves large arbitrariness in the solution. Moreover, this is by no means an effective way.

These problems are resolved by imposing the algebraic constraint

$$U(\lambda)^2 = I. \tag{5}$$

This amounts to finding $U(\lambda)$ in such a form as

$$U(\lambda) = \phi(\lambda) J \phi(\lambda)^{-1}, \tag{6}$$

where $\phi(\lambda)$ is another matrix of generating function

$$\phi(\lambda) = \sum_{n=0}^{\infty} \phi_n \lambda^{-n}, \quad \phi_0 = I,$$

that satisfies the differential equation

$$\partial_x \phi(\lambda) = A(\lambda)\phi(\lambda) - \phi(\lambda)J\lambda. \tag{7}$$

Remarkably, if the *existence* of a solution of this equation is ensured, one can uniquely determine U_n's by a set of recurrence relations as follows.

Note that the algebraic constraint implies the algebraic relations

$$0 = JU_n + U_n J + \sum_{m=1}^{n-1} U_m U_{n-m}, \quad n > 0, \tag{8}$$

that hold for U_n's. Combining this with the differential equations

$$\partial_x U_{n-1} = JU_n - U_n J + [A^{(1)}, U_{n-1}],$$

one obtains the relations

$$2JU_n = \partial_x U_{n-1} - [A^{(1)}, U_{n-1}] - \sum_{m=1}^{n-1} U_m U_{n-m}. \tag{9}$$

These relations take the form of recurrence relations, which enables one to calculate U_n's successively. The first few terms read

$$U_1 = \begin{pmatrix} 0 & u \\ v & 0 \end{pmatrix}, \quad U_2 = \begin{pmatrix} -\frac{1}{2}uv & \frac{1}{2}\partial_x u \\ -\frac{1}{2}\partial_x v & \frac{1}{2}uv \end{pmatrix}, \quad \text{etc.}$$

The matrix elements of U_n thus obtained are "local" quantities, namely, polynomials of derivatives of u, v.

What is left is to prove that the second generating function $\phi(\lambda)$ does exist. The equations for the Laurent coefficients of $\phi(\lambda)$ read

$$\partial_x \phi_n = [J, \phi_{n+1}] + A^{(1)}\phi_n. \tag{10}$$

One can split this matrix equation into the diagonal and off-diagonal part. This results in the two equations

$$\partial_x (\phi_n)_{\mathrm{diag}} = (A^{(1)}\phi_n)_{\mathrm{diag}}$$

and

$$\partial_x (\phi_n)_{\mathrm{off-diag}} = [J,\ (\phi_{n+1})_{\mathrm{off-diag}}] + (A^{(1)}\phi_n)_{\mathrm{off-diag}}$$

for the diagonal part $(\phi_n)_{\mathrm{diag}}$ and the off-diagonal parts $(\phi_n)_{\mathrm{off-diag}}$ of ϕ_n. The first equation determines the diagonal part of ϕ_n up to integration constants. The second equation is rather an algebraic equation that determines the off-diagonal part of ϕ_{n+1} from the lower coefficients $\phi_1, \cdots, \phi_n$. One can thus construct a solution of these equations.

Note that, unlike the aforementioned construction of U_n's, the construction of $\phi(\lambda)$ is not purely algebraic (the outcome is accordingly "nonlocal") and leaves large arbitrariness. It is, however, $\phi(\lambda)$ rather than $U(\lambda)$ that plays a more fundamental role in the passage to the Grassmannian perspective.

2.2. Formulation of hierarchy

Let $t = (t_1, t_2, \ldots)$ be a sequence of "time" variables; the first one t_1 is to be identified with the spatial variable x. The nth time evolution is generated by the matrix

$$A_n(\lambda) = \sum_{m=0}^{n} U_m \lambda^{n-m} = \big(U(\lambda)\lambda^n\big)_+. \tag{11}$$

Here $(\cdot)_+$ denotes the polynomial part of a Laurent series of λ. Having introduced these matrices, one can define the nonlinear Schrödinger hierarchy as the system of the Lax equations

$$[\partial_{t_n} - A_n(\lambda),\ U(\lambda)] = 0 \tag{12}$$

for $n = 1, 2, \ldots$. Since $A_1(\lambda) = A(\lambda)$, one can identify t_1 with x. In many aspects, this formulation of the nonlinear Schrödinger hierarchy resembles the formulation of the KP hierarchy [20]. For instance, as known in the case of the KP hierarchy, the system of Lax equations is equivalent to the system of zero-curvature equations

$$[\partial_{t_m} - A_m(\lambda),\ \partial_{t_n} - A_n(\lambda)] = 0 \tag{13}$$

for $m, n = 1, 2, \ldots$, namely, one can derive one from the other. The zero-curvature equation for $m = 1$ and $n = 2$ gives the equations

$$\partial_t u - \frac{1}{2}\partial_x^2 u + u^2 v = 0, \quad \partial_t v + \frac{1}{2}\partial_x^2 v - uv^2 = 0,$$

which turn into the usual nonlinear Schrödinger equation by rescaling the variables as $u \to e^{at}u$, $v \to e^{-at}v$, $t \to it$ $(a = \pm 1)$ and imposing the reality condition $v = \overline{u}$.

Yet another formulation of the hierarchy is achieved by the system of differential equations

$$\partial_{t_n}\phi(\lambda) = A_n(\lambda)\phi(\lambda) - \phi(\lambda)J\lambda^n. \tag{14}$$

If one introduces the co called "formal Baker-Akhiezer function"

$$\psi(\lambda) = \phi(\lambda)\exp\Big(\sum_{n=1}^{\infty} t_n J\lambda^n\Big), \tag{15}$$

the foregoing equations turn into the auxiliary linear equations

$$\partial_{t_n}\psi(\lambda) = A_n(\lambda)\psi(\lambda). \tag{16}$$

The Frobenius integrability condition of these equations yields the aforementioned zero-curvature equations. On the other hand, if one rewrites the definition of $A_n(\lambda)$ as

$$A_n(\lambda) = \Big(\phi(\lambda)J\lambda^n\phi(\lambda)^{-1}\Big)_+ \tag{17}$$

and insert it into the differential equations for $\phi(\lambda)$, the outcome is a system of nonlinear evolution equations for $\phi(\lambda)$ of the form

$$\partial_{t_n}\phi(\lambda) = -\left(\phi(\lambda)J\lambda^n\phi(\lambda)^{-1}\right)_-\phi(\lambda), \tag{18}$$

where $(\cdot)_-$ denotes the negative power part of a Laurent series of λ. These equations may be thought of as the most fundamental because the Lax and zero-curvature equations can be derived from these equations.

3. Nonlinear Schrödinger hierarchy as dynamical system on Lie group of Laurent series

The nonlinear Schrödinger hierarchy can be interpreted as a dynamical system on an infinite-dimensional Lie group. It is customary to formulate such a statement in terms of a loop group, namely, the set of a suitable class of (smooth, real-analytic or square-integrable) mappings from $S^1 = \{\lambda \in \mathbf{C} \mid |\lambda^{-1}| = a\}$ to $\mathrm{SL}(2,\mathbf{C})$. From an aesthetic point of view, however, fixing a circle is not beautiful; the circle is a kind of artifact that did not exist in the formulation of the hierarchy itself. A better approach is to use a Lie group of Laurent series that converge in a neighborhood of $\lambda = \infty$ except at the point $\lambda = \infty$.

3.1. Lie algebras and groups of Laurent series

Let $\mathfrak{g}$ denote the Lie algebra of Laurent series of the form

$$X(\lambda) = \sum_{n=-\infty}^{\infty} X_n\lambda^n, \quad X_n \in \mathrm{sl}(2,\mathbf{C}), \tag{19}$$

that converge in a neighborhood of $\lambda = \infty$ except at $\lambda = \infty$. In other words, the coefficients are assumed to satisfy the conditions

$$\lim_{n\to\infty} |X_n|^{1/n} = 0, \quad \limsup_{n\to\infty} |X_{-n}|^{1/n} < \infty.$$

This Lie algebra has the direct sum decomposition

$$\mathfrak{g} = \mathfrak{g}_+ \oplus \mathfrak{g}_-, \tag{20}$$

where $\mathfrak{g}_\pm$ are subalgebras of the form

$$\mathfrak{g}_+ = \{X(\lambda) \in \mathfrak{g} \mid X_n = 0 \text{ for } n < 0\},$$
$$\mathfrak{g}_- = \{X(\lambda) \in \mathfrak{g} \mid X_n = 0 \text{ for } n \geq 0\}.$$

The direct sum decomposition induces a factorization of the associated Lie group $G = \exp\mathfrak{g}$ to the subgroups $G_\pm = \exp\mathfrak{g}_\pm$, namely, any element $g(\lambda)$ of G near the unit matrix I can be uniquely factorized as

$$g(\lambda) = g_+(\lambda)^{-1}g_-(\lambda), \quad g_\pm(\lambda) \in G_\pm. \tag{21}$$

Analytically, this is nothing but the so called Riemann-Hilbert problem. In geometric terms, $g(\lambda)$ is the transition function of a holomorphic $\mathrm{SL}(2,\mathbf{C})$ bundle P over

$\mathbf{P}^1$ obtained by gluing trivial bundles over two disks D_+ and D_-, $D_+ \cup D_- = \mathbf{P}^1$, as

$$P = D_+ \times \mathrm{SL}(2,\mathbf{C}) \sqcup D_- \times \mathrm{SL}(2,\mathbf{C}) \;/\; \sim, \tag{22}$$

where $(\lambda, g_+) \in D_+ \times \mathrm{SL}(2,\mathbf{C})$ and $(\lambda, g_-) \in D_- \times \mathrm{SL}(2,\mathbf{C})$ are identified if $g_+ = g(\lambda) g_-$. Factorizability of $g(\lambda)$ amounts to holomorphic triviality of P and of associated vector bundles.

3.2. Factorization method

Let $\phi(\lambda)$ denote an arbitrary element of G_- and consider the factorization problem

$$\phi(\lambda) \exp\Bigl(-\sum_{n=1}^{\infty} t_n J \lambda^n\Bigr) = \chi(t,\lambda)^{-1} \phi(t,\lambda), \tag{23}$$
$$\chi(t,\lambda) \in G_+, \quad \phi(t,\lambda) \in G_-.$$

Lemma 1. *If t is sufficiently small, the factorization problem* (23) *has a unique solution.*

Proof. Since $\phi(t,\lambda)$ is expected to be a small deformation of $\phi(\lambda)$, one can assume it in the form

$$\phi(t,\lambda) = \tilde{\chi}(t,\lambda)\phi(\lambda), \quad \tilde{\chi}(t,\lambda) \in G_-,$$

and convert the problem to the form

$$\phi(\lambda) \exp\Bigl(-\sum_{n=1}^{\infty} t_n J \lambda^n\Bigr) \phi(\lambda)^{-1} = \chi(t,\lambda)^{-1} \tilde{\chi}(t,\lambda).$$

If t is sufficiently small, the left-hand side is close to the unit matrix I, so that one can resort to the local factorizability of G. □

Suppose that the factorization problem (23) does have a unique solution. The second factor $\phi(t,\lambda)$ then turns out to give a solution of (18):

Theorem 1. *The second factor $\phi(t,\lambda)$ of the factorization problem satisfies the evolution equations* (18) *and the initial condition $\phi(0,\lambda) = \phi(\lambda)$.*

Proof. If one rewrites the factorization relation as

$$\chi(t,\lambda)\phi(\lambda) = \phi(t,\lambda) \exp\Bigl(\sum_{n=1}^{\infty} t_n J \lambda^n\Bigr)$$

and differentiate both hand sides by t_n, the outcome reads

$$\begin{aligned} &\partial_{t_n} \chi(t,\lambda) \cdot \phi(\lambda) \\ &= \partial_{t_n} \phi(t,\lambda) \cdot \exp\Bigl(\sum_{n=1}^{\infty} t_n J \lambda^n\Bigr) + \phi(t,\lambda) J \lambda^n \exp\Bigl(\sum_{n=1}^{\infty} t_n J \lambda^n\Bigr). \end{aligned}$$

One can use the previous relation once again to eliminate $\phi(\lambda)$ and the exponential from this relation. This yields the relation

$$\partial_{t_n} \chi(t,\lambda) \cdot \chi(t,\lambda)^{-1} = \partial_{t_n} \phi(t,\lambda) \cdot \phi(t,\lambda)^{-1} + \phi(t,\lambda) J \lambda^n \phi(t,\lambda)^{-1}.$$

Let $A_n(t,\lambda)$ denote the 2×2 matrix defined by both hand sides of the last equation. This leads to the two expressions

$$A_n(t,\lambda) = \partial_{t_n}\chi(t,\lambda)\cdot\chi(t,\lambda)^{-1}$$

and

$$A_n(t,\lambda) = \partial_{t_n}\phi(t,\lambda)\cdot\phi(t,\lambda)^{-1} + \phi(t,\lambda)J\lambda^n\phi(t,\lambda)^{-1}$$

of $A_n(t,\lambda)$. The first expression shows that $A_n(t,\lambda)$ takes values in $\mathfrak{g}_+$, so that one can replace the right-hand side of the second expression by its projection onto $\mathfrak{g}_+$. Since the first term $\partial_{t_n}\phi(t,\lambda)\cdot\phi(t,\lambda)^{-1}$ obviously disappears upon projection, one finds that

$$A_n(\lambda) = \Big(\phi(t,\lambda)J\lambda^n\phi(t,\lambda)^{-1}\Big)_+.$$

These results show that $\phi(t,\lambda)$ does satisfy (18) as expected. Uniqueness of the factorization implies that $\phi(0,\lambda) = \phi(\lambda)$. □

This result can be restated in geometric terms as follows. Evolution equations (18) define a dynamical system on G_-. Factorizability of G implies that

$$\begin{array}{ccc} G_- & \to & G_+\backslash G \\ \phi(\lambda) & \mapsto & G_+\phi(\lambda) \end{array} \tag{24}$$

is an injective mapping with open (and dense) image. The dynamical system on G_- is nothing but the pullback, by (24), of the exponential flows

$$G_-g(\lambda) \mapsto G_-g(\lambda)\exp\Big(-\sum_{n=1}^{\infty} t_n J\lambda^n\Big) \tag{25}$$

on the coset $G_+\backslash G$. Moreover, the coset $G_+\backslash G$ may be interpreted as the moduli space of holomorphic $\mathrm{SL}(2,\mathbf{C})$ bundles over $\mathbf{P}^1$ equipped with trivialization over D_-, elements of the form $G_+\phi(\lambda)$ being a representative of trivial bundles.

4. Nonlinear Schrödinger hierarchy as dynamical system on infinite-dimensional Grassmann variety

The forgoing dynamical system on G_- can be mapped to a dynamical system in an infinite-dimensional Grassmann variety. In the literature, two different models of Grassmann varieties (or Grassmann manifolds) have been used for this kind of description. One is Sato's algebraic or complex analytic model based on a vector space of Laurent series [20]; the other is Segal and Wilson's functional analytic model made from the Hilbert space of square-integrable functions on a circle [21]. One should obviously choose Sato's model in the present setting.

4.1. Formulation of Grassmann variety

The Grassmann variety Gr to be used below is constructed from the vector space V of 2×2 matrices $X(\lambda)$ of Laurent series of the form

$$X(\lambda) = \sum_{n=-\infty}^{\infty} X_n \lambda^n, \quad X_n \in \mathrm{gl}(2, \mathbf{C}), \tag{26}$$

that converge in a neighborhood of $\lambda = \infty$ except at the point $\lambda = \infty$. This is almost the same thing as $\mathfrak{g}$ but the coefficients X_n are now an arbitrary matrix; recall that $\mathrm{gl}(2, \mathbf{C})$ denotes the vector space (or matrix Lie algebra) of arbitrary 2×2 complex matrices. This vector space is a matrix version of the vector space $V^{\mathrm{ana}(\infty)}$ in Sato's list of models [20]. As noted therein, this vector space has a natural linear topology. The Grassmann variety Gr consists of closed vector subspaces $W \subset V$ satisfying an additional condition as follows:

$$\begin{aligned} \mathrm{Gr} = \{W \subset V \mid \\ \dim \mathrm{Ker}(W \to V/V_-) = \dim \mathrm{Coker}(W \to V/V_-) < \infty\}. \end{aligned} \tag{27}$$

Here V_- denotes the vector subspace of V consisting of $X(\lambda)$'s that contain only negative powers of λ:

$$V_- = \{X(\lambda) \in V \mid X_n = 0 \text{ for } n \geq 0\}. \tag{28}$$

The map $W \to V/V_-$ is the composition of the inclusion $W \hookrightarrow V$ and the canonical projection $V \to V/V_-$. The so called "big cell" of Gr consists of subspaces $W \subset V$ for which this linear map is an isomorphism:

$$\mathrm{Gr}^\circ = \{W \in \mathrm{Gr} \mid W \simeq V/V_-\}. \tag{29}$$

This is an open subset of Gr, namely, sufficiently small deformations of any element of Gr° remains in Gr°.

4.2. Vacuum and dressing

Let W_0 be the subspace of V spanned by nonnegative powers of λ:

$$W_0 = \{X(\lambda) \in V \mid X_n = 0 \text{ for } n < 0\}. \tag{30}$$

The linear map $W_0 \to V/V_-$ is obviously isomorphic in view of the basis $\{E_{ij}\lambda^n \mid n \geq 0,\ i, j = 1, 2\}$ for both vector spaces (E_{ij} are the standard basis of $\mathrm{gl}(2, \mathbf{C})$). Hence W_0 is an element of the big cell Gr°. This special element of the big cell plays the role of "vacuum," which corresponds to the vacuum solution $u = v = 0$ of the nonlinear Schrödinger equation.

One can "dress" W_0 by an arbitrary element of G_-:

$$W = W_0 \phi(\lambda), \quad \phi(\lambda) = I + \sum_{n=1}^{\infty} \phi_n \lambda^{-n} \in G_-. \tag{31}$$

Lemma 2. *W is an element of the big cell* Gr°.

Proof. W is spanned by $E_{ij}\lambda^n\phi(\lambda)$, $n \geq 0$, $i,j = 1,2$. By a triangular linear transformation, one can modify this basis of W to another basis $\{w_{n,ij}(\lambda) \mid n \geq 0,\ i,j = 1,2\}$ such that

$$w_{n,ij}(\lambda) = E_{ij}\lambda^n + O(\lambda^{-1}).$$

More explicitly,

$$w_{n,ij}(\lambda) = \Big(\phi(\lambda)E_{ij}\lambda^n\phi(\lambda)^{-1}\Big)_+\phi(\lambda).$$

The linear map $W \to V/V_-$ sends this basis to the standard basis $\{E_{ij}\lambda^n \mid n \geq 0,\ i,j = 1,2\}$ of V/V_-, thereby turns out to be an isomorphism. □

The phase space G_- of the dynamical system of the last section can be thus mapped, by the correspondence

$$\phi(\lambda) \mapsto W = W_0\phi(\lambda), \tag{32}$$

to the set

$$\mathcal{M} = \{W \in \mathrm{Gr}^\circ \mid W = W_0\phi(\lambda),\ \phi(\lambda) \in G_-\} \tag{33}$$

of these "dressed vacua" in (the big cell of) the infinite-dimensional Grassmann variety Gr. The problem to be addressed next is to describe the dynamical motion on this new phase space.

Actually, the foregoing mapping $G_- \xrightarrow{\sim} \mathcal{M}$ can be understood in a slightly more general form. Namely, the mapping can be extended to

$$\begin{array}{ccc} G_+\backslash G & \to & \mathrm{Gr} \\ G_+g(\lambda) & \mapsto & W_0g(\lambda) \end{array} \tag{34}$$

that sends the coset $G_+\backslash G$ into Gr. Note that this mapping is well defined and injective because

$$g(\lambda) \in G_+ \iff W_0g(\lambda) = W_0$$

(cf. Lemma 3). Thus, combined with the open embedding (24) of G_- into $G_+\backslash G$, the mapping $G_- \xrightarrow{\sim} \mathcal{M}$ is substantially the well known embedding of the "affine Grassmannian" $G_+\backslash G$ into the Sato Grassmannian [21].

4.3. Dynamical system on space of dressed vacua

For simplicity, the following consideration is limited to small values of t. The factorization problem (23) is thereby ensured to have a unique solution. The goal is to elucidate the motion of $W(t) = W_0\phi(t,\lambda) \in \mathcal{M}$. A clue to the answer is the following.

Lemma 3. $W_0\chi(t,\lambda) = W_0$.

Proof. W_0 is obviously closed under multiplication of two element. By construction, $\chi(t,\lambda)$ is obviously an element of W_0. Therefore $W_0\chi(t,\lambda) \subseteq W_0$. On the other hand, the inverse $\chi(t,\lambda)^{-1}$ is also an element of G_+ as far as t is sufficiently small, so that the same reasoning leads to the conclusion that $W_0\chi(t,\lambda) \subseteq W_0$. Thus the equality follows. □

If one rewrites the factorization relation (23) as

$$\phi(t,\lambda) = \chi(t,\lambda)\phi(0,\lambda)\exp\Big(-\sum_{n=1}^{\infty} t_n J\lambda^n\Big)$$

and insert it into the definition $W(t) = W_0\phi(t,\lambda)$ of $W(t)$, one finds that

$$\begin{aligned} W(t) &= W_0\chi(t,\lambda)\phi(0,\lambda)\exp\Big(-\sum_{n=1}^{\infty} t_n J\lambda^n\Big) \\ &= W_0\phi(0,\lambda)\exp\Big(-\sum_{n=1}^{\infty} t_n J\lambda^n\Big) \\ &= W(0)\exp\Big(-\sum_{n=1}^{\infty} t_n J\lambda^n\Big). \end{aligned}$$

Note that the lemma has been used in the first stage; the first factor $\chi(t,z)$ of the factorization pair is absorbed by W_0. Thus the motion of the point $W(t)$ of $\mathcal{M}$ turns out to obey the simple exponential law

$$W(t) = W(0)\exp\Big(-\sum_{n=1}^{\infty} t_n J\lambda^n\Big). \tag{35}$$

One thus arrives at the following fundamental picture, which is an example of the Grassmannian perspective of soliton equations due to Sato [20] and Segal and Wilson [21].

Theorem 2. *The nonlinear Schrödinger hierarchy can be mapped, by the correspondence $W(t) = W_0\phi(t,\lambda)$, to a dynamical system on the set $\mathcal{M}$ of dressed vacua in the Grassmann variety* Gr. *The motion of $W(t)$ obeys the exponential law* (35).

Conversely, given an arbitrary element $\phi(\lambda)$ of G_-, one can derive a solution of the factorization problem (23) from this dynamical system. By Lemma 2, $W(0) = W_0\phi(\lambda)$ is an element of the big cell. If t is sufficiently small, the point $W(t)$ on the trajectory of the exponential flows (35) still remains in the big cell, because the big cell is an open subset of Gr. This means that the linear map $W(t) \to V/V_-$, i.e., the composition of the inclusion $W(t) \hookrightarrow V$ and the canonical projection $V \to V/V_-$, is an isomorphism. Let $\phi(t,\lambda) \in W(t)$ be the inverse image of $I \in V/V_-$ by this isomorphism. Being equal to I modulo V_-, $\phi(t,\lambda)$ is a Laurent series of the form

$$\phi(t,\lambda) = I + \sum_{n=1}^{\infty} \phi_n(t)\lambda^{-n}, \quad \phi_n(t) \in \mathrm{gl}(2,\mathbf{C}).$$

On the other hand, as an element of

$$W(t) = W_0\phi(\lambda)\exp\Big(-\sum_{n=1}^{\infty} t_n J\lambda^n\Big),$$

$\phi(t, \lambda)$ can also be expressed as

$$\phi(t, \lambda) = \chi(t, \lambda)\phi(\lambda) \exp\Big(-\sum_{n=1}^{\infty} t_n J \lambda^n\Big)$$

with an element $\chi(t, \lambda)$ of W_0. Taking the determinant of both hand sides of the last equality yields the equality

$$\det \phi(t, \lambda) = \det \chi(t, \lambda),$$

in which $\phi(\lambda)$ and the exponential disappear because they are known to be unimodular. Notice here that

$$\begin{aligned} \det \phi(t, \lambda) &= 1 + (\text{negative powers of } \lambda), \\ \det \chi(t, \lambda) &= (\text{nonnegative powers of } \lambda). \end{aligned}$$

Consequently, both hand sides of the determinant equality is actually equal to 1. This implies that $\phi(t, \lambda) \in G_-$ and $\chi(t, \lambda) \in G_+$, so that they give a solution of the factorization problem (23).

This shows another aspect of the Grassmannian perspective. Namely, the Grassmann variety can be used as a tool for solving a factorization or Riemann-Hilbert problem. This point of view turns out to be useful later.

5. Elliptic analogue of nonlinear Schrödinger hierarchy

We now turn to examples with an elliptic spectral parameter. The first example is based on an example of Krichever's general construction [12]. Let us briefly recall the background of Krichever's work.

It is well known, after the work of Zakharov and Mikhailov [27], that a naive attempt at the construction of a zero-curvature equation

$$[\partial_x - A(P),\ \partial_t - B(P)] = 0, \quad P \in \Gamma,$$

on an arbitrary algebraic curve Γ is confronted with a serious difficulty that stems from the Riemann-Roch theorem. If the construction for $\Gamma = \mathbf{P}^1$ also works in the general case, $A(P)$ and $B(P)$ are matrices of meromorphic functions on Γ with fixed poles, say, $Q_1, \ldots, Q_s$ of order $m_1, \ldots, m_s$ for $A(P)$ and $n_1, \ldots, n_s$ for $B(P)$. Choosing a suitable linearly independent set of meromorphic functions $f_j(P)$, $h = 1, \ldots, M$, and $g_k(P)$, $k = 1, \ldots, N$, one can expand $A(P)$ and $B(P)$ as

$$A(P) = \sum_{j=1}^{M} A_j f_j(P), \quad B(P) = \sum_{k=1}^{N} B_k g_k(P).$$

The (matrix-valued) coefficients A_j, B_k are interpreted as the field variables $A_j = A_j(x, t)$, $B_k = B_k(x, t)$, for which the zero-curvature equation induces a set of PDE's. Part of these field variables can be eliminated by gauge transformations $A_j \to g^{-1} A_j g - g_x g^{-1}$, $B_k \to g^{-1} B_k g - g_t g^{-1}$. In the case where $\Gamma = \mathbf{P}^1$, suitable gauge fixing leads to a determined system of PDE's (i.e., a system of evolution equations) for the reduced field variables. In contrast, if the genus of Γ is not zero, the Riemann-Roch theorem implies that the zero-curvature equation in a "general position" is an overdetermined system for A_j's and B_k's. This means that one has to assume some special structure in $A(P)$ and $B(P)$ to obtain a consistent system of evolution equations. An example is the Landau-Lifshitz equation (for which Γ is an elliptic curve).

Krichever [12] pointed out that this difficulty can be avoided by allowing $A(P), B(P)$ to have extra "movable" poles γ_s at which the solutions of the auxiliary linear system $\partial_x\psi(P) = A(P)\psi(P)$, $\partial_t\psi(P) = B(P)\psi(P)$ remain regular. This is reminiscent of the notion of "apparent singularities" in the theory of ordinary differential equations. The number of necessary movable poles turns out to be equal to rg, where r is the size of the matrices $A(P), B(P)$ and g the genus of Γ. Moreover, to each movable pole is assigned a directional vector $\boldsymbol{\alpha}_s \in \mathbf{P}^{r-1}$ as an extra parameter. These pairs $(\gamma_s, \boldsymbol{\alpha}_s)$, $s = 1, \ldots, rg$, are called "Tyurin parameters" and now join the game as new dynamical variables. The elliptic analogue of the nonlinear Schrödinger hierarchy amounts to the case where $g = 1$ and $r = 2$.

5.1. Matrix of elliptic functions parametrized by Tyurin parameters

The first stage of construction is to choose a suitable counterpart $A(z)$ of $A(\lambda)$. This is a 2×2 matrix of elliptic functions on a nonsingular elliptic curve. The spectral parameter z is now understood to be the standard complex coordinate on the torus $\Gamma = \mathbf{C}/(2\omega_1\mathbf{Z} + 2\omega_3\mathbf{Z})$ that realizes the elliptic curve.

In addition to a pole at $z = 0$ (which corresponds to $\lambda = \infty$ in the case of the nonlinear Schrödinger hierarchy), this matrix has two extra poles γ_1, γ_2, $\gamma_1 \neq \gamma_2$, that depend on x and t_n's. Two directional vectors $\boldsymbol{\alpha}_1, \boldsymbol{\alpha}_2 \in \mathbf{P}^1$ are introduced as the other half of the Tyurin parameters. These directional vectors can be normalized as $\boldsymbol{\alpha}_s = {}^{\mathrm{t}}(\alpha_s, 1)$. The two fields u, v in the usual nonlinear Schrödinger hierarchy also appear here. Thus one has altogether six dynamical variables $u, v, \gamma_1, \gamma_2, \alpha_1, \alpha_2$ in the formulation of this elliptic analogue.

The matrix $A(z)$ is defined, indirectly, by the following properties:

1. $A(z)$ has poles at $z = 0, \gamma_1, \gamma_2$ and is holomorphic at other points.
2. As $z \to 0$,
$$A(z) = \begin{pmatrix} z^{-1} & u \\ v & -z^{-1} \end{pmatrix} + O(z).$$
3. As $z \to \gamma_s$, $s = 1, 2$,
$$A(z) = \frac{\boldsymbol{\beta}_s {}^{\mathrm{t}}\boldsymbol{\alpha}_s}{z - \gamma_s} + O(1),$$
where $\boldsymbol{\alpha}_s$ and $\boldsymbol{\beta}_s$ are two-dimensional column vectors that do not depend on z. $\boldsymbol{\alpha}_s$ is normalized as $\boldsymbol{\alpha}_s = {}^{\mathrm{t}}(\alpha_s, 1)$.

Lemma 4. *If $\alpha_1 \neq \alpha_2$, a matrix $A(z)$ of meromorphic functions on Γ with these properties does exists. It is unique and can be written explicitly in terms of the*

Weierstrass zeta function $\zeta(z)$ *as*

$$A(z) = \sum_{s=1,2} \boldsymbol{\beta}_s {}^{\mathrm{t}}\boldsymbol{\alpha}_s(\zeta(z-\gamma_s)+\zeta(\gamma_s)) + \begin{pmatrix} \zeta(z) & u \\ v & -\zeta(z) \end{pmatrix}, \tag{36}$$

where

$$\boldsymbol{\beta}_1 = \frac{1}{\alpha_1-\alpha_2}\begin{pmatrix} -1 \\ -\alpha_2 \end{pmatrix}, \quad \boldsymbol{\beta}_2 = \frac{1}{\alpha_1-\alpha_2}\begin{pmatrix} 1 \\ \alpha_1 \end{pmatrix}.$$

Proof. One can express $A(z)$ as

$$A(z) = \sum_{s=1,2} \boldsymbol{\beta}_s {}^{\mathrm{t}}\boldsymbol{\alpha}_s\zeta(z-\gamma_s) + J\zeta(z) + C,$$

where C is a constant matrix. By the residue theorem, the coefficients of $\zeta(z-\gamma_1)$, $\zeta(z-\gamma_s)$ and $\zeta(z)$ have to satisfy the linear equation

$$\sum_{s=1,2} \boldsymbol{\beta}_s {}^{\mathrm{t}}\boldsymbol{\alpha}_s + J = 0$$

that ensures that $A(z)$ is single valued on Γ. Solving this equation for $\boldsymbol{\beta}_s$ leads to the formula stated in the lemma. On the other hand, matching with the Laurent expansion of $A(z)$ at $z=0$ leads to the relation

$$A^{(1)} = \sum_{s=1,2} \boldsymbol{\beta}_s {}^{\mathrm{t}}\boldsymbol{\alpha}_s\zeta(-\gamma_s) + C,$$

which determines C as shown in the formula above. □

The final task is to fulfill the requirement on the auxiliary linear system. By Krichever's lemma [12, Lemma 5.2], the auxiliary linear system $\partial_x\psi(z) = A(z)\psi(z)$ has a 2×2 matrix solution that is holomorphic at $z=\gamma_s$ and invertible except at these points if and only if γ_s and α_s satisfy the equations

$$\partial_x\gamma_s + \operatorname{Tr}\boldsymbol{\beta}_s {}^{\mathrm{t}}\boldsymbol{\alpha}_s = 0, \tag{37}$$

$$\partial_x {}^{\mathrm{t}}\boldsymbol{\alpha}_s + {}^{\mathrm{t}}\boldsymbol{\alpha}_s A^{(s,1)} = \kappa_s {}^{\mathrm{t}}\boldsymbol{\alpha}_s, \tag{38}$$

where $A^{(s,1)}$ denotes the constant term of the Laurent expansion of $A(z)$ at $z=\gamma_s$,

$$A^{(s,1)} = \lim_{z\to\gamma_s}\left(A(z) - \frac{\boldsymbol{\beta}_s {}^{\mathrm{t}}\boldsymbol{\alpha}_s}{z-\gamma_s}\right),$$

and κ_s is a constant to be determined by the equation itself.

5.2. Generating functions

The second stage is to introduce two generating functions

$$\phi(z) = I + \sum_{n=1}^{\infty}\phi_n z^n, \quad U(z) = J + \sum_{n=1}^{\infty}U_n z^n$$

as counterparts of $U(\lambda)$ and $\phi(\lambda)$ in the case of the usual nonlinear Schrödinger hierarchy. Recall that the point $\lambda=\infty$ of $\mathbf{P}^1$ corresponds to the origin $z=0$ of the torus Γ.

The first generating function $\phi(z)$ is a Laurent series that satisfies the differential equation

$$\partial_x \phi(z) = A(z)\phi(z) - \phi(z)Jz^{-1}, \tag{39}$$

where $A(z)$ is understood to be its Laurent expansion

$$A(z) = Jz^{-1} + \sum_{n=1}^{\infty} A^{(n)} z^{n-1} \tag{40}$$

at $z = 0$. One can construct a solution of this differential equation by essentially the same (but slightly more complicated) procedure as mentioned in the case of the usual nonlinear Schrödinger hierarchy.

The second generating function $U(z)$ can be obtained from $\phi(z)$ as

$$U(z) = \phi(z)J\phi(z)^{-1}, \tag{41}$$

which satisfies the equations

$$[\partial_x - A(z),\, U(z)] = 0, \quad U(z)^2 = I. \tag{42}$$

The Laurent coefficients are again determined by a set of recurrence relations:

$$2JU_{n+1} = \partial_x U_n - \sum_{m=1}^{n+1} [A^{(m)}, U_{n+1-m}] - \sum_{m=1}^{n} U_m U_{n+1-m}. \tag{43}$$

5.3. Construction of hierarchy

The third stage is to construct a set of generators $A_n(z)$, $n = 1, 2, \ldots$, of time evolutions. Just like $A(z)$, they are 2×2 matrices of elliptic functions and characterized by the following properties.

1. $A_n(z)$ has poles at $z = 0, \gamma_1, \gamma_2$ and is holomorphic at other points.
2. As $z \to 0$,

$$A_n(z) = U(z)z^{-n} + O(z).$$

3. As $z \to \gamma_s$, $s = 1, 2$,

$$A_n(z) = \frac{\boldsymbol{\beta}_{n,s}\, {}^{\mathrm{t}}\boldsymbol{\alpha}_s}{z - \gamma_s} + O(1),$$

where $\boldsymbol{\beta}_{n,s}$ is a two-dimensional column vector that does not depend on z.

Lemma 5. *If $\alpha_1 \neq \alpha_2$, a matrix $A_n(z)$ of meromorphic functions on Γ with these properties does exist. It is unique and can be written explicitly as*

$$\begin{aligned} A_n(z) \;&=\; \sum_{s=1,2} \boldsymbol{\beta}_{n,s}\, {}^{\mathrm{t}}\boldsymbol{\alpha}_s(\zeta(z - \gamma_s) + \zeta(\gamma_s)) \\ &\quad + \sum_{m=0}^{n-1} \frac{(-1)^m}{m!} \partial_z^m \zeta(z) U_{n-1-m} + U_n. \end{aligned} \tag{44}$$

The vectors $\boldsymbol{\beta}_{n,s}$ *are determined by the linear equation*

$$\sum_{s=1,2} \boldsymbol{\beta}_{n,s}\,{}^{\mathrm{t}}\boldsymbol{\alpha}_s + U_{n-1} = 0 \tag{45}$$

that ensures the single-valuedness of $A_n(z)$ *on* Γ.

In the following, the genericity condition

$$\alpha_1 \neq \alpha_2 \tag{46}$$

is always assumed; the matrices $A(z)$ and $A_n(z)$ are thereby determined. The elliptic analogue of the nonlinear Schrödinger hierarchy is defined by the Lax equations

$$[\partial_{t_n} - A_n(z),\ U(z)] = 0 \tag{47}$$

for the generating function $U(z)$ and the differential equations

$$\partial_{t_n}\gamma_s + \operatorname{Tr}\boldsymbol{\beta}_{n,s}\,{}^{\mathrm{t}}\boldsymbol{\alpha}_s = 0, \tag{48}$$

$$\partial_{t_n}\,{}^{\mathrm{t}}\boldsymbol{\alpha}_n + {}^{\mathrm{t}}\boldsymbol{\alpha}_s A_n^{(s,1)} = \kappa_{n,s}\boldsymbol{\alpha}_s \tag{49}$$

for the Tyurin parameters. Here $A_n^{(s,1)}$ denotes the constant term of the Laurent expansion of $A_n(z)$ at $z = \gamma_s$, i.e.,

$$A_n^{(s,1)} = \lim_{z\to\gamma_s}\left(A_n(z) - \frac{\boldsymbol{\beta}_{n,s}\,{}^{\mathrm{t}}\boldsymbol{\alpha}_s}{z-\gamma_s}\right),$$

and $\kappa_{n,s}$ is a constant determined by the differential equation itself. (48) and (49) are the necessary and sufficient conditions for the auxiliary linear system $\partial_{t_n}\psi(z) = A_n(z)\psi(z)$ to have a 2×2 matrix solution that is holomorphic at $z = \gamma_s$ and invertible except at these points.

One can confirm that the zero-curvature equations

$$[\partial_{t_m} - A_m(z),\ \partial_{t_n} - A_n(z)] = 0 \tag{50}$$

are satisfied by any solution of the three equations (47), (48) and (49). This implies, in particular, the commutativity of flows generated by $A_n(z)$. Actually, the following stronger statement holds as in the case of the usual nonlinear Schrödinger hierarchy.

Theorem 3. *The system of Lax equations* (47) *and the system of zero-curvature equations* (50) *are equivalent under the equations* (48) *and* (49) *for the Tyurin parameters.*

As regards the status of (48) and (49), one can derive them from the zero-curvature equations

$$[\partial_{t_n} - A_n(z),\ \partial_x - A(z)] = 0 \tag{51}$$

assuming that (37) and (38) are satisfied. In this respect, (37) and (38) should be understood as part of the definition of $A(z)$. Krichever's construction of a hierarchy is rather based on these zero-curvature equations [12].

6. Elliptic analogue of nonlinear Schrödinger hierarchy in Grassmannian perspective

A technical clue to the Grassmannian perspective of the elliptic analogue of the nonlinear Schrödinger hierarchy is again a factorization or Riemann-Hilbert problem. The situation is, however, far more complicated. First of all, the present case is concerned with the torus rather than the sphere. Moreover, whereas the usual Riemann-Hilbert problem on the sphere is based on the triviality of a holomorphic bundle (cf. Section 3.1), the present case is, by construction, related to a *nontrivial* holomorphic bundle in the Tyurin parameterization. As it turns out, what is relevant to the present setting is a Riemann-Hilbert problem with *degeneration points*; Tyurin parameters are nothing but the geometric data of those points. This kind of Riemann-Hilbert problems also appear in the work of Krichever and Novikov [9, 10, 11] on commutative rings of differential operators.

Another clue can be found in the paper of Previato and Wilson [18]. They demonstrate therein a "dressing method" based on an infinite-dimensional Grassmann variety to solve a Riemann-Hilbert problem of the Krichever-Novikov type. Moreover, their paper shows what should be the "vacuum" that corresponds to a holomorphic vector bundle in the Tyurin parametrization.

These ideas lead to a Grassmannian perspective of the elliptic analogue [24].

6.1. Riemann-Hilbert problem with degeneration points

In the following, t denotes the full set of time variables $(t_1, t_2, \ldots)$, in which x is identified with t_1. Moreover, any quantity that depends on t is written with its t-dependence indicated explicitly as $A(t,z)$, $A_n(t,z)$, $\gamma_s(t)$, $\alpha_s(t)$, etc.

The Lax equations (47) and the zero-curvature equations (50) are associated with the auxiliary linear system

$$\partial_{t_n}\psi(t,z) = A_n(t,z)\psi(t,z).$$

The Riemann-Hilbert problem is concerned with two distinct solutions of this linear system.

One solution is the Laurent series solution of the form

$$\psi(t,z) = \phi(t,z)\exp\Bigl(\sum_{n=1}^{\infty} t_n J z^{-n}\Bigr), \tag{52}$$

where the prefactor $\phi(t,z)$ is a Laurent series of the form

$$\phi(t,z) = I + \sum_{n=1}^{\infty} \phi_n(t) z^n.$$

This prefactor is nothing but the generating function introduced previously, but it is now required to satisfy the differential equations

$$\partial_{t_n}\phi(t,z) = A_n(t,z)\phi(t,z) - \phi(t,z)Jz^{-n} \tag{53}$$

for $n = 1, 2, \ldots$ as well. Note that this Laurent series solution, by its nature, carries no information on the global structure of $A_n(t,z)$'s.

Another solution $\chi(t,z)$ is characterized by the initial condition

$$\chi(0,z) = I. \tag{54}$$

This solution $\chi(t,z)$ turns out to carry global information. To avoid delicate problems, suppose that the solutions of the hierarchy under consideration are (real or complex) analytic in a neighborhood of the initial point $t = 0$. One can then expand it to a Taylor series in t. The Taylor coefficients of $\chi(t,z)$ at $t = 0$ can be evaluated by successively differentiating the differential equations as

$$\begin{aligned}
\partial_{t_n}\chi(t,z) &= A_n(t,z)\chi(t,z),\\
\partial_{t_m}\partial_{t_n}\chi(t,z) &= (\partial_{t_m}A_n(t,z) + A_n(t,z)A_m(t,z))\chi(t,z),\\
\partial_{t_k}\partial_{t_m}\partial_{t_n}\chi(t,z) &= \Big(\partial_{t_k}\partial_{t_m}A_n(t,z) + \partial_{t_k}(A_n(t,z)A_m(t,z))\\
&\quad + (\partial_{t_m}A_n(t,z))A_k(t,z) + A_n(t,z)A_m(t,z)A_k(t,z)\Big)\chi(t,z),
\end{aligned}$$

etc. Letting $t = 0$, we are left with a polynomial of derivatives of A_n's. One can deduce from these calculations that all Taylor coefficients of $\chi(t,z)$ at $t = 0$ are matrices of meromorphic functions of z on Γ with poles at $z = 0, \gamma_1(0), \gamma_2(0)$ and holomorphic at other points. Since the order of poles at $z = 0$ is unbounded for higher orders of the Taylor expansion, the Taylor series of $\chi(t,z)$ has an essential singularity at $z = 0$. On the other hand, the poles at $z = \gamma_1(0), \gamma_2(0)$ remain to be of the first order. More careful analysis [24] shows that the detailed structure of these first order poles:

Lemma 6. *As $z \to \gamma_s(0)$, $s = 1, 2$, $\chi(t,z)$ behaves as*

$$\chi(t,z) = \frac{\boldsymbol{\beta}_{\chi,s}(t)\,{}^{\mathrm{t}}\boldsymbol{\alpha}_s(0)}{z - \gamma_s(0)} + O(1),$$

where $\boldsymbol{\beta}_{\chi,s}(t)$ is a two-dimensional vector.

Another important property of $\chi(t,z)$ can be seen from the the linear system

$$\partial_x\chi(t,z) = A(t,z)\chi(t,z).$$

Taking the residue at $z = \gamma_s(t)$ yields the relation

$$0 = \boldsymbol{\beta}_s(t)\,{}^{\mathrm{t}}\boldsymbol{\alpha}_s(t)\chi(t,\gamma_s(t)),$$

which implies that

$${}^{\mathrm{t}}\boldsymbol{\alpha}_s(t)\chi(t,\gamma_s(t)) = \mathbf{0}.$$

Thus one finds the following.

Lemma 7. $\det\chi(t,z)$ *has zeroes at $z = \gamma_s(t)$, $s = 1, 2$. ${}^{\mathrm{t}}\boldsymbol{\alpha}_s(t)$ is a left null vector of $\chi(t,\gamma_s(t))$.*

This result shows that $\chi(t,z)$ is exactly the solution mentioned in Krichever's lemma [12, Lemma 5.2], namely a matrix solution of the auxiliary linear system that is holomorphic at the movable poles of $A(t,z)$.

Since $\psi(t,z)$ and $\chi(t,z)$ satisfy the same auxiliary linear system, their "matrix ratio" $\chi(t,z)^{-1}\psi(t,z)$ is independent of t, hence equal to its initial value at $t=0$. One thus obtains the relation

$$\psi(0,z) = \chi(t,z)^{-1}\psi(t,z) \tag{55}$$

or, equivalently,

$$\phi(0,z)\exp\Big(-\sum_{n=1}^{\infty} t_n J z^{-n}\Big) = \chi(t,z)^{-1}\phi(t,z). \tag{56}$$

This is the Riemann-Hilbert problem that plays the role of an intermediate step towards the Grassmannian perspective. The pair of $\phi(t,z)$ and $\chi(t,z)$ are referred to as a Riemann-Hilbert pair.

Note that this Riemann-Hilbert problem has a few unusual aspects. Firstly, in addition to the pole at $z=0$, $\chi(t,z)$ has extra poles at $z=\gamma_s(0)$, $s=1,2$. Secondly, $\chi(t,z)$ *degenerate* (namely, $\det\chi(t,z)$ has zeroes) at $z=\gamma_s(t)$, $s=1,2$. Moreover, these degeneration points are movable as t varies.

6.2. Grassmann variety and vacuum

Let V denote the vector space of all 2×2 matrices of Laurent series

$$X(z) = \sum_{n=-\infty}^{\infty} X_n z^n, \quad X_n \in \mathrm{gl}(2,\mathbf{C}), \tag{57}$$

that converges in a neighborhood of $z=0$ except at $z=0$, and V_+ the subspace

$$V_+ = \{X(z)\in V \mid X_n = 0 \text{ for } n\le 0\}. \tag{58}$$

of all $X(z)\in V$ that are holomorphic and vanish at $z=0$. Recalling that z amounts to λ^{-1}, this is essentially the same setting as the case of the nonlinear Schrödinger hierarchy. The Grassmann variety Gr and the big cell $\mathrm{Gr}^\circ \subset \mathrm{Gr}$ are defined as

$$\begin{aligned}\mathrm{Gr} = \{W\subset V \mid\ & \\ & \dim\mathrm{Ker}(W\to V/V_+) = \dim\mathrm{Coker}(W\to V/V_+) < \infty\}\end{aligned} \tag{59}$$

and

$$\mathrm{Gr}^\circ = \{W\in\mathrm{Gr} \mid W \simeq V/V_+\}. \tag{60}$$

The following lemma shows the construction of a special point $W_0(\gamma,\alpha)$ of Gr°, which plays the role of "vacuum" in the present setting. This is a matrix version of the vacuum that Previato and Wilson [18] suggest to use for a holomorphic vector bundle in the Tyurin parametrization.

Lemma 8. *Let $\gamma=(\gamma_1,\gamma_2)$ be a pair of distinct points of Γ, $\gamma_1\neq\gamma_2$, and $\alpha=(\alpha_1,\alpha_2)$ a pair of constants satisfying the genericity condition $\alpha_1\neq\alpha_2$. Then, for any integer $n\ge 0$ and the matrix indices $i,j=1,2$, there is a unique 2×2 matrix $w_{n,ij}(z)$ of meromorphic functions on Γ with the following properties:*

1. *$w_{n,ij}(z)$ has poles at $z = 0, \gamma_1, \gamma_2$ and is holomorphic at other points.*
2. *$w_{n,ij}(z) = E_{ij} z^{-n} + O(z)$ as $z \to 0$, where E_{ij}, $i, j = 1, 2$, are the standard basis of* $\mathrm{gl}(2, \mathbf{C})$.
3. *As $z \to \gamma_s$, $s = 1, 2$,*

$$w_{n,ij}(z) = \frac{\boldsymbol{\beta}_{n,ij,s}\, {}^{\mathrm{t}}\boldsymbol{\alpha}_s}{z - \gamma_s} + O(1),$$

where $\boldsymbol{\alpha}_s = {}^{\mathrm{t}}(\alpha_s, 1)$, and $\boldsymbol{\beta}_{n,ij,s}$ is another two-dimensional constant vector.

The subspace

$$W_0(\gamma, \alpha) = \langle w_{n,ij}(z) \mid n \geq 0,\ i, j = 1, 2 \rangle \tag{61}$$

spanned by (the Laurent series of) $w_{n,ij}(z)$'s is an element of the big cell.

This vacuum $W_0(\gamma, \alpha)$ is "dressed" by a Laurent series to become a dressed vacuum:

$$W = W_0(\gamma, \alpha)\phi(z), \quad \phi(z) = I + \sum_{n=1}^{\infty} \phi_n z^n, \quad \phi_n \in \mathrm{gl}(2, \mathbf{C}).$$

The set

$$\begin{aligned} \mathcal{M} = \{ W \in \mathrm{Gr}^o \mid W = W_0(\gamma, \alpha)\phi(z),\ \phi_n \in \mathrm{gl}(2, \mathbf{C}), \\ \gamma = (\gamma_1, \gamma_2) \in \Gamma^2,\ \alpha = (\alpha_1, \alpha_2) \in \mathbf{C}^2,\ \gamma_1 \neq \gamma_2,\ \alpha_1 \neq \alpha_2 \} \end{aligned} \tag{62}$$

of these dressed vacua is the phase space for the Grassmannian perspective of the elliptic analogue of the nonlinear Schrödinger hierarchy.

6.3. Interpretation of Riemann-Hilbert problem

For technical reasons, the following consideration is limited to a small neighborhood of $t = 0$. The goal is to translate the Riemann-Hilbert problem to the language of the set $\mathcal{M}$ of dressed vacua. A clue is the the following.

Lemma 9. $W_0(\gamma(t), \alpha(t))\chi(t, z) = W_0(\alpha(0), \gamma(0))$.

Proof. The following is an outline of the proof; see the paper [24] for details. Let $w_{n,ij}(t, z)$, $n \geq 0$, $i, j = 1, 2$, denote the elements of the basis of $W_0(\gamma(t), \alpha(t))$ defined in Lemma 8. $w_{n,ij}(t, z)$ has poles at $z = 0, \gamma_1(t), \gamma_2(t)$, and behaves as

$$w_{n,ij}(t, z) = \frac{\boldsymbol{\beta}_{n,ij,s}(t)\, {}^{\mathrm{t}}\boldsymbol{\alpha}_s(t)}{z - \gamma_s(t)} + O(1)$$

as $z \to \gamma_s(t)$. Upon multiplication with $\chi(t, z)$, the poles at $z = \gamma_s(t)$ are cancelled out because ${}^{\mathrm{t}}\boldsymbol{\alpha}_s(t)$ is a left null vector of $\chi(t, \gamma_s(t))$. Thus $w_{n,ij}(t, z)\chi(t, z)$ turns out to have an essential singularity at $z = 0$, first order poles at $z = \gamma_s(0)$, $s = 1, 2$, and is holomorphic at other points. The leading part of the Laurent expansion at $z = \gamma_s(0)$ takes the form

$$w_{n,ij}(t, z)\chi(t, z) = \frac{w_{n,ij}(t, \gamma_s(0))\boldsymbol{\beta}_{\chi,s}(t)\, {}^{\mathrm{t}}\boldsymbol{\alpha}_s(0)}{z - \gamma_s(0)} + O(1).$$

These results show that $w_{n,ij}(t,z)\chi(t,z)$ is an element of $W_0(\gamma(0),\alpha(0))$. One can thus see that

$$W_0(\gamma(t),\alpha(t))\chi(t,z) \subseteq W_0(\gamma(0),\alpha(0)).$$

A few more steps of consideration on the analytic properties of $\chi(t,z)$ lead to the conclusion that these two vector subspaces of V are equal. □

Thanks to this lemma, one can readily convert the Riemann-Hilbert problem to the language of dressed vacua. The Riemann-Hilbert relation yields the relation

$$W_0(\gamma(t),\alpha(t))\phi(t,z) = W_0(\gamma(t),\alpha(t))\chi(t,z)\phi(0,z)\exp\Bigl(-\sum_{n=1}^{\infty} t_n J z^{-n}\Bigr).$$

The lemma shows that $W_0(\gamma(t),\alpha(t))$ absorbs $\chi(t,z)$ to become $W_0(\gamma(0),\alpha(0))$. The outcome is the relation

$$W_0(\gamma(t),\alpha(t))\phi(t,z) = W_0(\gamma(0),\alpha(0))\phi(0,z)\exp\Bigl(-\sum_{n=1}^{\infty} t_n J z^{-n}\Bigr),$$

which means that the dressed vacuum $W(t) = W_0(\gamma(t),\alpha(t))\phi(t,z) \in \mathcal{M}$ obeys the exponential law

$$W(t) = W(0)\exp\Bigl(-\sum_{n=1}^{\infty} t_n J z^{-n}\Bigr). \tag{63}$$

Conversely, one can derive a solution of the Riemann-Hilbert problem from these exponential flows as follows. (This is a variation of the dressing method of Previato and Wilson [18].) Given a set of initial values $\gamma(0), \alpha(0)$ and $\phi(0,z)$, one can consider the exponential flows sending $W(0) = W_0(\gamma(0),\alpha(0))\phi(0,z)$ to $W(t)$. If t is sufficiently small, $W(t)$ remains in the big cell. This means that the linear map $W(t) \to V/V_+$ is an isomorphism. Let $\phi(t,z) \in W(t)$ be the inverse image of $I \in V/V_+$ by this isomorphism. Being equal to I modulo V_+, $\phi(t,z)$ is a Laurent series of the form

$$\phi(t,z) = 1 + \sum_{n=1}^{\infty} \phi_n(t) z^n.$$

On the other hand, as an element of

$$W(t) = W_0(\gamma(0),\alpha(0))\phi(0,z)\exp\Bigl(-\sum_{n=1}^{\infty} t_n J z^{-n}\Bigr),$$

$\phi(t,z)$ can also be expressed as

$$\phi(t,z) = \chi(t,z)\phi(0,z)\exp\Bigl(-\sum_{n=1}^{\infty} t_n J z^{-n}\Bigr)$$

with an element $\chi(t,z)$ of $W_0(\gamma(0),\alpha(0))$. Thus one obtains a Riemann-Hilbert pair. The associated Tyurin parameters $(\gamma_s(t), \boldsymbol{\alpha}_s(t))$ are determined as the position of zeros of $\chi(t,z)$ and the normalized left null vector of $\chi(t,z)$ at those degeneration points.

One thus eventually arrives at the following Grassmannian perspective in the present setting.

Theorem 4. *The elliptic analogue of the nonlinear Schrödinger hierarchy can be mapped, by the correspondence* $W(t) = W_0(\gamma(t), \alpha(t))\phi(t, z)$, *to a dynamical system on the set* $\mathcal{M}$ *of dressed vacua in the Grassmann variety* Gr. *The motion of* $W(t)$ *obeys the exponential law. Conversely, the exponential flows on* $\mathcal{M}$ *yield a solution of the Riemann-Hilbert problem.*

As a final remark, it should be stressed that the main characters of this story are all related to the geometry of holomorphic vector bundles over Γ. The Tyurin parameters $(\gamma(t), \alpha(t))$ correspond to a holomorphic vector bundle that deforms as t varies. The subspace $W_0(\gamma, \alpha) \subset V$ can be identified with the space of holomorphic sections of the associated $\mathrm{sl}(2, \mathbf{C})$ bundle over the punctured torus $\Gamma \setminus \{z = 0\}$. $\phi(t, z)$ is related to changing local trivialization of this bundle at $z = 0$. Note, in particular, that the primary role (as a dynamical variable) is now played by the data of local trivialization. This differs decisively from the work of Previato and Wilson [18]; they take, in place of the data of local trivialization, a set of functions in Krichever's "algebraic spectral data" [9] as main parameters. In this respect, the present setting is rather close to Li and Mulase's approach [17, 15] to commutative rings of differential operators; they treat the choice of local trivialization as an independent data.

7. Landau-Lifshitz hierarchy in Grassmannian perspective

The last example with an elliptic spectral parameter is the Landau-Lifshitz equation in $1 + 1$ dimensions and the associated hierarchy (Landau-Lifshitz hierarchy) of higher time evolutions. This is one of the classical examples of soliton equations with an elliptic zero-curvature representation [22, 3].

As regards the Grassmannian perspective of this equation, studies from a very close point of view have been done by Date, Jimbo, Kashiwara and Miwa [4] and Carey, Hannabuss, Mason and Singer [2]. Actually, Date et al. developed a free fermion formalism rather than a Grassmannian formalism. Carey et al. presented two approaches to a factorization method for solving the Landau-Lifshitz equation. The first approach uses an infinite-dimensional Grassmann manifold (rather than a "variety", because this is a functional analytic model). The second one is based on the geometry of a holomorphic vector bundle over the torus $\Gamma = \mathbf{C}/(2\omega_1\mathbf{Z} + 2\omega_3\mathbf{Z})$. This work is yet unsatisfactory because their usage of the Grassmann manifold fails to incorporate the bundle structure.

The lessons in the preceding examples show that a clue is always the choice of a suitable "vacuum" (and of course a Grassmann variety that accommodates that vacuum). As the paper of Previato and Wilson suggests [18], a correct choice of vacuum is somehow related to the structure of a holomorphic vector bundle. The Grassmannian perspective of the Landau-Lifshitz equation (and hierarchy), too, can be reached along the same lines [25].

7.1. Geometric and algebraic structures behind Landau-Lifshitz equation

The zero-curvature representation of the Landau-Lifshitz equation [22, 3] is based on the first order matrix differential operator $\partial_x - A(z)$ with the A-matrix of the form

$$A(z) = \sum_{a=1,2,3} w_a(z) S_a \sigma_a, \tag{64}$$

where S_a's are dynamical variables (spin fields) and σ_a's denote the Pauli matrices. The weight functions $w_a(z)$ are defined by Jacobi's elliptic functions $\mathrm{sn}, \mathrm{cn}, \mathrm{dn}$ as

$$w_1(z) = \frac{\alpha \mathrm{cn}(\alpha z)}{\mathrm{sn}(\alpha z)}, \quad w_2(z) = \frac{\alpha \mathrm{dn}(\alpha z)}{\mathrm{sn}(\alpha z)}, \quad w_3(z) = \frac{\alpha}{\mathrm{sn}(\alpha z)}, \tag{65}$$

where $\alpha = \sqrt{e_1 - e_3}$, $e_a = \wp(\omega_a)$.

The matrix $A(z)$ has the twisted double periodicity

$$A(z + 2\omega_a) = \sigma_a A(z) \sigma_a, \quad a = 1, 2, 3, \tag{66}$$

where ω_2 denotes the third half period $\omega_2 = -\omega_1 - \omega_3$. This is a manifestation of the structure of a nontrivial holomorphic $\mathrm{sl}(2, \mathbf{C})$ bundle over the torus; $A(z)$ is a meromorphic section of that bundle. The same bundle is known to play a fundamental role in the elliptic Gaudin model and an associated conformal field theory [13].

Compared with the equations formulated by Tyurin parameters, the Landau-Lifshitz equation is rather close to classical soliton equations with a rational zero-curvature representation, because one can treat this system by a factorization method based on a Lie group of Laurent series (or a loop group) with factorization structure [19, 2]. Geometrically, this fact is related to *rigidity* of the aforementioned holomorphic $\mathrm{sl}(2, \mathbf{C})$ bundle or of an associated $\mathrm{SL}(2, \mathbf{C})$ bundle [8].

To formulate the factorization structure, one starts from a Lie algebra with direct sum decomposition to two subalgebras. Let $\mathfrak{g}$ be the Lie algebra of Laurent series

$$X(z) = \sum_{n=-\infty}^{\infty} X_n z^n, \quad X_n \in \mathrm{sl}(2, \mathbf{C}), \tag{67}$$

that converge in a neighborhood of $z = 0$ except at $z = 0$. This Lie algebra has a direct sum decomposition of the form

$$\mathfrak{g} = \mathfrak{g}_{\mathrm{out}} \oplus \mathfrak{g}_{\mathrm{in}}, \tag{68}$$

where $\mathfrak{g}_{\mathrm{in}}$ and $\mathfrak{g}_{\mathrm{out}}$ are the following subalgebras:

1. $\mathfrak{g}_{\mathrm{in}}$ consists of all $X(z) \in \mathfrak{g}$ that are also holomorphic at $z = 0$, i.e., $X_n = 0$ for $n < 0$.
2. $\mathfrak{g}_{\mathrm{out}}$ consists of all $X(z) \in \mathfrak{g}$ that can be extended to a holomorphic mapping $X : \mathbf{C} \backslash (2\omega_1 \mathbf{Z} + 2\omega_3 \mathbf{Z} \to \mathrm{sl}(2, \mathbf{C})$ with singularity at each point of $2\omega_1 \mathbf{Z} + 2\omega_3 \mathbf{Z}$ and satisfy the twisted double periodicity condition
$$X(z + 2\omega_a) = \sigma_a X(z) \sigma_a \quad a = 1, 2, 3.$$

Note that constant matrices are excluded from $\mathfrak{g}_{\mathrm{out}}$, so that $\mathfrak{g}_{\mathrm{out}} \cap \mathfrak{g}_{\mathrm{in}} = \{0\}$. One can choose $\{\partial_z^n w_a(z)\sigma_a \mid n \geq 0,\ a = 1,2,3\}$ as a basis of $\mathfrak{g}_{\mathrm{out}}$; the projection $(\cdot)_{\mathrm{out}} : \mathfrak{g} \to \mathfrak{g}_{\mathrm{out}}$ thereby takes the simple form

$$\left(z^{-n-1}\sigma_a\right)_{\mathrm{out}} = \frac{(-1)^n}{n!}\partial_z^n w_a(z)\sigma_a, \quad \left(z^n\sigma_a\right)_{\mathrm{out}} = 0, \quad n \geq 0. \tag{69}$$

The direct sum decomposition of the Lie algebra $\mathfrak{g}$ induces the factorization of the associated Lie group $G = \exp \mathfrak{g}$ to the subgroups $G_{\mathrm{out}} = \exp \mathfrak{g}_{\mathrm{out}}$ and $G_{\mathrm{in}} = \exp \mathfrak{g}_{\mathrm{in}}$, namely, any element $g(z)$ of G near the unit matrix I can be uniquely factorized as

$$g(z) = g_{\mathrm{out}}(z)^{-1} g_{\mathrm{in}}(z), \quad g_{\mathrm{out}}(z) \in G_{\mathrm{out}}, \quad g_{\mathrm{in}}(z) \in G_{\mathrm{in}}. \tag{70}$$

7.2. Construction of hierarchy

The Landau-Lifshitz hierarchy can be obtained by the projection of the exponential flows

$$g(\lambda) \mapsto g(\lambda)\exp\Big(-\sum_{n=1}^{\infty} t_n J\lambda^n\Big) \tag{71}$$

on G to G_{in} with regard to the foregoing factorization [6, 2]. The fundamental dynamical variable is thus a Laurent series of the form

$$\phi(z) = \sum_{n=0}^{\infty} \phi_n z^n, \quad \det \phi(z) = 1,$$

that converges in a neighborhood of $z = 0$. The time evolution $\phi(0,z) \mapsto \phi(t,z)$ is achieved by the factorization

$$\phi(0,z)\exp\Big(-\sum_{n=1}^{\infty} t_n z^{-n}\sigma_3\Big) = \chi(t,z)^{-1}\phi(t,z), \tag{72}$$

where $\chi(t,z)$ is an element of G_{out} that also depends on t. As demonstrated in the case of the usual nonlinear Schrödinger hierarchy, one can derive the equations

$$\partial_{t_n}\phi(t,z) = A_n(t,z)\phi(t,z) - \phi(t,z)z^{-n}\sigma_3, \tag{73}$$

$$\text{where} \quad A_n(t,z) = \Big(\phi(t,z)z^{-n}\sigma_3\phi(t,z)^{-1}\Big)_{\mathrm{out}}, \tag{74}$$

$$\text{or} \quad \partial_{t_n}\phi(t,z) = -\Big(\phi(t,z)z^{-n}\sigma_3\phi(t,z)^{-1}\Big)_{\mathrm{in}}\phi(t,z) \tag{75}$$

as equations of motion of $\phi(t,z) \in G_{\mathrm{in}}$. $(\cdot)_{\mathrm{in}}$ denotes the projection $\mathfrak{g} \to \mathfrak{g}_{\mathrm{in}}$.

The zero-curvature equations

$$[\partial_{t_m} - A_m(t,z),\ \partial_{t_n} - A_n(t,z)] = 0 \tag{76}$$

follow from the auxiliary linear system

$$(\partial_{t_n} - A_n(t,z))\chi(t,z) = 0 \tag{77}$$

as the Frobenius integrability condition.

7.3. Grassmann variety and vacuum

It will be reasonable to use the same pair (V, V_+) of vector spaces as those for the elliptic analogues of the nonlinear Schrödinger hierarchy. Actually, already at this stage, the present approach differs from that of Carey et al. [2]. Carey et al. use a vector space of two-component vectors rather than 2×2 matrices; this is not suited for treating the aforementioned $\mathrm{sl}(2, \mathbf{C})$ bundle structure.

The next problem is the choice of a suitable subspace $W_0 \subset V$ that plays the role of "vacuum." In view of the previous examples, W_0 should be a vector subspace that absorbs the first factor $\chi(t, z)$ of the factorization pair. This will be the case if W_0 consists of matrix-valued functions of z with the same analytic properties as $\chi(t, z)$. As a t-dependent element of G_{out}, $\chi(t, z)$ is a matrix-valued holomorphic function on $\mathbf{C} \setminus (2\omega_1 \mathbf{Z} + 2\omega_3 \mathbf{Z})$ with twisted double periodicity.

For this reason, let W_0 be the subspace of V that consists of all $X(z) \in V$ with the following properties:

1. $X(z)$ can be extended to a holomorphic mapping
$$X : \mathbf{C} \setminus (2\omega_1 \mathbf{Z} + 2\omega_3 \mathbf{Z}) \to \mathrm{gl}(2, \mathbf{C}).$$
2. $X(z)$ has the twisted double periodicity
$$X(z + 2\omega_a) = \sigma_a X(z) \sigma_a, \quad a = 1, 2, 3.$$

This resembles the definition of $\mathfrak{g}_{\mathrm{out}}$; the difference is, firstly, that $X(z)$ now takes values in $\mathrm{gl}(2, \mathbf{C})$ rather than $\mathrm{sl}(2, \mathbf{C})$, and secondly, that $X(z)$ can be a constant matrix.

As it turns out, this subspace W_0 does *not* satisfy the condition in the definition of the Grassmann variety Gr that has been used in the previous case:

Lemma 10. *The following hold for the linear map $W_0 \to V/V_+$:*

1. $\mathrm{Im}(W_0 \to V/V+) \oplus \mathrm{sl}(2, \mathbf{C}) \oplus \mathbf{C}z^{-1} I = V/V_+$.
2. $\mathrm{Ker}(W_0 \to V/V_+) = \{0\}$.

Proof. It is a (slightly advanced) exercise of linear algebra and complex function theory to confirm that W_0 is spanned by I, $\partial_z^n w_a(z)\sigma_a$, $a = 1, 2, 3$, and $\partial_z^n \wp(z) I$ for $n \geq 0$. $\partial_z^n w_a(z)$ and $\partial_z^n \wp(z)$ have the Laurent expansion

$$\partial_z^n w_a(z)\sigma_a = (-1)^n n! z^{-n-1} \sigma_a + O(z)$$

and

$$\partial_z^n \wp(z) I = (-1)^n (n+1)! z^{-n-2} I + O(z)$$

at $z = 0$. This implies that these generators of W_0 are linearly independent, and that the image of $W \to V/V_+$ are spanned by I, $z^{-n-1}\sigma_a$, $a = 1, 2, 3$, and $z^{-n-2} I$ for $n \geq 0$ among the standard basis $\{z^{-n}\sigma_a,\ z^{-n} I \mid n \geq 0,\ a = 1, 2, 3\}$ of V/V_+. What is missing are σ_a, $a = 1, 2, 3$, and $z^{-1} I$, which respectively span the subspaces $\mathrm{sl}(2, \mathbf{C})$ and $\mathbf{C}z^{-1}$ of V/V_+. Thus the assertion on $\mathrm{Im}(W_0 \to V/V_+)$ follows. On the other hand, one has $\mathrm{Ker}(W_0 \to V/V_+) = W_0 \cap V_+$. Any element $X(z)$ of $W_0 \cap V_+$ has the twisted double periodicity and a zero at all points of $2\omega_1 \mathbf{Z} + 2\omega_3 \mathbf{Z}$; by Liouville's theorem, such a matrix-valued function is identically zero. □

This lemma implies that

$$\dim \operatorname{Ker}(W_0 \to V/V_+) = 0, \quad \dim \operatorname{Coker}(W_0 \to V/V_+) = 4. \tag{78}$$

Consequently, the Grassmann variety to accommodate W_0 is not Gr but the following one:

$$\begin{aligned} \mathrm{Gr}_{-4} = \{W \subset V \mid \\ \dim \operatorname{Ker}(W \to V/V_+) = \dim \operatorname{Coker}(W \to V/V_+) - 4 < \infty\}. \end{aligned} \tag{79}$$

The subset

$$\mathrm{Gr}^\circ_{-4} = \{W \in \mathrm{Gr}_{-4} \mid W \simeq V/(V_+ \oplus \mathrm{sl}(2, \mathbf{C}) \oplus \mathbf{C}z^{-1}I)\} \tag{80}$$

of Gr_{-4} is an open subset, in fact, the open cell (or "big cell") of a cell decomposition of Gr_{-4}. The foregoing lemma shows that W_0 is actually an element of this open subset:

$$W_0 \in \mathrm{Gr}^\circ_{-4}. \tag{81}$$

The set

$$\mathcal{M} = \{W \in \mathrm{Gr}^\circ_{-4} \mid W = W_0\phi(z),\ \phi(z) \in G_{\mathrm{in}}\} \tag{82}$$

of dressed vacua becomes the phase space of a dynamical system to which the Landau-Lifshitz hierarchy is mapped.

7.4. Interpretation of factorization problem

The following consideration is, again, limited to a small neighborhood of $t = 0$. In this situation, one can prove the following in the same way as the case of the nonlinear Schrödinger hierarchy.

Lemma 11. $W_0\chi(t, z) = W_0$.

Using this lemma, one can repeat the calculations done for the previous cases to show that the motion of the dressed vacuum $W(t) = W_0\phi(t, z) \in \mathcal{M}$ obeys the exponential law

$$W(t) = W(0) \exp\Big(-\sum_{n=1}^{\infty} t_n J z^{-n}\Big). \tag{83}$$

The converse, namely, deriving a solution of the factorization problem from the exponential flows needs an extra effort because the definition of the big cell is different from the previous cases. This is also related to the fact that the leading term $\phi_0(t)$ of $\phi(t, z)$ is generally not equal to I. A clue here is the fact that $W(t)$, as an element of the big cell, satisfies the condition that

$$\dim \operatorname{Im}(W(t) \to V/V_+) \cap \mathrm{gl}(2, \mathbf{C}) = 1. \tag{84}$$

The leading term $\phi_0(t)$ is picked out from this one-dimensional subspace; if t is sufficiently small, $\phi_0(t)$ is an invertible matrix. The rest of the construction is almost parallel to the case of the nonlinear Schrödinger hierarchy; see the paper [25] for details.

The conclusion is that the Grassmannian perspective also holds for this case, but with a different Grassmann variety:

Theorem 5. *The Landau-Lifshitz hierarchy can be mapped, by the correspondence* $W(t) = W_0\phi(t,\lambda)$*, to a dynamical system on the set* $\mathcal{M}$ *of dressed vacua in the Grassmann variety* Gr_{-4}*. The motion of* $W(t)$ *obeys the exponential law. Conversely, the exponential flows on* $\mathcal{M}$ *yield a solution of the factorization problem.*

8. Conclusion

A main conclusion of this case study is that the structure of a holomorphic vector bundle is the most important clue to the Grassmannian perspective of soliton equations with a zero-curvature representation constructed on an algebraic curve. This is also the case for classical soliton equations with a rational spectral parameter; the relevant holomorphic vector bundle therein is a trivial bundle. The two examples with an elliptic spectral parameters examined here are respectively accompanied by a bundle of its own particular type. The bundle for the elliptic nonlinear Schrödinger hierarchy is naturally the one in the Tyurin parametrization. The bundle for the Landau-Lifshitz hierarchy is a rigid bundle.

It is remarkable that the mapping to an infinite-dimensional Grassmann variety can be constructed in a fully parallel, almost universal way. Namely, the first thing to do is to choose a special base point W_0, called "vacuum," of the Grassmann variety. This is determined by the relevant holomorphic vector bundle E. More precisely, one has to choose a marked point P_0 of Γ, a local coordinate z in a neighborhood of P_0 and a local trivialization of E in a neighborhood of P_0 as extra geometric data. W_0 consists of Laurent series that represent (via the local trivialization of E) a holomorphic section of E over $\Gamma \setminus \{P_0\}$. The vacuum W_0 is then "dressed" by a Laurent series $\phi(z)$, which is related to changing the local trivialization of E. These geometric data are familiar stuff in the theories of algebro-geometric solutions of soliton equations, commutative rings of differential operators, etc. [9, 10, 11, 15, 16, 17, 18, 21].

This geometric point of view is already enough to tackle more general cases. It is rather straightforward to generalize the result for the elliptic analogue of the nonlinear Schrödinger hierarchy to higher genera, though explicit formulas of the A-matrices are not available therein. The work of Li and Mulase [17, 15], too, provides valuable material to this issue.

Acknowledgements

I would like to thank the organizers of the workshop, in particular, Nenad Manojlovic and Henning Samtleben, for invitation and hospitality. This work was partly supported by the Grant-in-Aid for Scientific Research (No. 14540172) from the Ministry of Education, Culture, Sports and Technology.

References

[1] D. Ben-Zvi and E. Frenkel, Spectral curves, opers and integrable systems, Publ. Math. Inst. Hautes Études Sci. **94** (2001), 87–159.

[2] A.L. Carey, K.C. Hannabuss, L.J. Mason and M.A. Singer, The Landau-Lifshitz equation, elliptic curve, and the Ward transform, Commun. Math. Phys. **154** (1993), 25–47.

[3] I.V. Cherednik, On integrability of the equation of a two-dimensional asymmetric $O(3)$-field and its quantum analogue, Yad. Fiz. **33** (1) (1981), 278–282.

[4] E. Date, M. Jimbo, M. Kashiwara and T. Miwa, Landau-Lifshitz equation: solitons, quasi-periodic solutions and infinite-dimensional Lie algebras, J. Phys. A: Math. Gen. **16** (1983), 221–236.

[5] E. Frenkel, Five lectures on soliton equations, Surveys in Differential Geometry, vol. 4, pp. 27–60 (International Press, 1998).

[6] F. Guil and M. Mañas, Loop algebras and the Krichever-Novikov equation, Phys. Lett. 153A (1991), 90–94.

[7] N. Hitchin, Stable bundles and integrable systems, Duke Math. J. **54** (1990), 91–114.

[8] J. C. Hurtubise and E. Markman, Surfaces and the Sklyanin bracket, Commun. Math. Phys. **230** (2002), 485–502.

[9] I.M. Krichever, Commutative rings of ordinary linear differential operators, Funct. Anal. Appl. **12** (1978), no. 3, 175–185.

[10] I.M. Krichever and S.P. Novikov, Holomorphic vector bundles over Riemann surfaces and the Kadomtsev-Petviashvili equation. I, Funct. Anal. Appl. **12** (1978), no. 4, 276–286.

[11] I.M. Krichever and S.P. Novikov, Holomorphic bundles over algebraic curves, and nonlinear equations, Russian Math. Surveys 35 (1980), no. 6, 53–80.

[12] I.M. Krichever, Vector bundles and Lax equations on algebraic curves, Commun. Math. Phys. **229** (2002), 229–269.

[13] G. Kuroki and T. Takebe, Twisted Wess-Zumino-Witten models on elliptic curves, Commun. Math. Phys. **190** (1997), 1–56.

[14] A.M. Levin, M.A. Olshanetsky and A. Zotov, Hitchin systems – symplectic Hecke correspondence and two-dimensional version, Commun. Math. Phys. **236** (2003), 93–133.

[15] Y. Li and M. Mulase, Prym varieties and integrable systems, Commun. Anal. Geom. **5** (1997), 279–332.

[16] M. Mulase, Cohomological structure in soliton equations and jacobian varieties, J. Differential Geom. **19** (1984), 403–430.

[17] M. Mulase, Category of vector bundles on algebraic curves and infinite-dimensional Grassmannians, Intern. J. Math. **1** (1990), 293–342.

[18] E. Previato and G. Wilson, Vector bundles over curves and solutions of the KP equations, Proc. Symp. Pure Math. vol. 49, part I, pp. 553–569 (American Mathematical Society, 1989).

[19] A.G. Reyman and M.A. Semenov-Tian-Shansky, Lie algebras and Lax equations with spectral parameter on an elliptic curve, Zap. Nauchn. Sem. LOMI **150** (1986), 104–118; J. Soviet Math. **46** (1989), 1631–1640.

[20] M. Sato and Y. Sato, Soliton equations as dynamical systems on an infinite-dimensional Grassmannian manifold, Lecture Notes in Num. Appl. Anal., vol. 5, pp. 259–271 (Kinokuniya, Tokyo, 1982).

[21] G.B. Segal and G. Wilson, Loop groups and equations of KdV type, Publ. Math. IHES **61** (1985), 5–65.

[22] E.K. Sklyanin, On complete integrability of the Landau-Lifshitz equation, Steklov Mathematical Institute Leningrad Branch preprint LOMI, E-3-79, 1979.

[23] K. Takasaki, A new approach to the self-dual Yang-Mills equations, Commun. Math. Phys. **94** (1984), 35–59.

[24] K. Takasaki, Tyurin parameters and elliptic analogue of nonlinear Schrödinger hierarchy, J. Math. Sci. Univ. Tokyo **11** (2004), 91–131.

[25] K. Takasaki, Landau-Lifshitz hierarchy and infinite-dimensional Grassmann variety, Lett. Math. Phys. **67** (2004), 141–152.

[26] A. Tyurin, Classification of vector bundles over an algebraic curve of arbitrary genus, AMS Translations II, Ser. 63, pp. 245–279. (American Mathematical Society, 1967).

[27] V.E. Zakharov and A.V. Mikhailov, Method of the inverse scattering problem with spectral parameter on an algebraic curve, Func. Anal. Appl. **17** (1982), No. 4, 247–251.

Kanehisa Takasaki
Graduate School of Human and Environmental Studies
Kyoto University
Yoshida, Sakyo
Kyoto 606-8501, Japan
e-mail: takasaki@math.h.kyoto-u.ac.jp

Progress in Mathematics, Vol. 237, 205–224

Trigonometric Degeneration and Orbifold Wess-Zumino-Witten Model. II

Takashi Takebe

Dedicated to Professor Akihiro Tsuchiya on his 60th birthday.

Abstract. The sheaves of conformal blocks and conformal coinvariants of the twisted WZW model have a factorisation property and are locally free even at the boundary of the moduli space, where the elliptic KZ equations and the Baxter-Belavin elliptic r-matrix degenerate to the trigonometric KZ equations and the trigonometric r-matrix, respectively. Etingof's construction of the elliptic KZ equations is geometrically interpreted.

Mathematics Subject Classification (2000). Primary 81T40; Secondary 14H15, 17B67, 17B81, 32G15.

Keywords. Trigonometric degeneration; twisted WZW model; orbifold WZW model; factorisation.

1. Introduction

This is a continuation of the paper [T]. We showed there that the trigonometric WZW model is factorised into the orbifold WZW models. Using this result, we show in the present article that the trigonometric WZW model is indeed the degenerate twisted WZW models on elliptic curves defined in [KT]. More precisely, we prove that there are locally free sheaves over the partially compactified family of elliptic curves, the fibre of which are the space of conformal blocks or the space of conformal coinvariants of the twisted WZW model at a generic point and those of the trigonometric WZW model at the discriminant locus, when all inserted modules are either Weyl modules (Proposition 4.2) or integrable highest weight modules (Theorem 6.1).

This work is partly supported by the Grant-in-Aid for Scientific Research © of the Japan Society for the Promotion of Science, No. 15540014.

Since the elliptic r-matrix (cf. [BD], [E]) describing the elliptic KZ equations degenerates to the trigonometric r-matrix, the above fact is naturally expected, though rigorous proof requires careful algebro-geometric arguments as in [TUY].

The paper is organised as follows. After reviewing the twisted WZW model on elliptic curves in §2 to recall basic notions and notations, we define family of elliptic curves with a singular fibre and a twisted Lie algebra bundle over it in §3. In §4 main objects of this paper, the sheaves of conformal coinvariants and conformal blocks, are defined and their coherence is proved. In particular when all the modules inserted to the curve are Weyl modules, they are locally free. To prove the locally freeness of the sheaf of conformal coinvariants for integrable highest weight modules, we examine its behavior at the discriminant locus. In this case the factorisation theorem, Theorem 7 of [T], is refined in §5. The proof of locally freeness in §6 follows the strategy of [TUY] and [TK].

Notations

We use the following notations besides other ordinary conventions in mathematics.

- N, L: fixed integers. $N \geqq 2$ will be the matrix size and $L \geqq 1$ will be the number of the marked points on a curve.
- $C_N := \mathbb{Z}/N\mathbb{Z}$: the cyclic group of order N.
- When X is an algebraic variety, $\mathcal{O}_X$ denotes the structure sheaf of X. When P is a point on X and $\mathcal{F}$ is an $\mathcal{O}_X$-sheaf, $\mathcal{F}_P$ denotes the stalk of $\mathcal{F}$ at P. $\mathfrak{m}_P$ is the maximal ideal of the local ring $\mathcal{O}_{X,P}$. $\mathcal{F}|_P := \mathcal{F}_P/\mathfrak{m}_P\mathcal{F}_P$, $\mathcal{F}_P^\wedge = \operatorname{proj\,lim}_{n\to\infty} \mathcal{F}_P/\mathfrak{m}_P^n\mathcal{F}_P$ are the fibre of $\mathcal{F}$ at P and the $\mathfrak{m}_P$-adic completion of $\mathcal{F}_P$, respectively.
- We shall use the same symbol for a vector bundle and for a locally free $\mathcal{O}_X$-module consisting of its local holomorphic sections.

2. Twisted WZW model on elliptic curves

In this section we briefly review the twisted WZW model on elliptic curves. See [KT] for details.

We fix an invariant inner product of $\mathfrak{g} = sl_N(\mathbb{C})$ by

$$(A|B) := \operatorname{tr}(AB) \quad \text{for } A, B \in \mathfrak{g}. \tag{1}$$

Define matrices β and γ by

$$\beta := \begin{pmatrix} 0 & 1 & & 0 \\ & 0 & \ddots & \\ & & \ddots & 1 \\ 1 & & & 0 \end{pmatrix}, \quad \gamma := \begin{pmatrix} 1 & & & 0 \\ & \varepsilon^{-1} & & \\ & & \ddots & \\ 0 & & & \varepsilon^{1-N} \end{pmatrix}, \tag{2}$$

where $\varepsilon = \exp(2\pi i/N)$. Then we have $\beta^N = \gamma^N = 1$ and $\gamma\beta = \varepsilon\beta\gamma$.

Let $E = E_\tau$ be the elliptic curve with modulus τ: $E_\tau := \mathbb{C}/\mathbb{Z}+\tau\mathbb{Z}$. We define a Lie algebra bundle $\mathfrak{g}^{\mathrm{tw}}$ with fibre $\mathfrak{g} = sl_N(\mathbb{C})$ over E by

$$\mathfrak{g}^{\mathrm{tw}} := (\mathbb{C} \times \mathfrak{g})/\approx, \tag{3}$$

where the equivalence relations $\approx$ are defined by

$$(z, A) \approx (z+1, \operatorname{Ad}\gamma(A)) \approx (z+\tau, \operatorname{Ad}\beta(A)). \tag{4}$$

Let $J_{ab} = \beta^a\gamma^{-b}$, which satisfies

$$\operatorname{Ad}\gamma(J_{ab}) = \varepsilon^a J_{ab}, \qquad \operatorname{Ad}\beta(J_{ab}) = \varepsilon^b J_{ab}. \tag{5}$$

Global meromorphic sections of $\mathfrak{g}^{\mathrm{tw}}$ are linear combinations of $J_{ab}f(z)$ $(a,b = 0,\dots,N-1,\ (a,b) \neq (0,0))$, where $f(z)$ is a meromorphic function with quasi-periodicity,

$$f(z+1) = \varepsilon^a f(z), \qquad f(z+\tau) = \varepsilon^b f(z). \tag{6}$$

For each point P on E, we define a Lie algebra,

$$\mathfrak{g}^P := (\mathfrak{g}^{\mathrm{tw}} \otimes_{\mathcal{O}_E} \mathcal{K}_E)^\wedge_P, \tag{7}$$

where $\mathcal{K}_E$ is the sheaf of meromorphic functions on E and $(\cdot)^\wedge_P$ means the completion of the stalk at P with respect to the maximal ideal $\mathfrak{m}_P$ of $\mathcal{O}_{E,P}$, the stalk of the structure sheaf. The Lie algebra $\mathfrak{g}^P$ is (non-canonically) isomorphic to the loop Lie algebra $\mathfrak{g}((z-z_0))$, where z_0 is the coordinate of P. The subspace

$$\mathfrak{g}^P_+ := (\mathfrak{g}^{\mathrm{tw}})^\wedge_P \cong \mathfrak{g}[[z-z_0]] \tag{8}$$

of $\mathfrak{g}^P$ is a Lie subalgebra.

Let us fix mutually distinct points $Q_1,\dots,Q_L$ on E whose coordinates are $z = z_1,\dots,z_L$ and put $D := \{Q_1,\dots,Q_L\}$. We shall also regard D as a divisor on E (i.e., $D = Q_1 + \cdots + Q_L$). The Lie algebra $\mathfrak{g}^D := \bigoplus_{i=1}^L \mathfrak{g}^{Q_i}$ has a 2-cocycle defined by

$$\mathrm{c_a}(A,B) := \sum_{i=1}^L \mathrm{c}_{\mathrm{a},i}(A_i,B_i), \qquad \mathrm{c}_{\mathrm{a},i}(A_i,B_i) := \operatorname{Res}_{Q_i}(dA_i|B_i), \tag{9}$$

where $A = (A_i)_{i=1}^L, B = (B_i)_{i=1}^L \in \mathfrak{g}^D$, Res_{Q_i} is the residue at Q_i and d is the exterior derivation. (The symbol "$\mathrm{c_a}$" stands for "Cocycle defining the Affine Lie algebra".) We denote the central extension of $\mathfrak{g}^D$ with respect to this cocycle by $\hat{\mathfrak{g}}^D$:

$$\hat{\mathfrak{g}}^D := \mathfrak{g}^D \oplus \mathbb{C}\hat{k}, \tag{10}$$

where $\hat{k}$ is a central element. Explicitly the bracket of $\hat{\mathfrak{g}}^D$ is represented as

$$[A,B] = ([A_i,B_i]^\circ)_{i=1}^L \oplus \mathrm{c_a}(A,B)\hat{k} \quad \text{for } A,B \in \mathfrak{g}^D, \tag{11}$$

where $[A_i,B_i]^\circ$ are the natural bracket in $\mathfrak{g}^{Q_i}$. The Lie algebra $\hat{\mathfrak{g}}^P$ for a point P is nothing but the affine Lie algebra $\hat{\mathfrak{g}}$ of type $A^{(1)}_{N-1}$ (a central extension of the loop algebra $\mathfrak{g}((t-z)) = sl_N(\mathbb{C}((t-z)))$).

The affine Lie algebra $\hat{\mathfrak{g}}^{Q_i}$ can be regarded as a subalgebra of $\hat{\mathfrak{g}}^D$. The subalgebra $\mathfrak{g}^{Q_i}_+$ of $\mathfrak{g}^{Q_i}$ (cf. (8)) can be also regarded as a subalgebra of $\hat{\mathfrak{g}}^{Q_i}$ and $\hat{\mathfrak{g}}^D$.

Let $\mathfrak{g}_{\text{out}}$ be the space of global meromorphic sections of $\mathfrak{g}^{\text{tw}}$ which are holomorphic on E except at D:

$$\mathfrak{g}_{\text{out}} := \Gamma(E, \mathfrak{g}^{\text{tw}}(*D)). \tag{12}$$

The residue theorem implies that we can regard $\mathfrak{g}_{\text{out}}$ as a Lie subalgebra of $\hat{\mathfrak{g}}^D$ by mapping an element of $\mathfrak{g}_{\text{out}}$ to its germs at Q_i's.

Definition 2.1. The space of *conformal coinvariants* $\mathrm{CC}_E(M)$ and that of *conformal blocks* $\mathrm{CB}_E(M)$ over E associated to $\hat{\mathfrak{g}}^{Q_i}$-modules M_i with the same level $\hat{k} = k$ are defined by

$$\mathrm{CC}_E(M) := M/\mathfrak{g}_{\text{out}}M, \quad \mathrm{CB}_E(M) := \mathrm{Hom}_{\mathbb{C}}(M/\mathfrak{g}_{\text{out}}M, \mathbb{C}), \tag{13}$$

where $M := \bigotimes_{i=1}^L M_i$. The module M_i is referred to as "a module inserted at the point Q_i".

3. Family of elliptic curves

In this section we construct a family of elliptic curves $\mathcal{E}$ with a singular fibre and its covering $\tilde{\mathcal{E}}$. A twisted Lie algebra bundle $\mathfrak{g}^{\text{tw}}$ over the generic fibres of $\mathcal{E}$ is one of the main representation theoretical data in the twisted WZW model but it does not directly extends to the singular fibre. Hence we pull it back to $\tilde{\mathcal{E}}$ and trivialise it. The sections of $\mathfrak{g}^{\text{tw}}$ on the singular fibre is understood as sections of the trivial bundle on $\tilde{\mathcal{E}}$ invariant under the action of the covering transformation group.

The construction of $\tilde{\mathcal{E}}$ is almost the same as that of the analytic fibre space of elliptic curves in [Wo]. We use N patches U_k ($k \in \mathbb{Z}/N\mathbb{Z}$):

$$U_k := \{(q, x_k, y_k) \mid |q| < 1, x_k y_k = q, |x_k| < |q|^{-1}, |y_k| < |q|^{-1}\}. \tag{14}$$

We denote $(q, x, y) \in U_k$ by $(q, x, y)_k$. The universal curve $\tilde{\mathcal{E}}$ is defined by

$$\tilde{\mathcal{E}} := \bigsqcup_{k \in \mathbb{Z}/N\mathbb{Z}} U_k \Big/ \sim, \tag{15}$$

where the equivalence relation $\sim$ is defined by

$$(q, x_k, y_k)_k \sim (q, x'_{k+1}, y'_{k+1})_{k+1} \text{ when } x_k y'_{k+1} = 1. \tag{16}$$

We have an analytic fibre space $\tilde{\pi} : \tilde{\mathcal{E}} \to \Delta$ over $\Delta = \{q \mid |q| < 1\}$. The fibre over $q \neq 0$ is an elliptic curve $\mathbb{C}^\times / q^{N\mathbb{Z}}$ and the fibre over $q = 0$ is singular with ordinary double points $(0, 0, 0)_k \in U_k$ ($k \in \mathbb{Z}/N\mathbb{Z}$).

Let $C_N = \{\bar{0}, \bar{1}, \dots, \overline{N-1}\}$ be the cyclic group of order N. The group $C_N^2 = C_N \times C_N$ acts on $\tilde{\mathcal{E}}$ from the right as follows: it is enough to define the action of the generators $(1, 0)$ and $(0, 1)$ of C_N^2.

$$(q, x_k, y_k)_k \cdot (1, 0) = (q, \varepsilon^{-1} x_k, \varepsilon y_k)_k, \qquad (q, x_k, y_k)_k \cdot (0, 1) = (q, x_k, y_k)_{k-1}. \tag{17}$$

The universal curve $\mathcal{E}$ is defined set-theoretically as the quotient space of $\tilde{\mathcal{E}}$ by this action:

$$\mathcal{E} := \tilde{\mathcal{E}} / C_N^2. \tag{18}$$

The canonical projection to Δ is denoted by $\pi : \mathcal{E} \to \Delta$. The fibre $\pi^{-1}(0)$ is the singular curve with one ordinary double point.

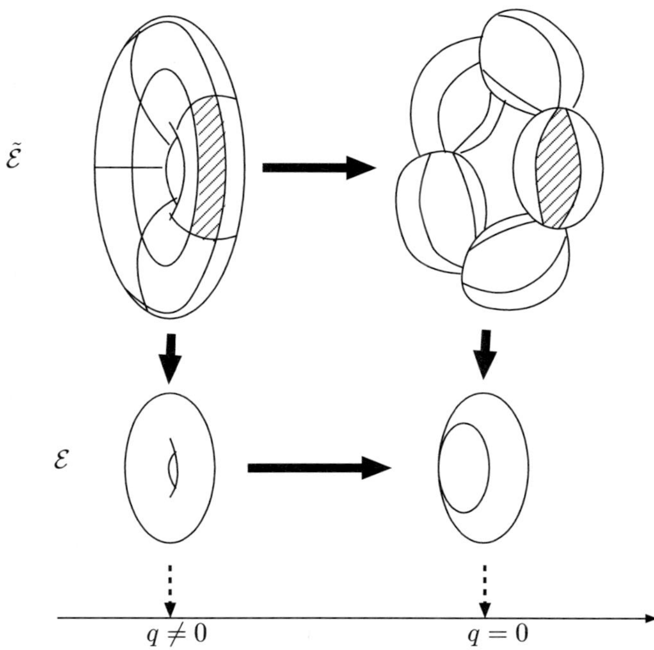

FIGURE 1. Degeneration of an elliptic curve and its N^2-covering.

The family of L-pointed elliptic curves and its covering are defined by

$$\mathfrak{X} := \mathcal{E} \times_\Delta S, \qquad \tilde{\mathfrak{X}} := \tilde{\mathcal{E}} \times_\Delta S. \tag{19}$$

Here the base space S is the fibre product of $\mathcal{E}$'s without diagonals:

$$\begin{aligned} S :=&\{(q; Q_1, \dots, Q_L) \in \overbrace{\mathcal{E} \times_\Delta \cdots \times_\Delta \mathcal{E}}^{L} \\ &\mid Q_i \neq Q_j \, (i \neq j), Q_i \neq [(0,0,0)_k] \text{ for any k.}\}, \end{aligned} \tag{20}$$

where Q_i is the point of the ith $\mathcal{E}$ in the fibre product with $\pi(Q_i) = q$. We exclude the degeneration of the types $Q_i \to Q_j$ and $Q_i \to$ (node). We denote the canonical projections $\mathfrak{X} \to S$, $\tilde{\mathfrak{X}} \to S$ and $\tilde{\mathfrak{X}} \to \mathfrak{X}$ by $\pi_{\mathfrak{X}/S}$, $\pi_{\tilde{\mathfrak{X}}/S}$ and $\pi_{\tilde{\mathfrak{X}}/\mathfrak{X}}$ respectively. The fibres of $\tilde{\mathfrak{X}}$ and $\mathfrak{X}$ over $S_0 := \{(0; Q_1, \dots, Q_L) \in S\}$ are singular curves.

The section q_i of $\mathfrak{X} \to S$ is defined by

$$q_i((q; Q_1, \dots, Q_L)) := (q; Q_i; Q_1, \dots, Q_L). \tag{21}$$

We denote the divisor $[q_1(S)] + \cdots + [q_L(S)]$ by D.

The group C_N^2 acts on $\tilde{\mathfrak{X}}$ naturally as covering transformation and on $\mathfrak{g}$ by

$$(m,n)\cdot A = (\gamma^m \beta^n) A (\gamma^m \beta^n)^{-1}. \tag{22}$$

The twisted Lie algebra bundle $\mathfrak{g}_{\mathfrak{X}}^{\mathrm{tw}}$ on $\overset{\circ}{\mathfrak{X}} := \mathfrak{X} \setminus \{\text{singular points}\}$ is defined as the associated bundle to the C_N^2-principal bundle $\overset{\circ}{\tilde{\mathfrak{X}}} := \tilde{\mathfrak{X}} \setminus \{\text{singular points}\} \to \overset{\circ}{\mathfrak{X}}$:

$$\mathfrak{g}_{\mathfrak{X}}^{\mathrm{tw}} := \overset{\circ}{\tilde{\mathfrak{X}}} \times_{C_N^2} \mathfrak{g}. \tag{23}$$

It is obvious that the restriction of $\mathfrak{g}_{\mathfrak{X}}^{\mathrm{tw}}$ to a fibre of $\mathfrak{X}$ at a point $(q; Q_1, \dots, Q_L)$ $(q \neq 0)$ is the bundle $\mathfrak{g}^{\mathrm{tw}}$ on the elliptic curve $\mathbb{C}^\times / q^{\mathbb{Z}}$ defined by (3).

Sheaf version of affine Lie algebras $\mathfrak{g}^P$, $\mathfrak{g}_+^P$, $\mathfrak{g}^D$ and $\hat{\mathfrak{g}}^D$ (cf. (7), (8), (10)) are $\mathcal{O}_S$-Lie algebras defined by

$$\begin{aligned}
&\mathfrak{g}_S^{Q_i} := \pi_{\mathfrak{X}/S,*}(\mathfrak{g}_{\mathfrak{X}}^{\mathrm{tw}}(*Q_i))^{\wedge}_{Q_i}, &&\mathfrak{g}_{S,+}^{Q_i} := \pi_{\mathfrak{X}/S,*}(\mathfrak{g}_{\mathfrak{X}}^{\mathrm{tw}})^{\wedge}_{Q_i}, \\
&\mathfrak{g}_S^{D} := \pi_{\mathfrak{X}/S,*}(\mathfrak{g}_{\mathfrak{X}}^{\mathrm{tw}}(*D))^{\wedge}_{D} = \bigoplus_{i=1}^{L} \mathfrak{g}_S^{Q_i}, &&\mathfrak{g}_{S,+}^{D} := \pi_{\mathfrak{X}/S,*}(\mathfrak{g}_{\mathfrak{X}}^{\mathrm{tw}})^{\wedge}_{D} = \bigoplus_{i=1}^{L} \mathfrak{g}_{S,+}^{Q_i}.
\end{aligned} \tag{24}$$

Since we assume that sections Q_i do not touch the singular point of the singular fibre, the definitions are the same as those for the non-singular case, (3.11) of [KT]. The central extensions of $\mathfrak{g}_S^D$ and $\mathfrak{g}_{S,+}^D$ are defined by the cocycle (9) with the coefficients in $\mathcal{O}_S$:

$$\hat{\mathfrak{g}}_S^D := \mathfrak{g}_S^D \oplus \mathcal{O}_S \hat{k}, \qquad \hat{\mathfrak{g}}_{S,+}^D := \mathfrak{g}_{S,+}^D \oplus \mathcal{O}_S \hat{k}. \tag{25}$$

The Lie subalgebra of meromorphic sections $\mathfrak{g}_{\mathrm{out}} \subset \hat{\mathfrak{g}}^D$, (12), would be replaced by $\pi_{\mathfrak{X}/S,*}(\mathfrak{g}_{\mathfrak{X}}^{\mathrm{tw}}(*D))$ if there were no singularity, as was the case in [KT]. Taking the singular fibre into account, we modify this naive definition as follows:

$$\mathfrak{g}_{\mathfrak{X},\mathrm{out}} = \left(\pi_{\tilde{\mathfrak{X}}/S,*}(\mathfrak{g} \otimes \mathcal{O}_{\tilde{\mathfrak{X}}}(*\tilde{D}))\right)^{C_N^2}, \tag{26}$$

where $\tilde{D}$ is the C_N^2-orbit of the divisor D and $(\cdot)^{C_N^2}$ denotes the C_N^2-invariant section of the equivariant locally free sheaf $\mathfrak{g} \otimes \mathcal{O}_{\tilde{\mathfrak{X}}}$. In other words, a $\mathfrak{g}$-valued meromorphic function $f(s,P)$ $(s = (q; Q_1, \dots, Q_L) \in S,\ P \in \tilde{\mathcal{E}}|_s)$ belongs to $\mathfrak{g}_{\mathfrak{X},\mathrm{out}}$ if and only if it satisfies

$$f(s,(1,0)\cdot P) = \operatorname{Ad}\gamma(f(s,P)), \qquad f(s,(0,1)\cdot P) = \operatorname{Ad}\beta(f(s,P)), \tag{27}$$

where $(m,n)\cdot$ $(m,n \in \mathbb{Z})$ is the left action of the generators of C_N^2 on the fibre of $\tilde{\mathcal{E}}$ defined by $(m,n)\cdot P := P \cdot (-m,-n)$. (See (17).)

Our construction is so explicit that we have an explicit basis of $\mathfrak{g}_{\mathfrak{X},\mathrm{out}}$. In [KT] meromorphic functions $w_{ab}(\tau;t)$ $(a,b = 0,\dots,N-1,\ \operatorname{Im}\tau > 0,\ t \in \mathbb{C})$ characterized by the following properties were introduced:

- Additive quasi-periodicity:

$$w_{ab}(\tau;t+1)=\varepsilon^a w_{ab}(\tau;t),\quad w_{ab}(\tau;t+\tau)=\varepsilon^b w_{ab}(\tau;t);$$

- As a function of $t\in\mathbb{C}$, $w_{ab}(\tau;t)$ has a simple pole with residue 1 at $\mathbb{Z}+\mathbb{Z}\tau$.

Let us denote this function by $w_{ab}^{\mathrm{add}}(\tau;t)$. (The superscript "add" stands for "additive".) Let us rewrite it to a multiplicatively quasi-periodic function $w_{ab}^{\mathrm{mul}}(q;u)$ as follows:

$$\begin{aligned} w_{ab}^{\mathrm{mul}}(q;u) :=& \frac{2\pi i u^a}{u^N-1} \\ &\times \frac{(q^{N-a}\varepsilon^b u^{-N};q^N)_\infty (q^a\varepsilon^{-b}u^N;q^N)_\infty}{(q^{N-a}\varepsilon^b;q^N)_\infty (q^N u^{-n};q^N)_\infty (q^a\varepsilon^{-b};q^N)_\infty (q^N u^N;q^N)_\infty}, \end{aligned} \tag{28}$$

where $(x;q)_\infty=\prod_{n=0}^\infty(1-xq^n)$ is the standard infinite product symbol. The function $w_{ab}^{\mathrm{mul}}(q;u)$ is related to w_{ab}^{add} by $w_{ab}^{\mathrm{mul}}(q;u)=w_{ab}^{\mathrm{add}}(N\log u/2\pi i, N\log q/2\pi i)$ when $q\neq 0$, that is, we replaced the arguments of w_{ab}^{add} by $e^{2\pi i z}=u^N$, $e^{2\pi i\tau}=q^N$ and used the product formula for the theta function. When $q=0$, w_{ab}^{mul} becomes a rational function of u:

$$w_{ab}^{\mathrm{mul}}(0;u)=\begin{cases}2\pi i u^a(u^N-1)^{-1}, & (a\neq 0),\\ 2\pi i(1-\varepsilon^b)^{-1}(u^N-\varepsilon^b)(u^N-1)^{-1}, & (a=0).\end{cases} \tag{29}$$

The important property of w_{ab}^{mul} is that it inherits the quasi-periodicity of w_{ab}^{add}:

$$w_{ab}^{\mathrm{mul}}(q;\varepsilon u)=\varepsilon^a w_{ab}^{\mathrm{mul}}(q;u),\qquad w_{ab}^{\mathrm{mul}}(q;qu)=\varepsilon^b w_{ab}^{\mathrm{mul}}(q;u). \tag{30}$$

We define the function $w_{ab,i}(P)$ $(i=1,\dots,L)$ on $\tilde{\mathfrak{X}}$ in terms of w_{ab}^{mul}: for $(q,x_k,y_k)_k\in U_k$ (cf. (14)), $q\neq 0$,

$$w_{ab,i}(((q,x_k,y_k)_k,s)) := w_{ab}^{\mathrm{mul}}(q;q^{k-k'}x_k/x'_{k'}). \tag{31}$$

Here $s=(q;Q_1,\dots,Q_L)\in S$ and we fix an index k' to express Q_i as a point $(q,x'_{k'},y'_{k'})_{k'}$ in $U_{k'}$. (The function $w_{ab,i}$ is determined up to this choice.) This function is extended to the points with $q=0$. (For example, $w_{ab,i}((0,x_0,y_0)_0)$ is the rational function (29) of $x=x_0/x'_0$ when Q_i is represented as a point $(0;x'_0,y'_0)_0$ in U_0.) The main properties of this function are

- Quasi-periodicity with respect to C_N^2-action:

$$w_{ab,i}((1,0)\cdot P)=\varepsilon^a w_{ab,i}(P),\qquad w_{ab,i}((0,1)\cdot P)=\varepsilon^b w_{ab,i}(P), \tag{32}$$

- All poles are simple and located at Q_i modulo C_N^2 action.

It is easy to see that any section of $\mathfrak{g}_{\mathfrak{X},\mathrm{out}}$ is a linear combination of $J_{ab}\otimes w_{ab,i}(P)$'s and their derivatives along the fibre.

Lemma 3.1. $\hat{\mathfrak{g}}_S^D=\mathfrak{g}_{\mathfrak{X},\mathrm{out}}\oplus\hat{\mathfrak{g}}_{S,+}^D$.

Proof. The singular part of an element of $\hat{\mathfrak{g}}_S^D$ can be expressed by a linear combination of $J_{ab}\otimes w_{ab,i}(P)$ and derivatives in a unique way. Subtracting such combination which belongs to $\mathfrak{g}_{\mathfrak{X},\mathrm{out}}$, we end up with a regular element of $\hat{\mathfrak{g}}_{S,+}^D$. □

4. Sheaves of conformal coinvariants and conformal blocks

In this section we introduce the sheaf $\mathcal{CC}$ of conformal coinvariants and the sheaf $\mathcal{CB}$ of conformal blocks and show their basic properties.

Definitions of $\mathcal{CC}$ and $\mathcal{CB}$ are literally the same as those for the non-singular case, Definition 3.3 of [KT].

Definition 4.1. For any $\hat{\mathfrak{g}}^D_S$-module $\mathcal{M}$ of level k (i.e., $\hat{k}$ acts as $k \cdot \mathrm{id}$), we define the *sheaf* $\mathcal{CC}(\mathcal{M})$ *of conformal coinvariants* and the *sheaf* $\mathcal{CB}(\mathcal{M})$ *of conformal blocks* by

$$\mathcal{CC}(\mathcal{M}) := \mathcal{M}/\mathfrak{g}_{\mathfrak{X},\mathrm{out}}\mathcal{M}, \tag{33}$$

$$\mathcal{CB}(\mathcal{M}) := \mathcal{H}om_{\mathcal{O}_S}(\mathcal{CC}(\mathcal{M}), \mathcal{O}_S). \tag{34}$$

We can regard $\mathcal{CC}(\cdot)$ as a covariant right exact functor from the category of $\hat{\mathfrak{g}}^D_S$-modules to that of $\mathcal{O}_S$-modules and similarly $\mathcal{CB}(\cdot)$ as a contravariant left exact functor.

The goal of this paper is to prove that $\mathcal{CC}$ and $\mathcal{CB}$ are locally free. Since $\mathcal{CB}$ is the dual of $\mathcal{CC}$, we mainly discuss about $\mathcal{CC}$ and briefly mention on $\mathcal{CB}$ when it is necessary.

We assume that the $\hat{\mathfrak{g}}^D_S$-module $\mathcal{M}$ are of the following type:

$$\mathcal{M} = \mathcal{O}_S \otimes \bigotimes_{i=1}^{L} M_i = \bigotimes_{i=1}^{L} (\mathcal{O}_S \otimes M_i), \tag{35}$$

where each M_i is a quotient of a $\hat{\mathfrak{g}}$-Weyl module $M(V_i) := \mathrm{Ind}_{\hat{\mathfrak{g}}^{Q_i}_{+}}^{\mathfrak{g}^{Q_i} \oplus \mathbb{C}\hat{k}} V_i$ of level k for a finite dimensional irreducible $\mathfrak{g}$-module V_i. (See §6 of [T] or §2.4 of [KL].) To endow $\mathcal{M}$ in (35) with the $\hat{\mathfrak{g}}^D_S$-module structure, we need to fix the coordinate of $\mathcal{E}$ and the trivialisation of $\mathfrak{g}^{\mathrm{tw}}_{\mathfrak{X}}$, which is irrelevant to the statements of theorems below. In the concrete computations, we use the coordinates and the trivialisation obtained naturally from the construction in §3.

Proposition 4.2. *$\mathcal{CC}(\mathcal{M})$ is a coherent $\mathcal{O}_S$-sheaf. When all M_i's are Weyl modules, $\mathcal{CC}(\mathcal{M})$ is (and hence $\mathcal{CB}(\mathcal{M})$ is) locally free.*

Proof. Lemma 3.1 makes it possible to apply the same argument as the proof for the non-singular case, Corollary 3.5 of [KT]. In fact, if each M_i in (35) is a Weyl module $M(V_i)$, $\mathcal{M}$ is expressed as

$$\mathcal{M} = U_S(\hat{\mathfrak{g}}^D_S) \otimes_{U_S(\hat{\mathfrak{g}}^D_{S,+})} (V \otimes \mathcal{O}_S) \xleftarrow{\sim} U_S(\mathfrak{g}_{\mathfrak{X},\mathrm{out}}) \otimes_{\mathcal{O}_S} (V \otimes \mathcal{O}_S), \tag{36}$$

by the Poincaré-Birkhoff-Witt theorem ($V = \otimes_{i=1}^{L} V_i$). Here $U_S(\cdot)$ denotes the universal $\mathcal{O}_S$-enveloping algebra. Modding out by $\mathfrak{g}_{\mathfrak{X},\mathrm{out}}\mathcal{M}$, we have

$$\mathcal{CC}(\mathcal{M}) = \mathcal{M}/\mathfrak{g}_{\mathfrak{X},\mathrm{out}}\mathcal{M} \xleftarrow{\sim} V \otimes \mathcal{O}_S, \tag{37}$$

which means that $\mathcal{CC}(\mathcal{M})$ is locally free, in particular, coherent.

If M_i's are quotients of Weyl modules, $\mathcal{M}$ is a quotient of $\bigotimes M(V_i) \otimes \mathcal{O}_S$. Hence by the right exactness of the functor $\mathcal{CC}$, $\mathcal{CC}(\mathcal{M})$ is a quotient of a coherent sheaf, and therefore coherent. □

In §6 we prove the locally freeness of $\mathcal{CC}(\mathcal{M})$ for integrable M_i's, examining the behavior of $\mathcal{CC}(\mathcal{M})$ at the boundary of the moduli space ($S_0 = \{q = 0\} \subset S$) carefully.

5. Sheaf version of trigonometric and orbifold WZW model

Everything in previous two sections can be restricted on S_0, namely on the configuration space of points on a singular rational curve with one ordinary double point. (As we have mentioned, the restriction of the functions $w_{ab,i}$ needs special care.) Hence we can define the corresponding sheaves $\mathcal{CC}(\mathcal{M})$ and $\mathcal{CB}(\mathcal{M})$ which we denote by $\mathcal{CC}_{\mathrm{trig}}(\mathcal{M})$ and $\mathcal{CB}_{\mathrm{trig}}(\mathcal{M})$. The subscript "trig" is put here because, as we shall see below, there are connections on them expressed in terms of the trigonometric r-matrix.

In the proof of Theorem 6.1 we shall use the sheaf of conformal coinvariants of the orbifold WZW model, $\mathcal{CC}_{\mathrm{orb}}(\mathcal{M}_0 \otimes \mathcal{M} \otimes \mathcal{M}_\infty)$ on S_0, where $\mathcal{M}_* = M_* \otimes \mathcal{O}_{S_0}$ for a $\hat{\mathfrak{g}}^{(*)}$-module M_* ($* = 0, \infty$). We shall recall the definition of the twisted affine algebras $\hat{\mathfrak{g}}^{(0)}$ and $\hat{\mathfrak{g}}^{(\infty)}$ and the details of $\mathrm{CC}_{\mathrm{orb}}$ soon later. Here we only say that $\mathcal{CC}_{\mathrm{orb}}$ is defined exactly in the same way as $\mathcal{CC}_{\mathrm{trig}}(\mathcal{M})$ if we replace the degenerate elliptic curve (the fibre of $\mathcal{E}$ at $q = 0$) by the orbifold $\mathbb{P}^1/C_N$. Note that S_0 can be regarded as the configuration space of points on $\mathbb{P}^1/C_N$. (See §3 of [T].)

Let $\mathcal{M}$ be as in (35) and M_* be a $\hat{\mathfrak{g}}^{(*)}$-Verma module. Then Proposition 4.2 holds for $\mathcal{CC}_{\mathrm{trig}}$, $\mathcal{CB}_{\mathrm{trig}}$, $\mathcal{CC}_{\mathrm{orb}}$ and $\mathcal{CB}_{\mathrm{orb}}$ as well. In fact, we can prove locally freeness under this assumption.

Proposition 5.1.

(i) *The sheaf $\mathcal{CC}_{\mathrm{trig}}(\mathcal{M})$ and the sheaf $\mathcal{CB}_{\mathrm{trig}}(\mathcal{M})$ are locally free $\mathcal{O}_{S_0}$-sheaves.*

(ii) *The sheaf $\mathcal{CC}_{\mathrm{orb}}(\mathcal{M}_0 \otimes \mathcal{M} \otimes \mathcal{M}_\infty)$ and the sheaf $\mathcal{CB}_{\mathrm{orb}}(\mathcal{M}_0 \otimes \mathcal{M} \otimes \mathcal{M}_\infty)$ are locally free $\mathcal{O}_{S_0}$-sheaves.*

Proof. (i) The proof of the locally freeness for the non-singular case, Corollary 5.3 of [KT] is true also in this case: $\mathcal{CC}_{\mathrm{trig}}(\mathcal{M})$ is coherent as shown at the end of §4 and there is a connection and D-module structure on it, which implies that it is locally free $\mathcal{O}_{S_0}$-sheaf. The only difference is that we do not change the curve itself (the modulus q is fixed to 0) in the present case, and hence there is nothing corresponding to the connection in the direction of $\partial/\partial\tau$ in [KT]. The connection in the direction of $\partial/\partial z_i$ (z_i is the coordinate of Q_i) is $\nabla_i = \partial/\partial z_i - \rho_i(T[-1])$ (cf. (5.14) of [KT]) as is well known, where ρ_i is the representation of the Virasoro algebra on M_i constructed via the Sugawara construction and $T[-1]$ is one of the Virasoro generator, usually denoted by L_{-1}.

(ii) We might proceed as the proof of (i) from the beginning but the short cut is to use the result of (i). The coherence of $\mathcal{CC}_{\mathrm{orb}}(\mathcal{M}_0 \otimes \mathcal{M} \otimes \mathcal{M}_\infty)$ having been proved, we have only to check that the above connection operators ∇_i ($i = 1, \dots, L$) on $\mathcal{CC}_{\text{trig}}(\mathcal{M})$ also define the flat connection on $\mathcal{CC}_{\text{orb}}$. What we need to check is

- $[\nabla_i, \mathfrak{g}^{\text{orb}}_{\text{out}}] \subset \mathfrak{g}^{\text{orb}}_{\text{out}}$,
- $[\nabla_i, \nabla_j] = 0$ on $\mathcal{CC}_{\text{orb}}$,

which is proved in the same way as in the case of the ordinary WZW model, e.g., Lemma 4 of [FFR]. □

The connection on $\mathcal{CC}_{\text{trig}}(\mathcal{M})$ mentioned in the proof of (i) is obtained by the degeneration $q \to 0$ of the elliptic Knizhnik-Zamolodchikov connection in [E] and [KT]. They are expressed as the first order differential operators on $V \otimes \mathcal{O}_{S_0}$ in terms of the *trigonometric r-matrix.* In fact, by tracing the argument which leads to the explicit form (Theorem 5.9 in [KT]) of the connection, we have only to replace the functions $w_{ab}(z_j - z_i)$ there with $w_{ab,i}(Q_j)$ which is expressed by the rational function of the form (29) on S_0. Hence the KZ equation for the WZW model on the degenerate elliptic curve is the *trigonometric KZ equation.*

Lemma 5.2.
(i) *The fibre of $\mathcal{CC}_{\text{trig}}(\mathcal{M})$ at $s \in S_0$, $\mathcal{CC}_{\text{trig}}(\mathcal{M})|_s$, is isomorphic to $\mathrm{CC}_{\text{trig}}(M)$, the space of conformal coinvariants of the trigonometric WZW model for the geometric data corresponding to s.*
(ii) *The fibre of $\mathcal{CC}_{\text{orb}}(\mathcal{M})$ at s, $\mathcal{CC}_{\text{orb}}(\mathcal{M})|_s$, is isomorphic to $\mathrm{CC}_{\text{orb}}(M)$, the space of conformal coinvariants of the orbifold WZW model for the geometric data corresponding to s.*

See Definition 3.2 of [T] for the definition of $\mathrm{CC}_{\text{trig}}$ and CC_{orb}.

Proof. We can modify the proof of the corresponding statement for the non-singular case, Corollary 5.4 of [KT]. For example, for the case (i), the isomorphism $\mathfrak{g}_{\mathfrak{X},\text{out}}|_s \cong \mathfrak{g}^{\text{trig}}_{\text{out}}$ is a consequence of the existence of the sections $w_{ab,i}$ and their derivatives (cf. the end of §3) which span both $\mathfrak{g}_{\mathfrak{X},\text{out}}|_s$ and $\mathfrak{g}^{\text{trig}}_{\text{out}}$. The rest of the proof can be translated to the present case without change.

The proof of (ii) is similar. □

Combining Proposition 5.1, Lemma 5.2 and Theorem 5.1 of [T], we have the following isomorphism:

$$\iota : \mathcal{CC}_{\text{trig}}(\mathcal{M}) \xrightarrow{\sim} \bigoplus_{\lambda \in \mathrm{wt}(V)} \mathcal{CC}_{\text{orb}}(\mathcal{M}^{(0)}_{\tilde{\lambda}} \otimes \mathcal{M} \otimes \mathcal{M}^{(\infty)}_{\tilde{\lambda}'}), \tag{38}$$

where $V = \bigotimes_{i=1}^L V_i$ is the $\mathfrak{g}$-module generating $M = \bigotimes_{i=1}^L M_i$ (recall M_i is a quotient of Weyl module $M(V_i)$), $\mathrm{wt}(V)$ is the set of its weights,

$$\begin{aligned} \tilde{\lambda} &:= -\lambda \circ (1 - \mathrm{Ad}\,\beta^{-1})^{-1} = \lambda \circ (1 - \mathrm{Ad}\,\beta)^{-1} \circ \mathrm{Ad}\,\beta, \\ \tilde{\lambda}' &:= -\lambda \circ (1 - \mathrm{Ad}\,\beta)^{-1}, \end{aligned} \tag{39}$$

$\mathcal{M}_\mu^{(*)} = M_\mu^{(*)} \otimes \mathcal{O}_{S_0}$ for a Verma module $M_\mu^{(*)}$ of $\hat{\mathfrak{g}}^{(*)}$ with the highest weight μ (cf. Definition 4.1 (i) of [T]).

In §6 the modules M_i are assumed to be integrable highest weight modules. (cf. Chapter 10 of [K].) In this case the above result can be refined. For this purpose we recall the details of the orbifold WZW model defined in §3 of [T].

Let us denote the standard coordinate of $\mathbb{P}^1(\mathbb{C})$ by t. The cyclic group C_N acts as $t \mapsto \varepsilon^a t$ $(a \in C_N)$ and the quotient $E_{\mathrm{orb}} = \mathbb{P}^1/C_N$ is an ordinary orbifold.

The definition of the space of conformal coinvariants/blocks of the orbifold WZW model on E_{orb} is almost the same as that on elliptic curves, Definition 2.1, except that we also insert modules to the singular points 0 and ∞. The Lie algebra $\mathfrak{g}_{\mathrm{out}}$ in (13) is replaced by $\mathfrak{g}_{\mathrm{out}}^{\mathrm{orb}}$ which consists of $\mathfrak{g}$-valued meromorphic functions $f(t)$ on $\mathbb{P}^1$ such that: (1) poles belong to $\{0, Q_1, \dots, Q_L, \infty\}$; (2) $f(\varepsilon t) = \operatorname{Ad}\gamma(f(t))$. Accordingly, the module inserted at 0 is the $\hat{\mathfrak{g}}^{(0)}$-module and the module inserted at ∞ is the $\hat{\mathfrak{g}}^{(\infty)}$-module, where

$$\hat{\mathfrak{g}}^{(0)} = \bigoplus_{\substack{a,b=0,\dots,N-1\\(a,b)\neq(0,0)}} \bigoplus_{m\in\mathbb{Z}} \mathbb{C}J_{a,b} \otimes t^{a+mN} \oplus \mathbb{C}\hat{k}, \tag{40}$$

$$\begin{aligned}\mathfrak{g}^{(\infty)} &= \bigoplus_{\substack{a,b=0,\dots,N-1\\(a,b)\neq(0,0)}} \bigoplus_{m\in\mathbb{Z}} \mathbb{C}J_{a,b} \otimes t^{a+mN} \oplus \mathbb{C}\hat{k} \\ &= \bigoplus_{\substack{a,b=0,\dots,N-1\\(a,b)\neq(0,0)}} \bigoplus_{m\in\mathbb{Z}} \mathbb{C}J_{a,b} \otimes s^{-a+mN} \oplus \mathbb{C}\hat{k}. \qquad (s = t^{-1})\end{aligned} \tag{41}$$

The cocycles which defines the central extension of $\mathfrak{g}^{(0)}$ and $\mathfrak{g}^{(\infty)}$ are:

$$\mathrm{c}_{\mathrm{a},0}(A,B) := \frac{1}{N}\operatorname{Res}_{t=0}(dA|B), \qquad \mathrm{c}_{\mathrm{a},\infty}(A,B) := \frac{1}{N}\operatorname{Res}_{s=0}(dA|B), \tag{42}$$

for $A, B \in \mathfrak{g}^{(0)}$ and $A, B \in \mathfrak{g}^{(\infty)}$ respectively. ($\mathfrak{g}^{(*)}$ is the loop algebra part of $\hat{\mathfrak{g}}^{(*)}$.)

As Etingof showed (Lemma 1.1 of [E]), $\hat{\mathfrak{g}}^{(0)}$ and $\hat{\mathfrak{g}}^{(\infty)}$ are isomorphic to the ordinary affine Lie algebra $\hat{\mathfrak{g}}$ of $A_{N-1}^{(1)}$ type. Explicitly the isomorphism $\phi_0 : \hat{\mathfrak{g}}^{(0)} \xrightarrow{\sim} \hat{\mathfrak{g}}$ is defined by

$$\begin{aligned} \phi_0(E_{i,i+1} \otimes t) &= e_i, & \phi_0(E_{i+1,i} \otimes t^{-1}) &= f_i, \\ \phi_0(E_{N,1} \otimes t) &= e_0, & \phi_0(E_{1,N} \otimes t^{-1}) &= f_0, \\ \phi_0(H_{i,i+1} \otimes 1) &= \alpha_i^\vee - \frac{\hat{k}}{N}, & \phi_0(\hat{k}) &= \hat{k}, \end{aligned} \tag{43}$$

for $i = 1, \dots, N-1$, where E_{ij} is the matrix unit, $H_{ij} = E_{ii} - E_{jj}$, e_i, f_i $(i = 0, \dots, N-1)$ are the Chevalley generators of $\hat{\mathfrak{g}}$ and $\alpha_i^\vee$ $(i = 0, \dots, N-1)$ are coroots of $\hat{\mathfrak{g}}$. (cf. §6.2 and §7.4 of [K].) Since the positive powers of s (cf. (41)) kill the highest weight vector of the Verma module $M_\mu^{(\infty)}$ inserted at ∞ (cf. Definition 4.1 (i) of [T]), we identify $\hat{\mathfrak{g}}^{(\infty)}$ and $\hat{\mathfrak{g}}$ through an isomorphism which is essentially a composition of ϕ_0 and the Chevalley involution:

$$\begin{aligned} \phi_\infty(E_{i,i+1}\otimes s^{-1}) &= -f_i, & \phi_\infty(E_{i+1,i}\otimes s) &= -e_i,\\ \phi_\infty(E_{N,1}\otimes s^{-1}) &= -f_0, & \phi_\infty(E_{1,N}\otimes s) &= -e_0,\\ \phi_\infty(H_{i,i+1}\otimes 1) &= -\alpha_i^\vee + \frac{\hat{k}}{N}, & \phi_\infty(\hat{k}) &= \hat{k}, \end{aligned} \tag{44}$$

for $i = 1, \dots, N-1$. Identified through ϕ_0 and ϕ_∞, the Verma modules of $\hat{\mathfrak{g}}^{(0)}$ and $\hat{\mathfrak{g}}^{(\infty)}$ are Verma modules of $\hat{\mathfrak{g}}$ in ordinary sense.

In this section, e_i and f_i denote the Chevalley generators of $\hat{\mathfrak{g}}$ identified with the elements in $\hat{\mathfrak{g}}^{(0)}$ or $\hat{\mathfrak{g}}^{(\infty)}$ by means of ϕ_0 or ϕ_∞.

Proposition 5.3. *Assume that all M_i $(i = 1, \dots, L)$ are integrable highest weight modules and that $M^{(*)}_{\mu_*}$ $(* = 0, \infty)$ is a Verma modules of $\hat{\mathfrak{g}}^{(*)}$.*

Then $\mathrm{CC}_{\mathrm{orb}}(M^{(0)}_{\mu_0} \otimes M \otimes M^{(\infty)}_{\mu_\infty})$ is 0 unless μ_0 and μ_∞ are dominant integral weights of $\hat{\mathfrak{g}}$ identified with $\hat{\mathfrak{g}}^{(0)}$ and $\hat{\mathfrak{g}}^{(\infty)}$. If μ_0 and μ_∞ are dominant integral weights,

$$\mathrm{CC}_{\mathrm{orb}}(M^{(0)}_{\mu_0} \otimes M \otimes M^{(\infty)}_{\mu_\infty}) \cong \mathrm{CC}_{\mathrm{orb}}(L^{(0)}_{\mu_0} \otimes M \otimes L^{(\infty)}_{\mu_\infty}), \tag{45}$$

where $L^{(*)}_{\mu_*}$ $(* = 0, \infty)$ is the irreducible quotient of $M^{(*)}_{\mu_*}$.

Remark 5.4. In physics context, this proposition is a consequence of the propagation of the null field. See §4 of [Z]. The author thanks Yasuhiko Yamada for this comment.

Remark 5.5. Proposition 5.3 is in sharp contrast to the Weyl module case. See §6 of [T].

Proof. The following lemma shall be proved later.

Lemma 5.6. *Let N_* $(* = 0, \infty)$ be a quotient of the Verma module $M^{(*)}_{\mu_*}$. Suppose $v_\kappa \in N_0$ is a singular vector of weight κ which is not a dominant integral weight. (The weight κ may possibly be the highest weight μ_0.) Then for any $v \in M$ and $v_\infty \in N_\infty$,*

$$v_\kappa \otimes v \otimes v_\infty \equiv 0 \mod \mathfrak{g}^{\mathrm{orb}}_{\mathrm{out}}(N_0 \otimes M \otimes N_\infty). \tag{46}$$

The same is true for a singular vector $v_\kappa \in N_\infty$.

The first statement of Proposition 5.3 is a consequence of Lemma 5.6. For example, assume μ_0 is not a dominant integral weight. Let us show

$$X_1[-n_1]\cdots X_l[-n_l]|\mu_0\rangle \otimes v \otimes v_\infty \equiv 0 \mod \mathfrak{g}^{\mathrm{orb}}_{\mathrm{out}}, \tag{47}$$

where $X_i[-n_i] \in \hat{\mathfrak{g}}^{(0)}$ $(X_i \in \mathfrak{g}, n_i > 0)$, $v \in M$, $v_\infty \in M^{(\infty)}_{\mu_\infty}$ and $\mod \mathfrak{g}^{\mathrm{orb}}_{\mathrm{out}}$ denotes $\mod \mathfrak{g}^{\mathrm{orb}}_{\mathrm{out}}(M^{(0)}_{\mu_0} \otimes M \otimes M^{(\infty)}_{\mu_\infty})$. (This abbreviation shall be used throughout this paper.) Let $f_1(t)$ be an element of $\mathfrak{g}^{\mathrm{orb}}_{\mathrm{out}}$ such that $f_1(t) \sim X_1 \otimes t^{-n_1} + O(t^n)$ for sufficiently large n. (Such f_1 exists due to the Riemann-Roch theorem. It is not difficult to construct such a function concretely.[1]) Then we may replace $X_1[-n_1]$ by $\rho_0(f_1(t))$:

$$\begin{aligned} & X_1[-n_1]\cdots X_l[-n_l]|\mu_0\rangle\otimes v\otimes v_\infty \\ =&\rho_0(f_1(t))X_2[-n_2]\cdots X_l[-n_l]|\mu_0\rangle\otimes v\otimes v_\infty \\ \equiv&-X_2[-n_2]\cdots X_l[-n_l]|\mu_0\rangle\otimes \\ &\otimes\left(\sum_{i=1}^{L}\rho_i(f_1(t))v\otimes v_\infty+v\otimes\rho_\infty(f_1(t))v_\infty\right)\quad \mathrm{mod}\ \mathfrak{g}^{\mathrm{orb}}_{\mathrm{out}}. \end{aligned} \tag{48}$$

By induction on l, the problem is reduced to showing

$$|\mu_0\rangle\otimes v\otimes v_\infty\equiv 0 \quad \mathrm{mod}\ \mathfrak{g}^{\mathrm{orb}}_{\mathrm{out}}, \tag{49}$$

which immediately follows from Lemma 5.6.

To prove the second statement of Proposition 5.3, assume that μ_0 and μ_∞ are dominant integral weight. Then the irreducible quotients of $M^{(*)}_{\mu_*}$ are expressed as

$$L^{(*)}_{\mu_*}=M^{(*)}_{\mu_*}/\sum_{i=0}^{N-1}U(\mathfrak{n}_-)f_i^{\langle\mu_*,\alpha_i^\vee\rangle+1}|\mu_*\rangle. \tag{50}$$

See (10.4.6) of [K]. Therefore to prove (45), it is enough to show

$$U(\mathfrak{n}_-)f_i^{\langle\mu_0,\alpha_i^\vee\rangle+1}|\mu_0\rangle\otimes M\otimes M^{(\infty)}_{\mu_\infty}\equiv 0\quad \mathrm{mod}\ \mathfrak{g}^{\mathrm{orb}}_{\mathrm{out}}, \tag{51}$$

and a similar statement with the indices "0" and "∞" for μ_*, $M^{(*)}_{\mu_*}$ etc. interchanged. They are proved as above, namely by the arguments like (48) and (49), because the weight of the singular vector $f_i^{\langle\mu_*,\alpha_i^\vee\rangle+1}|\mu_*\rangle$ is not a dominant integral weight. $\square$

Proof of Lemma 5.6. Since κ is not a dominant integral weights, there is an index i $(0\leqq i\leqq N-1)$ such that $\kappa(\alpha_i^\vee)$ is not a non-negative integer. By an easy calculation, we have

$$e_i^n f_i^n v_\kappa = c v_\kappa,\qquad c=n!\prod_{l=1}^{n}(\kappa(\alpha_i^\vee)-l+1), \tag{52}$$

for any $n\in\mathbb{N}$. Note that the constant c never vanishes.

Let $e(t)$ be an element of $\mathfrak{g}^{\mathrm{orb}}_{\mathrm{out}}$ such that: (1) $e(t)\sim\phi_0^{-1}(e_i)+O(t^n)$ for sufficiently large n; (2) $\rho_\infty(e(t))v_\infty=0$ (i.e., $e(t)$ has a zero of large order at $t=\infty$). Such an element can be constructed in the form $X\otimes F(t)$, where $X=E_{i,i+1}$ $(i=1,\dots,N-1)$ or $X=E_{N,1}$ $(i=0)$ and $F(t)$ is a rational function. Hence we

[1]A useful technique: for any $\mathfrak{g}$-valued function $f(t)$ with poles in $\{0,Q_1,\dots,Q_L,\infty\}$, $f(t)+\mathrm{Ad}\,\gamma(f(\varepsilon^{-1}t))+(\mathrm{Ad}\,\gamma)^2(f(\varepsilon^{-2}t))+\cdots+(\mathrm{Ad}\,\gamma)^{N-1}(f(\varepsilon^{-N+1}t))\in\mathfrak{g}^{\mathrm{orb}}_{\mathrm{out}}$.

can rewrite $v_\kappa \otimes v \otimes v_\infty$ modulo $\mathfrak{g}^{\mathrm{orb}}_{\mathrm{out}}$ as follows (cf. p.479 of [TUY]):

$$\begin{aligned} & v_\kappa \otimes v \otimes v_\infty \\ & = c^{-1} \rho_0(e_i^n f_i^n) v_\kappa \otimes v \otimes v_\infty \\ & = c^{-1} \rho_0(e(t)^n f_i^n) v_\kappa \otimes v \otimes v_\infty \\ & \equiv (-1)^n c^{-1} \sum_{n_1+\cdots+n_L=n} \frac{n!}{n_1! \cdots n_L!} \rho_0(f_i^n) v_\kappa \otimes \prod_{j=1}^{L} \rho_j(e(t))^{n_j} v \otimes v_\infty . \end{aligned} \tag{53}$$

Recall that $\rho_j(e(t)) = \rho_j(X \otimes F(t))$ is locally nilpotent on M_j (Corollary 1.4.6 of [TUY]). Thus the right-hand side of (53) is 0 for large n, which completes the proof of the lemma. □

Because of the difference of the sign in (43) and (44) and the fact that $\mathrm{Ad}\,\beta$ is a Dynkin automorphism, if $\tilde\lambda$ is a dominant integral weight for $\hat{\mathfrak{g}}^{(0)}$, $\tilde\lambda'$ is a dominant integral weight for $\hat{\mathfrak{g}}^{(\infty)}$ and vice versa. (See (39).)

Corollary 5.7. *Under the same assumption as Proposition* 5.3 *we have*

$$\mathrm{CC}_{\mathrm{trig}}(M) \cong \bigoplus_{\lambda \in \mathrm{wt}(V), \tilde\lambda:\mathrm{dom.\ int.}} \mathrm{CC}_{\mathrm{orb}}(L^{(0)}_{\tilde\lambda} \otimes M \otimes L^{(\infty)}_{\tilde\lambda'}), \tag{54}$$

where "dom. int." means "dominant integral weight". Similarly the decomposition (38) *becomes*

$$\iota : \mathcal{CC}_{\mathrm{trig}}(\mathcal{M}) \xrightarrow{\sim} \bigoplus_{\lambda \in \mathrm{wt}(V), \tilde\lambda:\mathrm{dom.\ int.}} \mathcal{CC}_{\mathrm{orb}}(\mathcal{L}^{(0)}_{\tilde\lambda} \otimes \mathcal{M} \otimes \mathcal{L}^{(\infty)}_{\tilde\lambda'}), \tag{55}$$

where $\mathcal{L}^{(*)}_\mu = L^{(*)}_\mu \otimes \mathcal{O}_{S_0}$ $(* = 0, \infty)$.

6. Locally freeness

The main theorem of this paper is proved in this section. We show that $\mathcal{CC}$ is locally free at the discriminant locus, $S_0 = \{q = 0\} \subset S$, provided that all modules inserted are *integrable highest weight modules*. Thus, combining the result in [KT], we have locally freeness of the sheaf of conformal coinvariants and consequently locally freeness of its dual, the sheaf of conformal blocks. Corresponding statement for the Weyl modules has been proved in Proposition 4.2.

The arguments in this section is parallel to those in §7.3 of [TK].

We assume the condition (35) for $\mathcal{M}$. Moreover we assume that all M_i's are integrable highest weight modules. In particular, the level k is a non-negative integer.

Theorem 6.1. *The sheaf* $\mathcal{CC}(\mathcal{M})$ *and hence the sheaf* $\mathcal{CB}(\mathcal{M})$ *are locally free* $\mathcal{O}_S$-*sheaves.*

The rest of the paper is devoted to the proof of this theorem. The main strategy of the proof is the same as that of [TK]. See also [SU], [NT] and [U]:

- Since $\mathcal{CC}(\mathcal{M})$ is coherent, it is sufficient to prove that each stalk $\mathcal{CC}(\mathcal{M})_s$ ($s \in S$) is a free $\mathcal{O}_{S,s}$-module.
- We proved the locally freeness of $\mathcal{CC}(\mathcal{M})$ on the non-singular part $S \smallsetminus S_0$ in [KT]. Thus we have only to prove the case $s \in S_0$.
- When $s \in S_0$, we prove that the stalk of the completion of $\mathcal{CC}(\mathcal{M})$ at s is isomorphic to $\mathcal{CC}_{\mathrm{trig}}(\mathcal{M})_s[[q]]$.
- $\mathcal{CC}(\mathcal{M})_s$ is a free $\mathcal{O}_{S,s}$-module because of Proposition 5.1 and the faithfully flatness of the completion functor.

We define completion of the sheaf $\mathcal{CC}(\mathcal{M})$ along the divisor S_0 of S by taking completion of each ingredient of the definition (33). Let $\hat{\mathcal{O}}_{S/S_0}$ be the completion of $\mathcal{O}_S$ along S_0:

$$\hat{\mathcal{O}}_{S/S_0} := \operatorname*{proj\,lim}_{n\to\infty} \mathcal{O}_S/\mathfrak{m}_{S_0}^n. \tag{56}$$

($\mathfrak{m}_{S_0}$ is the defining ideal of S_0.) As an $\mathcal{O}_{S_0}$-module, it is isomorphic to the ring of formal power series: $\hat{\mathcal{O}}_{S/S_0} \cong \mathcal{O}_{S_0}[[q]]$. The completion of $\mathcal{M} = M \otimes \mathcal{O}_S$ is obviously

$$\widehat{\mathcal{M}} := \mathcal{M} \otimes_{\mathcal{O}_S} \hat{\mathcal{O}}_{S/S_0} \cong \mathcal{M}_{S_0}[[q]], \tag{57}$$

where $\mathcal{M}_{S_0} := M \otimes \mathcal{O}_{S_0}$. The $\mathcal{O}_S$-Lie algebra $\mathfrak{g}_{\mathfrak{X},\mathrm{out}}$ acts on $\widehat{\mathcal{M}}$ naturally as follows: a germ $f(P)$ of $\mathfrak{g}_{\mathfrak{X},\mathrm{out}}$ at $(q = 0; Q_1, \dots, Q_L) \in S_0$ is expanded at $(q = 0; (0,0,0)_k; Q_1, \dots, Q_L) \in \tilde{\mathfrak{X}}$ in terms of the coordinates $(q; x_k, y_k)_k$ as

$$\begin{aligned} f(P) &= \sum_{m,n=0}^{\infty} f_{k,m,n}(s) x_k^m y_k^n \\ &= \sum_{m,n=0}^{\infty} f_{k,m,n}(s) x_k^{m-n} q^n = \sum_{m,n=0}^{\infty} f_{k,m,n}(s) y_k^{n-m} q^m \\ &= \sum_{n=0}^{\infty} f_{k,n,x}(s, x_k) q^n = \sum_{m=0}^{\infty} f_{k,m,y}(s, y_k) q^m, \end{aligned} \tag{58}$$

where $f_{k,m,n}(s) \in \mathfrak{g} \otimes \mathcal{O}_{S_0}$ ($s \in S_0$) and

$$f_{k,n,x}(s, x_k) = \sum_{m=0}^{\infty} f_{k,m,n}(s) x_k^{m-n}, \qquad f_{k,m,y}(s, y_k) = \sum_{n=0}^{\infty} f_{k,m,n}(s) y_k^{n-m}. \tag{59}$$

Note that the periodicity condition (27) implies

$$\operatorname{Ad}\gamma(f_{k,m,n}(s)) = \varepsilon^{m-n} f_{k,m,n}(s), \qquad \operatorname{Ad}\beta(f_{k,m,n}(s)) = f_{k+1,m,n}(s), \tag{60}$$

and hence $f_{k,n,x}(s, x_k)$ and $f_{k,m,y}(s, y_k)$ are meromorphic (rational) function on $\mathbb{P}^1$ with quasi-periodicity

$$\begin{aligned} f_{k,n,x}(s, \varepsilon x_k) &= \operatorname{Ad}\gamma(f_{k,n,x}(s, x_k)), \\ f_{k,m,y}(s, \varepsilon^{-1} y_k) &= \operatorname{Ad}\gamma(f_{k,m,y}(s, y_k)), \\ f_{k+1,n,x}(s, x_k) &= \operatorname{Ad}\beta(f_{k,n,x}(s, x_k)), \\ f_{k+1,m,y}(s, y_k) &= \operatorname{Ad}\beta(f_{k,m,y}(s, y_k)), \end{aligned} \tag{61}$$

the poles of which are in the divisor

$$*\tilde{D}_0 := \sum_{i=1}^{L} \sum_{j\in\mathbb{Z}/N\mathbb{Z}} *[\varepsilon^j Q_i] + *[0] + *[\infty]. \tag{62}$$

Therefore $\{f_{k,n,x}(s,x_k)\}_{k\in\mathbb{Z}/N\mathbb{Z}}$ and $\{f_{k,n,y}(s,y_k)\}_{k\in\mathbb{Z}/N\mathbb{Z}}$, namely the nth coefficients of the expansion

$$f(P) = \sum_{n=0}^{\infty} f_n(P) q^n, \tag{63}$$

define a section $f_n(P)$ of $\mathfrak{g}^{\mathrm{tw}}_{\mathfrak{X}}|_{\pi^{-1}_{\mathfrak{X}/S}(S_0)}$ with poles at $Q_1,\dots,Q_L,0,\infty$. See (23) and (27). The action of $f(P)$ on $v\otimes g \in \widehat{\mathcal{M}}$ ($v\in\mathcal{M}$, $g\in\hat{\mathcal{O}}_{S/S_0}$) is defined by

$$f(P)\cdot(v\otimes g) := \sum_{i=1}^{L}\sum_{n=0}^{\infty} \rho_i(f_n(P))v\otimes q^n g. \tag{64}$$

Here ρ_i denotes the usual action of the Laurent expansion of $f_n(P)$ at Q_i on M_i. The space of coinvariants of $\widehat{\mathcal{M}}$ with respect to this action is the completion of $\mathcal{CC}(\mathcal{M})$:

$$\widehat{\mathcal{CC}}(\mathcal{M}) := \widehat{\mathcal{M}}/\mathfrak{g}_{\mathfrak{X},\mathrm{out}}(\widehat{\mathcal{M}}). \tag{65}$$

Lemma 6.2.

$$\widehat{\mathcal{CC}}(\mathcal{M}) \cong \mathcal{CC}(\mathcal{M}) \otimes_{\mathcal{O}_S} \hat{\mathcal{O}}_{S/S_0}. \tag{66}$$

Proof. By definition, we have an exact sequence

$$\mathfrak{g}_{\mathfrak{X},\mathrm{out}}\otimes\mathcal{M} \to \mathcal{M} \to \mathcal{CC}(\mathcal{M}) \to 0. \tag{67}$$

Tensoring $\hat{\mathcal{O}}_{S/S_0}$ we obtain an exact sequence

$$(\mathfrak{g}_{\mathfrak{X},\mathrm{out}}\otimes\mathcal{M})[[q]] \to \widehat{\mathcal{M}} \to \mathcal{CC}(\mathcal{M})\otimes\hat{\mathcal{O}}_{S/S_0} \to 0. \tag{68}$$

It is sufficient to show that the image of the map $(\mathfrak{g}_{\mathfrak{X},\mathrm{out}}\otimes\mathcal{M})[[q]] \to \widehat{\mathcal{M}}$ is $\mathfrak{g}_{\mathfrak{X},\mathrm{out}}(\widehat{\mathcal{M}})$ defined by the action (64). This is almost trivial since $f(P)\cdot v$ for $f(P)\in\mathfrak{g}_{\mathfrak{X},\mathrm{out}}$ and $v\in\mathcal{M}$ is expressed as

$$f(P)\cdot v = \sum_{i=1}^{L}\sum_{n=0}^{\infty} q^n \rho_i(f_n(P))v,$$

because of the expansion (63). □

The next step is to make a completion of the isomorphism (55). For this purpose we need a lemma on Verma modules of $\hat{\mathfrak{g}}^{(0)}$ and $\hat{\mathfrak{g}}^{(\infty)}$. Note that the Verma module $M^{(*)}_\mu$ of $\hat{\mathfrak{g}}^{(*)}$ ($*=0,\infty$) is graded by the degree:

$$M^{(*)}_\mu = \bigoplus_{d\geq 0} M^{(*)}_\mu(d), \qquad M^{(*)}_\mu(d) := \sum_{n_1+\cdots+n_l=d} \mathbb{C}X_1[-n_1]\cdots X_l[-n_l]|\mu\rangle, \tag{69}$$

where $|\mu\rangle$'s are the highest weight vectors of $M^{(*)}_\mu$ and $X_i[-n_i]\in\hat{\mathfrak{g}}^{(*)}$ ($X_i\in\mathfrak{g}$, $n_i\in\mathbb{N}$).

Lemma 6.3.

(i) *For any* $\mu \in \mathfrak{h}^*$ *there exists a pairing between the Verma modules* $M^{(0)}_{\mu\circ\mathrm{Ad}\,\beta}$ *and* $M^{(\infty)}_{-\mu}$*:*

$$M^{(0)}_{\mu\circ\mathrm{Ad}\,\beta} \times M^{(\infty)}_{-\mu} : (u,v) \mapsto \langle u,v\rangle \in \mathbb{C}, \tag{70}$$

which satisfies $\big\langle |\mu\circ\mathrm{Ad}\,\beta\rangle, |-\mu\rangle\big\rangle = 1$ *and*

$$\langle X[n]u,v\rangle + \langle u, \mathrm{Ad}\,\beta(X)[-n]v\rangle = 0, \tag{71}$$

for any $u \in M^{(0)}_{\mu\circ\mathrm{Ad}\,\beta}$, $v \in M^{(\infty)}_{-\mu}$, $X \in \mathfrak{g}$ *and* $n \in \mathbb{Z}$.

(ii) $\left\langle M^{(0)}_{\mu\circ\mathrm{Ad}\,\beta}(n), M^{(\infty)}_{-\mu}(n')\right\rangle = 0$ *if* $n \neq n'$.

(iii) *The radical* $R^{(0)} = \{u \in M^{(0)}_{\mu\circ\mathrm{Ad}\,\beta} \mid \langle u,v\rangle = 0$ *for all* $v \in M^{(\infty)}_{-\mu}\}$ *is the largest proper submodule of* $M^{(0)}_{\mu\circ\mathrm{Ad}\,\beta}$. *Similarly the radical in* $M^{(\infty)}_{-\mu}$ *is the largest proper submodule. Hence the pairing descends to a non-degenerate pairing between the irreducible quotients* $L^{(0)}_{\mu\circ\mathrm{Ad}\,\beta}$ *and* $L^{(\infty)}_{-\mu}$.

Proof. Let ν be the anti-isomorphism

$$\nu : U\hat{\mathfrak{g}}^{(0)} \ni X[n] = X\otimes t^n \mapsto -X\otimes t^n = -X[-n] \in U\hat{\mathfrak{g}}^{(\infty)}, \qquad \nu(\hat{k}) = \hat{k}. \tag{72}$$

This induces a linear isomorphism $\nu_\beta : M^{(0)}_{\mu\circ\mathrm{Ad}\,\beta} \to \mathrm{Hom}_{\mathbb{C}}(M^{(\infty)}_{-\mu}, \mathbb{C})$ defined by

$$\nu_\beta(x|\mu\circ\mathrm{Ad}\,\beta\rangle) = \langle -\mu|\nu(\mathrm{Ad}\,\beta(x)), \tag{73}$$

where $\langle -\mu|$ is the generating vector of the right $\hat{\mathfrak{g}}^{(\infty)}$-module $\mathrm{Hom}_{\mathbb{C}}(M^{(\infty)}_{-\mu}, \mathbb{C})$, normalised by $\langle -\mu \mid -\mu\rangle = 1$. We define the pairing $\langle , \rangle$ by

$$\langle v, v'\rangle := \nu_\beta(v)v'. \tag{74}$$

Straightforward computation shows that for $x \in U\hat{\mathfrak{g}}^{(0)}$ we have

$$\langle xv, v'\rangle = \langle v, \nu(\mathrm{Ad}\,\beta(x))v'\rangle, \tag{75}$$

which means (71) for $x = X[n]$.

(ii) follows from the construction.

(iii) is proved in the same way as Proposition 3.26 of [Wa]. □

Let $\{e_{\lambda,d,i}\}$ be a basis of $L^{(0)}_{\tilde{\lambda}}(d)$ and $\{e^i_{\lambda,d}\}$ be its dual basis of $L^{(\infty)}_{\tilde{\lambda}'}(d)$ with respect to $\langle , \rangle$.

Proposition 6.4. *There exists an isomorphism*

$$\hat{\imath} : \widehat{\mathcal{CC}}(\mathcal{M}) \to \bigoplus_{\lambda\in\mathrm{wt}(V), \tilde{\lambda}:\text{dom. int.}} \mathcal{CC}_{\mathrm{orb}}(\mathcal{L}_{\tilde{\lambda}} \otimes \mathcal{M} \otimes \mathcal{L}_{\tilde{\lambda}'})[[q]], \tag{76}$$

of $\hat{\mathcal{O}}_{S/S_0}$*-modules defined by*

$$\hat{\imath}([v]) := \bigoplus_{\lambda} \left[\sum_{d=0}^{\infty}\sum_i e_{\lambda,d,i} \otimes v \otimes e^i_{\lambda,d}\right] q^d, \tag{77}$$

for $v \in \mathcal{M}$.

Proof. First we prove the well-definedness of (77), for which it is enough to show the well-definedness of its component $\iota_\lambda : \widehat{\mathcal{CC}}(\mathcal{M}) \to \mathcal{CC}_{\mathrm{orb}}(\mathcal{M}_{\tilde{\lambda}} \otimes \mathcal{M} \otimes \mathcal{M}_{\tilde{\lambda}'})[[q]]$, namely,

$$\sum_{d=0}^{\infty} \sum_{i} e_{\lambda,d,i} \otimes f(P) \cdot v \otimes e^{i}_{\lambda,d} q^d \in \mathfrak{g}^{\mathrm{orb}}_{\mathrm{out}}(\mathcal{M}_{\tilde{\lambda}} \otimes \mathcal{M} \otimes \mathcal{M}_{\tilde{\lambda}'}) \tag{78}$$

for $f(P) \in \mathfrak{g}_{\mathfrak{X},\mathrm{out}}$ and $v \in \mathcal{M}$. Since

$$f(P) \cdot \left(\sum_{d=0}^{\infty} \sum_{i} e_{\lambda,d,i} \otimes v \otimes e^{i}_{\lambda,d} \right) q^d \in \mathfrak{g}^{\mathrm{orb}}_{\mathrm{out}}(\mathcal{M}_{\tilde{\lambda}} \otimes \mathcal{M} \otimes \mathcal{M}_{\tilde{\lambda}'}), \tag{79}$$

we have only to show that the left-hand side of (78) is equal to the left-hand side of (79), which is equivalent to an equation in $L_{\tilde{\lambda}} \otimes L_{\tilde{\lambda}'}$:

$$\sum_{d=0}^{\infty} \sum_{i} \left(\rho_0(f(P)) \cdot e_{\lambda,d,i} \otimes e^{i}_{\lambda,d} + e_{\lambda,d,i} \otimes \rho_\infty(f(P)) \cdot e^{i}_{\lambda,d} \right) q^d = 0. \tag{80}$$

Recall that the germ of $f(P)$ at 0 and the germ at ∞ is related by $f(P)_\infty = \mathrm{Ad}\,\beta(f(P)_0)$. See (27) of this paper or (18) of [T]. Hence using the expansion (58) and the invariance (71) of the pairing, we can show (80) in the same way as the proof of Claim 3 in the proof of Theorem 6.2.1 in [TUY]. Thus $\hat{\iota}$ is well defined.

Obviously the $q = 0$ part of $\hat{\iota}$ is the isomorphism ι, (55). Therefore by termwise approximation (in analytic language) or, in other words, by Nakayama's lemma (in algebraic language), $\hat{\iota}$ is shown to be an isomorphism. □

With these preparations, the proof of the locally freeness of $\mathcal{CC}(\mathcal{M})$ goes as follows. As is mentioned after the statement of Theorem 6.1, it is enough to prove that the stalk $\mathcal{CC}(\mathcal{M})_s$ at $s \in S_0$ is a free $\mathcal{O}_{S,s}$-module. Since $\widehat{\mathcal{CC}}(\mathcal{M})_s$ is isomorphic to $\mathcal{CC}(\mathcal{M})_s \otimes_{\mathcal{O}_{S,s}} \otimes \hat{\mathcal{O}}_{S/S_0,s}$ (Lemma 6.2) and to $\bigoplus_\lambda \mathcal{CC}_{\mathrm{orb}}(\mathcal{L}_{\tilde{\lambda}} \otimes \mathcal{M} \otimes \mathcal{L}_{\tilde{\lambda}'})_s[[q]]$ (Proposition 6.4), we have an isomorphism

$$\mathcal{CC}(\mathcal{M})_s \otimes_{\mathcal{O}_{S,s}} \hat{\mathcal{O}}_{S/S_0,s} \cong \bigoplus_{\lambda} \mathcal{CC}_{\mathrm{orb}}(\mathcal{L}_{\tilde{\lambda}} \otimes \mathcal{M} \otimes \mathcal{L}_{\tilde{\lambda}'})_s[[q]]. \tag{81}$$

The right-hand side of (81) being a free $\hat{\mathcal{O}}_{S/S_0,s}$-module (Proposition 5.1 (ii) and Proposition 5.3), faithfully flatness of $\hat{\mathcal{O}}_{S/S_0,s}$ over $\mathcal{O}_{S,s}$ implies that $\mathcal{CC}(\mathcal{M})_s$ is a free $\mathcal{O}_{S,s}$-module. Thus Theorem 6.1 is proved.

7. Concluding comments

We have proved locally freeness of $\mathcal{CC}(\mathcal{M})$ in two cases; Weyl module case (Proposition 4.2) and integrable highest weight module case (Theorem 6.1). A few comments are in order:

- In the Weyl module case, $\mathcal{CC}(\mathcal{M}) \cong V \otimes \mathcal{O}_S$ as shown in the proof of Proposition 4.2 and the rank of $\mathcal{CC}(\mathcal{M})$ is $\dim V$.

- In the integrable highest module case, the rank is computed by further degenerating the orbifold. Degeneration of the type $Q_i \to Q_j$ should be considered in the same way as in [TUY] or [NT]. The final results of the degeneration is a combination of the three-punctured orbifold $\mathbb{P}^1/C_N$. In principle a Verlinde-type formula would be obtained in this way.
- In [KT] we have shown that $\mathcal{CC}(\mathcal{M})$ has a flat connection. It has a regular singularity along $S_0 = \{q = 0\}$, which is easily deduced from the explicit form of the connection, Theorem 5.9 of [KT]. Hence there is a one-to-one correspondence between flat sections around S_0 and its restriction to S_0 or, in other words, the "initial value" at S_0 because of the locally flatness.

Acknowledgments

The author expresses his gratitude to Akihiro Tsuchiya who explained details of [TUY] and [NT], Toshiro Kuwabara who showed the manuscript of [TK] (the best guide to [TUY] for $\mathbb{P}^1$ case) before publishing, Michio Jimbo, Gen Kuroki, Tetsuji Miwa, Hiroyuki Ochiai, Kiyoshi Ohba, Nobuyoshi Takahashi, Tomohide Terasoma and Yasuhiko Yamada for discussion and comments.

The atmosphere and environment of Institute for Theoretical and Experimental Physics (Moscow, Russia) and the conference "Infinite-Dimensional Algebras and Integrable Systems" (Faro, Portugal) were very important. The author thanks their hospitality.

References

[BD] A.A. Belavin, V.G. Drinfeld, Solutions of the classical Yang-Baxter equations for simple Lie algebras. *Funkts. Anal. i ego Prilozh.* **16-3**, 1–29 (1982) (in Russian); *Funct. Anal. Appl.* **16**, 159–180 (1982) (English transl.)

[E] P.I. Etingof, Representations of affine Lie algebras, elliptic r-matrix systems, and special functions. *Comm. Math. Phys.* **159**, 471–502 (1994).

[FFR] B. Feigin, E. Frenkel, N. Reshetikhin, Gaudin model, Bethe Ansatz and critical level. Commun. Math. Phys. **166**, 27–62 (1994)

[K] V.G. Kac, *Infinite-dimensional Lie algebras*, 3rd Edition, Cambridge University Press 1990.

[KL] D. Kazhdan, G. Lusztig, Tensor structures arising from affine Lie algebras. I, II, *J. Amer. Math. Soc.* **6**, 905–948, 949–1011 (1993).

[KT] G. Kuroki, T. Takebe, Twisted Wess-Zumino-Witten models on elliptic curves. *Comm. Math. Phys.* **190**, 1–56 (1997).

[NT] K. Nagatomo, A. Tsuchiya, Conformal field theories associated to regular chiral vertex operator algebras I: theories over the projective line, `math.QA/0206223`.

[SU] Y. Shimizu, K. Ueno, *Moduli theory III*, (Iwanami, Tokyo, 1999) Gendai Suugaku no Tenkai series (in Japanese); *Advances in moduli theory*, Translations of Mathematical Monographs, **206**, Iwanami Series in Modern Mathematics, American Mathematical Society, Providence, U.S.A. (2002) (English translation)

[T] T. Takebe, Trigonometric Degeneration and Orbifold Wess-Zumino-Witten Model. I In *the Proceedings of the 6th International workshop on Conformal and Integrable models, Chernogolovka, Sep. 2002, International Journal of Modern Physics, A*, **19**, Supplement, 418–435 (2004)

[TK] A. Tsuchiya, T. Kuwabara, Introduction to Conformal Field Theory, to appear as MSJ Suugaku Memoir of Mathematical Society of Japan.

[TUY] A. Tsuchiya, K. Ueno, Y. Yamada, Conformal field theory on universal family of stable curves with gauge symmetries. In *Integrable systems in quantum field theory and statistical mechanics, Adv. Stud. Pure Math.* **19**, 459–566 (1989).

[U] K. Ueno, On conformal field theory, In *Vector bundles in algebraic geometry (Durham, 1993)*, ed. by N. J. Hitchin, P. E. Newstead and W. M. Oxbury, *London Math. Soc. Lecture Note Ser.* **208**, (Cambridge Univ. Press, Cambridge, 1995) pp. 283–345,

[Wa] M. Wakimoto, *Infinite-dimensional Lie algebras*, (Iwanami, Tokyo, 1999) Gendai Suugaku no Tenkai series (in Japanese); Translations of Mathematical Monographs, **195**, Iwanami Series in Modern Mathematics, American Mathematical Society, Providence, U.S.A. (2001) (English translation by K. Iohara)

[Wo] S. Wolpert, On the homology of the moduli space of stable curves. *Ann. of Math.* **118**, 491–523 (1983).

[Z] A.B. Zamolodchikov, Exact solutions of conformal field theory in two dimensions and critical phenomena. *Rev. Math. Phys.* **1** 197–234 (1989). (Translated from the Russian by Y. Kanie.)

Takashi Takebe
Department of Mathematics
Ochanomizu University
Otsuka 2-1-1, Bunkyo-ku
Tokyo, 112-8610, Japan
e-mail: takebe@math.ocha.ac.jp

Progress in Mathematics, Vol. 237, 225–233

Weil-Petersson Geometry of the Universal Teichmüller Space

Leon A. Takhtajan and Lee-Peng Teo

Mathematics Subject Classification (2000). 32F60 (Primary) 32G15, 46E20, 58B20 (Secondary).

Keywords. Universal Teichmüller space, Bers embedding, Hilbert manifold, Velling-Kirillov metric, Weil-Petersson metric, Riemann curvature tensor.

1. Introduction

The universal Teichmüller space $T(1)$ is the simplest Teichmüller space that bridges spaces of univalent functions and general Teichmüller spaces. It was introduced by Bers [Ber65, Ber72, Ber73] and it is an infinite-dimensional complex Banach manifold. The universal Teichmüller space $T(1)$ contains Teichmüller spaces of Riemann surfaces as complex submanifolds.

The universal Teichmüller space $T(1)$ plays an important role in one of the approaches to non-perturbative bosonic closed string field theory based on Kähler geometry. Namely, in the "old approach" to string field theory as the Kähler geometry of the loop space [BR87a, BR87b], the loop space $\mathcal{L}(\mathbb{R}^d)$ is the configuration space for the closed strings,

$$\mathcal{L}(\mathbb{R}^d) = \mathbb{R}^d \times \Omega(\mathbb{R}^d).$$

The space $\Omega(\mathbb{R}^d)$ of based loops has a natural structure of an infinite-dimensional Kähler manifold. The space of all complex structures of $\Omega(\mathbb{R}^d)$ is

$$\mathcal{M} = S^1 \backslash \operatorname{Diff}_+(S^1).$$

The space $\mathcal{M}$ parameterizes vacuum states for Faddeev-Popov ghosts in the string field theory. The "flag manifolds" $\mathcal{M}$ and

$$\mathcal{N} = \text{Möb}(S^1) \backslash \operatorname{Diff}_+(S^1)$$

Talk given by the first author at the workshop "Infinite-Dimensional Algebras and Quantum Integrable Systems" in Faro, Portugal, July 21–25, 2003. Detailed exposition and proofs can be found in [TT03].

are infinite-dimensional complex Fréchet manifolds carrying a natural Kähler metrics [BR87a, BR87b, Kir87, KY87]. These manifolds also have an interpretation as coadjoint orbits of the Bott-Virasoro group, and the corresponding Kähler forms coincide with Kirillov-Kostant symplectic forms [Kir87, KY87]. Ricci tensor for $\mathcal{M}$ is related to the problem of constructing reparametrization-invariant vacuum for ghosts.

The natural inclusion $\mathcal{N} \hookrightarrow T(1)$ is holomorphic ($\mathcal{N}$ is a leaf of a holomorphic foliation of $T(1)$), and the Kirillov-Kostant symplectic form at the origin of $\mathcal{N}$ is a pull-back of a certain symplectic form on the subspace of the tangent space to $T(1)$ at the origin [NV90] (an avatar of the Weil-Petersson structure on $T(1)$).

2. Basic facts

2.1. Definitions

Let

$$\begin{aligned} \mathbb{D} &= \{z \in \mathbb{C} : |z| < 1\}, \\ \mathbb{D}^* &= \{z \in \mathbb{C} : |z| > 1\}. \end{aligned}$$

The complex Banach spaces $L^\infty(\mathbb{D}^*)$ and $L^\infty(\mathbb{D})$ are the spaces of bounded Beltrami differentials on $\mathbb{D}^*$ and $\mathbb{D}$ respectively. Let $L^\infty(\mathbb{D}^*)_1$ be the unit ball in $L^\infty(\mathbb{D}^*)$. Two classical models of Bers' universal Teichmüller space $T(1)$ are the following.

Model A. Extend every $\mu \in L^\infty(\mathbb{D}^*)_1$ to $\mathbb{D}$ by the reflection

$$\mu(z) = \overline{\mu\left(\frac{1}{\bar{z}}\right)\frac{z^2}{\bar{z}^2}}\,, \quad z \in \mathbb{D},$$

and consider the unique quasiconformal mapping $w_\mu : \mathbb{C} \to \mathbb{C}$, which fixes $-1, -i$ and 1, and satisfies the Beltrami equation

$$\frac{\partial w_\mu}{\partial \bar{z}} = \mu\,\frac{\partial w_\mu}{\partial z}\,.$$

The mapping w_μ satisfies

$$\frac{1}{w_\mu(z)} = \overline{w_\mu\left(\frac{1}{\bar{z}}\right)}$$

and fixes the domains $\mathbb{D}$, $\mathbb{D}^*$, and the unit circle S^1. For $\mu, \nu \in L^\infty(\mathbb{D}^*)_1$ set $\mu \sim \nu$ if

$$w_\mu|_{S^1} = w_\nu|_{S^1}\,.$$

The universal Teichmüller space $T(1)$ is defined as the set of equivalence classes of the mappings w_μ,

$$T(1) = L^\infty(\mathbb{D}^*)_1/\sim\,.$$

Model B. Extend every $\mu \in L^\infty(\mathbb{D}^*)_1$ to be zero outside $\mathbb{D}^*$ and consider the unique solution w^μ of the Beltrami equation

$$\frac{\partial w^\mu}{\partial \bar{z}} = \mu \frac{\partial w^\mu}{\partial z},$$

satisfying $f(0) = 0$, $f'(0) = 1$ and $f''(0) = 0$, where $f = w^\mu|_\mathbb{D}$ is holomorphic on $\mathbb{D}$. For $\mu, \nu \in L^\infty(\mathbb{D}^*)_1$ set $\mu \sim \nu$ if

$$w^\mu|_\mathbb{D} = w^\nu|_\mathbb{D}.$$

The universal Teichmüller space is defined as the set of equivalence classes of the mappings w^μ,

$$T(1) = L^\infty(\mathbb{D}^*)_1/\sim .$$

Since $w_\mu|_{S^1} = w_\nu|_{S^1}$ if and only if $w^\mu|_\mathbb{D} = w^\nu|_\mathbb{D}$, the two definitions of the universal Teichmüller space are equivalent. The set $T(1)$ is a topological space with the quotient topology induced from $L^\infty(\mathbb{D}^*)_1$.

2.2. Properties of $T(1)$

1. The universal Teichmüller space $T(1)$ has a unique structure of a complex Banach manifold such that the projection map

$$\Phi : L^\infty(\mathbb{D}^*)_1 \to T(1)$$

is a holomorphic submersion.

2. The holomorphic tangent space $T_0T(1)$ at the origin is identified with the Banach space $\Omega^{-1,1}(\mathbb{D}^*)$ of harmonic Beltrami differentials,

$$\Omega^{-1,1}(\mathbb{D}^*) = \{\mu \in L^\infty(\mathbb{D}^*) : \\ \mu(z) = (1-|z|^2)^2\overline{\phi(z)},\ \phi \in A_\infty(\mathbb{D}^*)\},$$

where

$$A_\infty(\mathbb{D}^*) = \{\phi \text{ holomorphic on } \mathbb{D}^* : \\ \|\phi\|_\infty = \sup_{z\in\mathbb{D}^*} \left|(1-|z|^2)^2\phi(z)\right| < \infty\}.$$

3. The universal Teichmüller space $T(1)$ is a group (not a topological group!) under the composition of the quasiconformal mappings. The group law on $L^\infty(\mathbb{D}^*)_1$

$$\lambda = \nu * \mu^{-1}$$

is defined through $w_\lambda = w_\nu \circ w_\mu^{-1}$ and projects to $T(1)$. Explicitly,

$$\lambda = \left(\frac{\nu - \mu}{1 - \bar{\mu}\nu} \frac{(w_\mu)_z}{(\overline{w}_\mu)_{\bar{z}}}\right) \circ w_\mu^{-1}.$$

For every $\mu \in L^\infty(\mathbb{D}^*)_1$ the right translations

$$R_{[\mu]} : T(1) \longrightarrow T(1), \quad [\lambda] \longmapsto [\lambda * \mu],$$

where $[\lambda] = \Phi(\lambda) \in T(1)$, are biholomorphic automorphisms of $T(1)$. The left translations, in general, are not even continuous mappings.

4. The group $T(1)$ is isomorphic to the subgroup of the group $\mathrm{Homeo}_{qs}(S^1)$ of quasisymmetric homeomorphisms of S^1 fixing -1, $-i$ and 1. By definition, $\gamma \in \mathrm{Homeo}_{qs}(S^1)$ if it is orientation preserving and satisfies

$$\frac{1}{M} \leq \left| \frac{\gamma\left(e^{i(\theta+t)}\right) - \gamma\left(e^{i\theta}\right)}{\gamma\left(e^{i\theta}\right) - \gamma\left(e^{i(\theta-t)}\right)} \right| \leq M$$

for all θ and all $|t| \leq \pi/2$ with some constant $M > 0$.

Remark 1. The closure of $\mathcal{N}$ in $T(1)$ is the subgroup of symmetric homeomorphisms in $\mathrm{Möb}(S^1)\backslash\mathrm{Homeo}_{qs}(S^1)$ satisfying the above inequality with M replaced by $1 + o(t)$ as $t \to 0$.

2.3. Bers embedding and the complex structure of $T(1)$

Let $A_\infty(\mathbb{D}) = \left\{ \phi \text{ holomorphic on } \mathbb{D} : \|\phi\|_\infty = \sup_{z\in\mathbb{D}} \left|(1-|z|^2)^2\phi(z)\right| < \infty \right\}.$

and let $\mathcal{S}(f)$ be the Schwarzian derivative,

$$\mathcal{S}(f) = \frac{f_{zzz}}{f_z} - \frac{3}{2}\left(\frac{f_{zz}}{f_z}\right)^2.$$

For every $\mu \in L^\infty(\mathbb{D}^*)_1$ the holomorphic function $\mathcal{S}(w^\mu)|_\mathbb{D} \in A_\infty(\mathbb{D})$ and, by Kraus-Nehari inequality, lies in the ball of radius 6. The Bers embedding $\beta : T(1) \hookrightarrow A_\infty(\mathbb{D})$ is defined by

$$\beta([\mu]) = \mathcal{S}(w^\mu|_\mathbb{D}),$$

and is a holomorphic map of complex Banach manifolds. Define the mapping $\Lambda : A_\infty(\mathbb{D}) \to \Omega^{-1,1}(\mathbb{D}^*)$ by

$$\Lambda(\phi)(z) = -\frac{1}{2}(1-|z|^2)^2\phi\left(\frac{1}{\bar{z}}\right)\frac{1}{\bar{z}^4}.$$

By Ahlfors-Weill theorem, the mapping Λ is inverse to the Bers embedding β over the ball of radius 2 in $A_\infty(\mathbb{D})$.

The complex structure of $T(1)$ is explicitly described as follows. For every $\mu \in L^\infty(\mathbb{D}^*)_1$ let $U_\mu \subset T(1)$ be the image of the ball of radius 2 in $A_\infty(\mathbb{D})$ under the map $h_\mu^{-1} = R_{[\mu]}^{-1} \circ \Lambda$. The inverse map $h_\mu = \beta \circ R_{[\mu]} : U_\mu \to A_\infty(\mathbb{D})$ and the maps $h_{\mu\nu} = h_\mu \circ h_\nu^{-1} : h_\mu(U_\mu) \bigcap h_\nu(U_\nu) \to h_\mu(U_\mu) \bigcap h_\nu(U_\nu)$ are biholomorphic (as functions in the Banach space $A_\infty(\mathbb{D})$). The open covering $T(1) = \bigcup_{\mu \in L^\infty(\mathbb{D}^*)_1} U_\mu$ with coordinate maps h_μ and transition maps $h_{\mu\nu}$ defines a complex-analytic atlas on $T(1)$ modelled on the Banach space $A_\infty(\mathbb{D})$.

The canonical projection $\Phi : L^\infty(\mathbb{D}^*)_1 \to T(1)$ is a holomorphic submersion and the Bers embedding $\beta : T(1) \to A_\infty(\mathbb{D})$ is a biholomorphic map with respect to this complex structure. Complex coordinates on $T(1)$ defined by the coordinate charts (U_μ, h_μ) are called Bers coordinates.

2.4. The universal Teichmüller curve

The universal Teichmüller curve $\mathcal{T}(1)$ is a complex fiber space over $T(1)$ with a holomorphic projection map

$$\pi : \mathcal{T}(1) \to T(1).$$

The fiber over each point $[\mu]$ is the quasi-disk $w^\mu(\mathbb{D}^*) \subset \hat{\mathbb{C}} = \mathbb{C} \cup \{\infty\}$ with the complex structure induced from $\hat{\mathbb{C}}$ and

$$\mathcal{T}(1) = \{([\mu], z) : [\mu] \in T(1),\ z \in w^\mu(\mathbb{D}^*)\}.$$

The fibration $\pi : \mathcal{T}(1) \longrightarrow T(1)$ has a natural holomorphic section given by

$$T(1) \ni [\mu] \mapsto ([\mu], \infty) \in \mathcal{T}(1)$$

which defines the embedding $T(1) \hookrightarrow \mathcal{T}(1)$. The universal Teichmüller curve is a complex Banach manifold modelled on $A_\infty(\mathbb{D}) \oplus \mathbb{C}$.

2.5. Velling-Kirillov metric on $\mathcal{T}(1)$

The Velling-Kirillov metric at the origin of $\mathcal{T}(1)$ is defined by

$$\| v \|_{VK}^2 = \sum_{n=1}^{\infty} n|c_n|^2, \qquad \text{where} \qquad v = \sum_{n\neq 0} c_n e^{in\theta} \frac{\partial}{\partial\theta} \in T_0 S^1\backslash\mathrm{Homeo}_{qs}(S^1)$$

– the tangent space at the origin of a real Banach manifold $S^1\backslash\mathrm{Homeo}_{qs}(S^1)$. (The series in the definition of $\| v \|_{VK}^2$ is always convergent.) At other points the Velling-Kirillov metric is defined by the right translations. The Velling-Kirillov metric on $\mathcal{T}(1)$ is Kähler with symplectic form ω_{VK}.

Remark 2. For the space $S^1\backslash \mathrm{Diff}_+(S^1)$ this metric was introduced by Kirillov [Kir87] and has been studied by Kirillov-Yuriev [KY87]. Velling [Vel] introduced a Hermitian metric for $\mathcal{T}(1)$ using geometric theory of functions, and in [Teo02] the second author extended Kirillov's metric to $\mathcal{T}(1)$ and proved that it coincides with the metric introduced by Velling. The Velling-Kirillov metric is the unique Kähler metric on $\mathcal{T}(1)$ invariant under the right translations [Kir87, Teo02].

3. Weil-Petersson metric on $T(1)$

As a Banach manifold, the universal Teichmüller space does not carry a natural Hermitian metric. However, it is possible (see [TT03] for detailed construction and proofs) to introduce a new Hilbert manifold structure on $T(1)$ such that it has a natural Hermitian metric. Namely, define the Hilbert space of harmonic Beltrami differentials on $\mathbb{D}^*$ by

$$H^{-1,1}(\mathbb{D}^*) = \Big\{\mu = \rho^{-1}\bar{\phi},\ \phi \text{ holomorphic on } \mathbb{D}^* : \\ \|\mu\|_2^2 = \iint_{\mathbb{D}^*} |\mu|^2 \rho(z) d^2 z < \infty\Big\},$$

where

$$\rho(z) = \frac{4}{(1-|z|^2)^2}$$

is the density of the hyperbolic metric on $\mathbb{D}^*$.

The natural inclusion map $H^{-1,1}(\mathbb{D}^*) \hookrightarrow \Omega^{-1,1}(\mathbb{D}^*)$ is bounded, and it can be shown that the family $\mathfrak{D}$, defined by

$$T(1) \ni [\mu] \mapsto D_0 R_{[\mu]} \left(H^{-1,1}(\mathbb{D}^*)\right) \subset T_{[\mu]}T(1),$$

is an integrable distribution on $T(1)$. Integral manifolds of the distribution $\mathfrak{D}$ are Hilbert manifolds modelled on the Hilbert space $H^{-1,1}(\mathbb{D}^*)$. Thus the universal Teichmüller space $T(1)$ carries a new structure of a Hilbert manifold. Similarly to the Banach manifold structure, the Hilbert manifold structure can be also described by a complex-analytic atlas. Let $T_0(1)$ be the component of origin of the Hilbert manifold $T(1)$, $\mathrm{M\ddot{o}b}(S^1)\backslash \mathrm{Diff}_+(S^1) \subset T_0(1)$.

As a Hilbert manifold, the universal Teichmüller space $T(1)$ has a natural Hermitian metric, defined by the Hilbert space inner product on tangent spaces. Thus the Weil-Petersson metric is a right-invariant metric on $T(1)$, defined at the origin of $T(1)$ by

$$g_{\mu\bar{\nu}} = \langle \mu, \nu \rangle = \iint\limits_{\mathbb{D}^*} \mu\bar{\nu}\rho(z)d^2z, \quad \mu, \nu \in H^{-1,1}(\mathbb{D}^*) = T_0T(1).$$

If

$$v = \sum_{n\neq -1,0,1} c_n e^{in\theta}\frac{\partial}{\partial\theta} \in T_0\, \mathrm{M\ddot{o}b}(S^1)\backslash \mathrm{Homeo}_{qs}(S^1)$$

– the tangent space to a real Hilbert manifold $\mathrm{M\ddot{o}b}(S^1)\backslash\mathrm{Homeo}_{qs}(S^1)$ at the origin – then

$$\| v \|^2_{WP} = \sum_{n=2}^{\infty} (n^3 - n)|c_n|^2,$$

The Weil-Petersson metric on $T(1)$ is Kähler with symplectic form ω_{WP}.

4. Riemann tensor of the Weil-Petersson metric

Let $G = \frac{1}{2}\left(\Delta_0 + \frac{1}{2}\right)^{-1}$ be (the one-half of) the resolvent kernel of the Laplace-Beltrami operator of the hyperbolic metric on $\mathbb{D}^*$ (acting on functions) at $\lambda = \frac{1}{2}$. Explicitly

$$G(z,w) = \frac{2u+1}{2\pi}\log\frac{u+1}{u} - \frac{1}{\pi}, \qquad \text{where} \qquad u(z,w) = \frac{|z-w|^2}{(1-|z|^2)(1-|w|^2)}.$$

Set

$$G(f)(z) = \iint\limits_{\mathbb{D}^*} G(z,w)f(w)\rho(w)d^2w.$$

Theorem A.

(i) *The Weil-Petersson metric is a Kähler metric on a Hilbert manifold* $T(1)$*, and the Bers coordinates are geodesic coordinates at the origin of* $T(1)$.

(ii) *Let* $\mu_\alpha, \mu_\beta, \mu_\gamma, \mu_\delta \in H^{-1,1}(\mathbb{D}^*) \simeq T_0T(1)$ *be orthonormal tangent vectors. Then the Riemann tensor at the origin of T(1) is given by*

$$R_{\alpha\bar{\beta}\gamma\bar{\delta}} = -\frac{\partial^2 g_{\alpha\bar{\beta}}}{\partial t_\gamma \partial \bar{t}_\delta} = -\langle G(\mu_\alpha\bar{\mu}_\delta), \mu_\beta\bar{\mu}_\gamma\rangle - \langle \mu_\alpha\bar{\mu}_\beta, G(\bar{\mu}_\gamma\mu_\delta)\rangle.$$

(iii) *The Hilbert manifold* $T_0(1)$ *is Kähler-Einstein with the negative definite Ricci tensor,*

$$Ric_{WP} = -\frac{13}{12\pi}\omega_{WP}.$$

5. Characteristic forms of $\mathcal{T}(1)$

Let $V = T_v\mathcal{T}(1)$ be the vertical tangent bundle of the fibration

$$\pi : \mathcal{T}(1) \to T(1).$$

The hyperbolic metric on $w^\mu(\mathbb{D}^*)$ defines a Hermitian metric on V, defining the first Chern form $c_1(V)$ – a $(1,1)$-form on $\mathcal{T}(1)$.

Mumford-Morita-Miller characteristic forms ("κ-forms") are (n,n)-forms on the Hilbert manifold $T(1)$, defined by

$$\kappa_n = (-1)^{n+1}\pi_*\left(c_1(V)^{n+1}\right),$$

where $\pi_* : \Omega^*(\mathcal{T}(1)) \to \Omega^{*-2}(T(1))$ is the operation of "integration over the fibers" of $\pi : \mathcal{T}(1) \to T(1)$, considered as a fibration of Hilbert manifolds.

Theorem B.

(i) *On* $\mathcal{T}(1)$*, considered as a Banach manifold,*

$$c_1(V) = -\frac{2}{\pi}\omega_{VK}.$$

(ii) *On* $T(1)$*, considered as a Hilbert manifold,*

$$\kappa_1 = \frac{1}{\pi^2}\omega_{WP}.$$

(iii) *The characteristic forms* κ_n *are right-invariant on the Hilbert manifold* $T(1)$ *and for* $\mu_1, \dots, \mu_n, \nu_1, \dots, \nu_n \in H^{-1,1}(\mathbb{D}^*) \simeq T_0T(1)$,

$$\begin{aligned}&\kappa_n(\mu_1,\dots,\mu_n,\bar{\nu}_1,\dots,\bar{\nu}_n)\\ &=\frac{i^n(n+1)!}{(2\pi)^{n+1}}\sum_{\sigma\in S_n} sgn(\sigma)\iint\limits_{\mathbb{D}^*} G\left(\mu_1\bar{\nu}_{\sigma(1)}\right)\dots G\left(\mu_n\bar{\nu}_{\sigma(n)}\right)\rho(z)d^2z.\end{aligned}$$

6. Applications

The Weil-Petersson properties of the universal Teichmüller space $T(1)$ are "universal" in the sense that all curvature properties of finite-dimensional Teichmüller spaces can be deduced from them. In particular, Wolpert explicit formulas [Wol86] follow from Theorems **A** and **B** by using an "averaging procedure", based on a uniform distribution of lattice points of a cofinite Fuchsian group in the hyperbolic plane (see [TT03] for details). The Kähler potential for the Weil-Petersson metric on the universal Teichmüller space $T(1)$ – "the universal Liouville action" – is constructed in [TT04].

Acknowledgments

The first author is grateful to the organizers of the workshop "Infinite-Dimensional Algebras and Quantum Integrable Systems" in Faro, Portugal, July 21-25, 2003, for their kind hospitality.

References

[Ber65] L Bers, *Automorphic forms and general Teichmüller spaces*, Proc. Conf. Complex Analysis (Minneapolis, 1964), Springer, Berlin, 1965, pp. 109–113.

[Ber72] Lipman Bers, *Uniformization, moduli and Kleinian groups*, Bull. London. Math. Soc. **4** (1972), 257–300.

[Ber73] ______, *Fiber spaces over Teichmüller spaces*, Acta. Math. **130** (1973), 89–126.

[BR87a] M.J. Bowick and S.G. Rajeev, *The holomorphic geometry of closed bosonic string theory and* Diff S^1/S^1, Nuclear Phys. B **293** (1987), no. 2, 348–384.

[BR87b] ______, *String theory as the Kähler geometry of loop space*, Phys. Rev. Lett. **58** (1987), no. 6, 535–538.

[Kir87] A.A. Kirillov, *Kähler structure on the K-orbits of a group of diffeomorphisms of the circle*, Funktsional. Anal. i Prilozhen. **21** (1987), no. 2, 42–45.

[KY87] A.A. Kirillov and D.V. Yur′ev, *Kähler geometry of the infinite-dimensional homogeneous space* $M = \mathrm{diff}_+(S^1)/\mathrm{rot}(S^1)$, Funktsional. Anal. i Prilozhen. **21** (1987), no. 4, 35–46.

[NV90] Subhashis Nag and Alberto Verjovsky, $\mathrm{diff}(S^1)$ *and the Teichmüller spaces*, Comm. Math. Phys. **130** (1990), no. 1, 123–138.

[Teo02] Lee-Peng Teo, *Velling-Kirillov metric on the universal Teichmüller curve*, J. Analyse Math. **93** (2004), 271–308.

[TT03] Leon A. Takhtajan and Lee-Peng Teo, *Weil-Petersson metric on the universal Teichmüller space I: Curvature properties and Chern forms*, Preprint arXiv: math.CV/0312172 (2003).

[TT04] Leon A. Takhtajan and Lee-Peng Teo, *Weil-Petersson metric on the universal Teichmüller space II: Kähler potential and period mapping*, Preprint arXiv: math.CV/0406408.

[Vel] John A. Velling, *A projectively natural metric on Teichmüller's spaces*, unpublished manuscript.

[Wol86] Scott A. Wolpert, *Chern forms and the Riemann tensor for the moduli space of curves*, Invent. Math. **85** (1986), no. 1, 119–145.

Leon A. Takhtajan
Department of Mathematics
SUNY at Stony Brook
Stony Brook
NY 11794-3651, USA
e-mail: `leontak@math.sunysb.edu`

Lee-Peng Teo
Department of Applied Mathematics
National Chiao Tung University
1001, Ta-Hsueh Road
Hsinchu City, 30050
Taiwan, R.O.C.
e-mail: `lpteo@math.nctu.edu.tw`

Progress in Mathematics, Vol. 237, 235–263

Duality for Knizhinik-Zamolodchikov and Dynamical Equations, and Hypergeometric Integrals

V. Tarasov

Mathematics Subject Classification (2000). 17B37, 17B80, 33C70, 33C80, 81R10.

Keywords. Knizhnik-Zamolodchikov equations, dynamical equations, $(\mathfrak{gl}_k, \mathfrak{gl}_n)$ duality, hypergeometric integrals.

1. Introduction

The Knizhnik-Zamolodchikov (KZ) equations is a holonomic system of differential equations for correlation functions in conformal field theory on the sphere [KZ]. The KZ equations play an important role in representation theory of affine Lie algebras and quantum groups, see for example [EFK]. There are rational, trigonometric and elliptic versions of KZ equations, depending on what kind of coefficient functions the equations have. In this paper we will consider only the rational and trigonometric versions of the KZ equations.

The rational KZ equations associated with a reductive Lie algebra $\mathfrak{g}$ is a system of equations for a function $u(z_1, \dots, z_n)$ of complex variables $z_1, \dots, z_n$, which takes values in a tensor product $V_1 \otimes \dots \otimes V_n$ of $\mathfrak{g}$-modules $V_1, \dots, V_n$. The equations depend on a complex parameter κ, and their coefficients are expressed in terms of the symmetric tensor $\Omega \in U(\mathfrak{g}) \otimes U(\mathfrak{g})$ corresponding to a nondegenerate invariant bilinear form on $\mathfrak{g}$. For example, if $\mathfrak{g} = \mathfrak{sl}_2$ and e, f, h are its standard generators such that $[e, f] = h$, then $\Omega = e \otimes f + f \otimes e + h \otimes h/2$.

The rational KZ equations are

$$\kappa \frac{\partial u}{\partial z_i} = \sum_{\substack{j=1 \\ j \neq i}}^{n} \frac{\Omega^{(ij)}}{z_i - z_j} \, u, \qquad i = 1, \dots, n, \tag{1.1}$$

where $\Omega^{(ij)} \in \operatorname{End}(V_1 \otimes \dots \otimes V_n)$ is the operator acting as Ω on $V_i \otimes V_j$ and as the identity on all other tensor factors; for instance,

$$\Omega^{(12)}(v_1 \otimes \dots \otimes v_n) = \bigl(\Omega(v_1 \otimes v_2)\bigr) \otimes v_3 \otimes \dots \otimes v_n .$$

All over the paper we will assume that κ is not a rational number. Properties of solutions of the *KZ* equations depend much on whether κ is rational or not.

Equations (1.1) can be generalized to a holonomic system of differential equations depending on an element $\lambda \in \mathfrak{g}$:

$$\kappa \frac{\partial u}{\partial z_i} = \lambda^{(i)} u + \sum_{\substack{j=1 \\ j \neq i}}^{n} \frac{\Omega^{(ij)}}{z_i - z_j} u, \qquad i = 1, \ldots, n. \tag{1.2}$$

Here $\lambda^{(i)} \in \mathrm{End}(V_1 \otimes \cdots \otimes V_n)$ acts as λ on V_i and as the identity on all other tensor factors: $\lambda^{(i)}(v_1 \otimes \cdots \otimes v_n) = v_1 \otimes \cdots \otimes \lambda v_i \otimes \cdots \otimes v_n$. System (1.2) is also called the *rational KZ equations.*

Further on we will assume that λ is a semisimple regular element of $\mathfrak{g}$. Let $\mathfrak{h} \subset \mathfrak{g}$ be the Cartan subalgebra containing λ, and let $e_\alpha \in \mathfrak{g}$ be a root vector corresponding to a root $\alpha \in \mathfrak{h}^*$. We normalize the root vectors by $(e_\alpha, e_{-\alpha}) = 1$, where $(\,,)$ is the bilinear form on $\mathfrak{g}$ corresponding to the tensor Ω.

In [FMTV] system (1.2) was extended to a larger system of holonomic differential equations for a function $u(z_1, \ldots, z_n; \lambda)$ on $\mathbb{C}^n \oplus \mathfrak{h}$. In addition to equations (1.2) the extended system includes the following equations with respect to λ:

$$\kappa D_\mu u = \sum_{i=1}^{n} z_i \mu^{(i)} u + \sum_{\alpha} \frac{(\mu, \alpha)}{2(\lambda, \alpha)} e_\alpha e_{-\alpha} u, \qquad \mu \in \mathfrak{h}, \tag{1.3}$$

where D_μ is the directional derivative: $D_\mu u(\lambda) = \bigl(\partial_t u(\lambda + t\mu)\bigr)\big|_{t=0}$. Equations (1.3) are called the *rational dynamical differential* (*DD*) *equations.*

A special case of equations (1.3), when $n = 1$ and $z_1 = 0$, was discovered for a completely different reason. Around 1995 studying hyperplanes arrangements De Concini and Procesi introduced in an unpublished work a connection on the set of regular elements of the Cartan subalgebra $\mathfrak{h}$. The equations for horizontal sections of the De Concini-Procesi connection coincide with the rational *DD* equations. The same connection also appeared later in [TL]. De Concini and Procesi conjectured that the monodromy of their connection is described in terms of the quantum Weyl group of type $\mathfrak{g}$. For $\mathfrak{g} = \mathfrak{sl}_n$ this conjecture was proved in [TL].

If all $\mathfrak{g}$-modules $V_1, \ldots, V_n$ are highest weight modules, solutions of the *KZ* equations (1.1) can be written in terms of multidimensional hypergeometric integrals [SV], [V]. The construction of hypergeometric solutions can be generalized in a straightforward way to the case of *KZ* equations (1.2), see [FMTV]. Moreover, it is shown in [FMTV] that the hypergeometric solutions of the *KZ* equations obey the *DD* equations (1.3) as well. Generically, hypergeometric solutions of the *KZ* and *DD* equations are complete, that is, they form a basis of solutions of those systems of differential equations.

An amusing fact about the hypergeometric solutions is that though systems (1.2) and (1.3) have rather similar look, and the variables $z_1, \ldots, z_n$ and λ seem to play nearly interchangeable roles, the formulae for the hypergeometric solutions of the *KZ* and *DD* equations involve $z_1, \ldots, z_n$ and λ in a highly nonsymmetric way.

While the variables $z_1, \dots, z_n$ determine singularities of integrands of the hypergeometric integrals and enter there in a rather complicated manner, λ appears in the integrands only in a very simple way via the exponential of a linear form. Such asymmetry suggests the following idea. Suppose that a certain holonomic system of differential equations can be viewed both as a special case of system (1.2) and as a special case of system (1.3), maybe not for the same Lie algebra $\mathfrak{g}$. Then one can get two types of integral formulae for solutions of that system, and solutions of one kind should be linear combinations of solutions of the other kind. Thus, this can lead to nontrivial relations between hypergeometric integrals of different dimensions.

It turns out that the mentioned idea indeed can be realized in the framework of the $(\mathfrak{gl}_k, \mathfrak{gl}_n)$ duality. This duality plays an important role in the representation theory and the classical invariant theory, see [Zh1], [Ho]. It was observed in [TL] that under the $(\mathfrak{gl}_k, \mathfrak{gl}_n)$ duality the *KZ* equations (1.1) for the Lie algebra $\mathfrak{sl}_k$ correspond to the *DD* equations (1.3) (with n replaced by k and all z's being equal to zero) for the Lie algebra $\mathfrak{sl}_n$. This fact was used in [TL] to compute the monodromy of the De Concini–Procesi connection in terms of the quantum Weyl group action.

Systems (1.2) and (1.3) are counterparts of each other under the $(\mathfrak{gl}_k, \mathfrak{gl}_n)$ duality in general as well, see [TV4]. Employing this claim for $k = n = 2$, after all one arrives to identities for hypergeometric integrals of different dimensions [TV6]. One can expect that there are similar identities for hypergeometric integrals for an arbitrary pair k, n.

There are various generalizations of the *KZ* equations. The function Ω/z, describing the coefficients of the *KZ* equations, is the simplest example of a classical r-matrix – a solution of the classical Yang-Baxter equation. Starting from any classical r-matrix with a spectral parameter one can write down a holonomic system of differential equations, see [Ch2]. The obtained system is called the *KZ* equations associated with the given r-matrix. For example, the standard trigonometric r-matrix is

$$r(z) = \frac{\Omega}{z-1} + \frac{1}{2} \sum_a \xi_a \otimes \xi_a + \sum_{\alpha>0} e_\alpha \otimes e_{-\alpha},$$

where $\{\xi_a\}$ is an orthonormal basis of the Cartan subalgebra, and the second sum is taken over all positive roots α, cf. (3.1) for the Lie algebra $\mathfrak{gl}_k$. The trigonometric r-matrix satisfies the classical Yang-Baxter equation

$$\bigl[r_{12}(z/w), r_{13}(z) + r_{23}(w)\bigr] + \bigl[r_{13}(z), r_{23}(w)\bigr] = 0.$$

The corresponding *KZ* equations are

$$\kappa z_i \frac{\partial u}{\partial z_i} = \lambda^{(i)} u + \sum_{\substack{j=1 \\ j \neq i}}^{n} r^{(ij)}(z_i/z_j)\, u, \qquad i = 1, \dots, n, \tag{1.4}$$

where λ is an element of the Cartan subalgebra. They are called the *trigonometric KZ* equations associated with the Lie algebra $\mathfrak{g}$. System (1.2) can be considered as a limiting case of system (1.4) by the following procedure: one replaces the variables $z_1, \dots, z_n$ by $e^{\varepsilon z_1}, \dots, e^{\varepsilon z_n}$ and λ by λ/ε, and then sends ε to 0.

The difference analogue of the *KZ* equations – the *quantized Knizhnik-Zamolodchikov (qKZ) equations* – were introduced in [FR]. Coefficients of the *qKZ* equations are given in terms of quantum *R*-matrices – solutions of the quantum Yang-Baxter equation:

$$R_{12}(z-w)\,R_{13}(z)\,R_{23}(w) \;=\; R_{23}(w)\,R_{13}(z)\,R_{12}(z-w)\,.$$

There are rational, trigonometric and elliptic versions of *KZ* equations, the corresponding *R*-matrices coming from the representation theory of Yangians, quantum affine algebra algebras and elliptic quantum groups, respectively. The *rational qKZ* equations associated with the Lie algebra $\mathfrak{g}$ is a holonomic system of difference equations for a function $u(z_1, \dots, z_n)$ with values in a tensor product $V_1 \otimes \dots \otimes V_n$ of modules over the Yangian $Y(\mathfrak{g})$:

$$\begin{aligned} u(z_1, \dots, z_i + \kappa, \dots, z_n) \;=\; & \big(R_{1i}(z_1 - z_i - \kappa) \dots R_{i-1,i}(z_{i-1} - z_i - \kappa)\big)^{-1} \\ & \times\, (e^{\mu})^{(i)}\, R_{in}(z_i - z_n) \dots R_{i,i+1}(z_i - z_{i+1})\, u(z_1, \dots, z_n)\,, \end{aligned} \tag{1.5}$$

$i = 1, \dots, n$. Here μ is an element of the Cartan subalgebra and $R_{ij}(z)$ is the *R*-matrix for the tensor product $V_i \otimes V_j$ of the Yangian modules.

There are also several generalizations of the rational differential dynamical equations. The difference analogue of the *DD* equations – the *rational difference dynamical (qDD) equations* – was suggested in [TV3]. The idea was to extend the trigonometric *KZ* equations (1.4) by equations with respect to λ similarly to the way in which system (1.3) extends the rational *KZ* equations (1.2), and to obtain a holonomic system of differential-difference equations for a function $u(z_1, \dots, z_n; \lambda)$ on $\mathbb{C}^n \oplus \mathfrak{h}$. The rational *qDD* equations have the form

$$u(z_1, \dots, z_n; \lambda + \kappa\omega) \;=\; Y_\omega(z_1, \dots, z_n; \lambda)\, u(z_1, \dots, z_n; \lambda) \tag{1.6}$$

where ω is an integral weight of $\mathfrak{g}$, and the operators Y_ω are written in terms of the extremal cocycle on the Weyl group of $\mathfrak{g}$. The extremal cocycles and their special values, the extremal projectors, are important objects in the representation theory of Lie algebras and Lie groups, see [AST], [Zh2], [Zh3], [ST].

The ideas used in [TV3] were further developed in [EV] where a new concept of the dynamical Weyl group was introduced, and the trigonometric version of the difference dynamical equations was suggested.

There is also the trigonometric version of the differential dynamical equations, which, in principle, can be obtained by degenerating the trigonometric difference dynamical equations. The explicit form of the trigonometric differential dynamical equations for the Lie algebras $\mathfrak{gl}_k$ and $\mathfrak{sl}_k$ was obtained in [TV4] by extending the rational *qKZ* equations (1.5) by equations with respect to μ in such a way that the result is a holonomic system of difference-differential equations for a function $u(z_1, \ldots, z_n; \mu)$ on $\mathbb{C}^n \oplus \mathfrak{h}$.

The $(\mathfrak{gl}_k, \mathfrak{gl}_n)$ duality naturally applies to the trigonometric and difference versions of the *KZ* and dynamical equations. Under the duality, the trigonometric *KZ* equations (1.4) for the Lie algebra $\mathfrak{gl}_k$ correspond to the trigonometric differential dynamical equations for the Lie algebra $\mathfrak{gl}_n$, and vice versa. At the same time the rational *qKZ* equations for $\mathfrak{gl}_k$ are counterparts of the rational *qDD* equations for $\mathfrak{gl}_n$. To relate the trigonometric *qKZ* and *qDD* equations, one has to employ the q-analogue of the $(\mathfrak{gl}_k, \mathfrak{gl}_n)$ duality: the $\big(U_q(\mathfrak{gl}_k), U_q(\mathfrak{gl}_n)\big)$ duality described in [B], [TL].

Hypergeometric solutions of the trigonometric *KZ* equations (1.4) can be written almost in the same manner as those of the rational *KZ* equations (1.1), see [Ch1], [MV]. Conjecturally, the hypergeometric solutions of the trigonometric *KZ* equations obey the corresponding rational *qDD* equations. For the Lie algebra $\mathfrak{sl}_k$ this claim was proved in [MV]. On the other hand, solutions of the rational *qKZ* equations can be written in terms of suitable q-hypergeometric Jackson integrals [TV1], or q-hypergeometric integrals of Mellin-Barnes type [TV2]. Thus, using the $(\mathfrak{gl}_k, \mathfrak{gl}_n)$ duality, one can obtain solutions of a certain system of differential-difference equations both in terms of ordinary hypergeometric integrals and q-hypergeometric integrals of Mellin-Barnes type, and establish nontrivial relations between those integrals. For $k = n = 2$ this has been done in [TV7]. The obtained relations are multidimensional analogues of the equality of two integral representations for the Gauss hypergeometric function ${}_2F_1$:

$$\begin{aligned} {}_2F_1(\alpha, \beta; \gamma; z) &= \frac{\Gamma(\gamma)}{\Gamma(\alpha)\Gamma(\gamma-\alpha)} \int_0^1 u^{\alpha-1}(1-u)^{\gamma-\alpha-1}(1-uz)^{-\beta}\, du \\ &= \frac{1}{2\pi i}\, \frac{\Gamma(\gamma)}{\Gamma(\alpha)\Gamma(\beta)} \int_{-i\infty-\varepsilon}^{+i\infty-\varepsilon} (-z)^s\, \frac{\Gamma(-s)\Gamma(s+\alpha)\Gamma(s+\beta)}{\Gamma(s+\gamma)}\, ds\,. \end{aligned}$$

As it was pointed out by J.Harnad, the duality between the *KZ* and *DD* equations in the rational differential case is essentially the "quantum" version of the duality for isomonodromic deformation systems [H1]. The relation of the differential *KZ* equations and the isomonodromic deformation systems is described in [R], [H2]. From this point of view the rational *qDD* equations can be considered as "quantum" analogues of the Schlesinger transformations, though the correspondence is not quite straightforward.

The paper is organized as follows. After introducing basic notation we subsequently describe the differential *KZ* and *DD* equations, and the rational difference *qKZ* and *qDD* equations, for the Lie algebra $\mathfrak{gl}_k$. This is done in Sections 2–5. Then we consider the $(\mathfrak{gl}_k, \mathfrak{gl}_n)$ duality in application to the *KZ* and dynamical equations. In the last two sections we describe the hypergeometric solutions of the equations, and use the duality relations to establish identities for hypergeometric and q-hypergeometric integrals of different dimensions.

2. Basic notation

Let n be a nonnegative integer. A partition $\lambda = (\lambda_1, \lambda_2, \dots)$ with at most k parts is an infinite nonincreasing sequence of nonnegative integers such that $\lambda_{k+1} = 0$. Denote by $\mathcal{P}_k$ the set of partitions with at most k parts and by $\mathcal{P}$ the set of all partitions. We often make use of the embedding $\mathcal{P}_k \to \mathbb{C}^k$ given by truncating the zero tail of a partition: $(\lambda_1, \dots, \lambda_k, 0, 0, \dots) \mapsto (\lambda_1, \dots, \lambda_k)$. Since obviously $\mathcal{P}_m \subset \mathcal{P}_k$ for $m \leqslant k$, in fact, one has a collection of embeddings $\mathcal{P}_m \to \mathbb{C}^k$ for any $m \leqslant k$. What particular embedding is used will be clear from the context.

Let e_{ab}, $a, b = 1, \dots, k$, be the standard basis of the Lie algebra $\mathfrak{gl}_k$: $[e_{ab}, e_{cd}] = \delta_{bc} e_{ad} - \delta_{ad} e_{cb}$. We take the Cartan subalgebra $\mathfrak{h} \subset \mathfrak{gl}_k$ spanned by $e_{11}, \dots, e_{kk}$, and the nilpotent subalgebras $\mathfrak{n}_+$ and $\mathfrak{n}_-$ spanned by the elements e_{ab} for $a < b$ and $a > b$, respectively. One has the standard Gauss decomposition $\mathfrak{gl}_k = \mathfrak{n}_+ \oplus \mathfrak{h} \oplus \mathfrak{n}_-$.

Let $\varepsilon_1, \dots, \varepsilon_k$ be the basis of $\mathfrak{h}^*$ dual to $e_{11}, \dots, e_{kk}$: $\langle \varepsilon_a, e_{bb} \rangle = \delta_{ab}$. We identify $\mathfrak{h}^*$ with $\mathbb{C}^k$ mapping $\lambda_1 \varepsilon_1 + \dots + \lambda_k \varepsilon_k$ to $(\lambda_1, \dots, \lambda_k)$. The root vectors of $\mathfrak{gl}_k$ are e_{ab} for $a \neq b$, the corresponding root being equal to $\alpha_{ab} = \varepsilon_a - \varepsilon_b$. The roots α_{ab} for $a < b$ are positive.

We choose the standard invariant bilinear form $(\,,)$ on $\mathfrak{gl}_k$: $(e_{ab}, e_{cd}) = \delta_{ad} \delta_{bc}$. It defines an isomorphism $\mathfrak{h} \to \mathfrak{h}^*$. The induced bilinear form on $\mathfrak{h}^*$ is $(\varepsilon_a, \varepsilon_b) = \delta_{ab}$.

For a $\mathfrak{gl}_k$-module W and a weight $\lambda \in \mathfrak{h}^*$ let $W[\lambda]$ be the weight subspace of W of weight λ.

For any $\lambda \in \mathcal{P}_k$ we denote by V_λ the irreducible $\mathfrak{gl}_k$-module with highest weight λ. By abuse of notation, for any $l \in \mathbb{Z}_{\geqslant 0}$ we write V_l instead of $V_{(l,0,\dots,0)}$. Thus, $V_0 = \mathbb{C}$ is the trivial $\mathfrak{gl}_k$-module, $V_1 = \mathbb{C}^k$ with the natural action of $\mathfrak{gl}_k$, and V_l is the l-th symmetric power of V_1.

Define a $\mathfrak{gl}_k$-action on the polynomial ring $\mathbb{C}[x_1, \dots, x_k]$ by differential operators: $e_{ab} \mapsto x_a \partial_b$, where $\partial_b = \partial / \partial x_b$, and denote the obtained $\mathfrak{gl}_k$-module by $\mathbb{V}$. Then

$$\mathbb{V} = \bigoplus_{l=0}^{\infty} V_l\,, \tag{2.1}$$

the submodule V_l being spanned by homogeneous polynomials of degree l. The highest weight vector of the submodule V_l is x_1^l.

3. Knizhnik-Zamolodchikov and differential dynamical equations

For any $g \in U(\mathfrak{gl}_k)$ set $g^{(i)} = \mathrm{id} \otimes \dots \otimes \underset{i\mathrm{th}}{g} \otimes \dots \otimes \mathrm{id} \in \bigl(U(\mathfrak{gl}_k)\bigr)^{\otimes n}$. We consider $U(\mathfrak{gl}_k)$ as a subalgebra of $\bigl(U(\mathfrak{gl}_k)\bigr)^{\otimes n}$, the embedding $U(\mathfrak{gl}_k) \hookrightarrow \bigl(U(\mathfrak{gl}_k)\bigr)^{\otimes n}$ being given by the n-fold coproduct, that is, $x \mapsto x^{(1)} + \dots + x^{(n)}$ for any $x \in \mathfrak{gl}_k$.

Let $\Omega = \sum_{a,b=1}^{k} e_{ab} \otimes e_{ba}$ be the Casimir tensor, and let

$$\Omega_+ = \frac{1}{2}\sum_{a=1}^{k} e_{aa} \otimes e_{aa} + \sum_{1 \leqslant a < b \leqslant k} e_{ab} \otimes e_{ba},$$

$$\Omega_- = \frac{1}{2}\sum_{a=1}^{k} e_{aa} \otimes e_{aa} + \sum_{1 \leqslant a < b \leqslant k} e_{ba} \otimes e_{ab},$$

so that $\Omega = \Omega_+ + \Omega_-$. The standard trigonometric r-matrix, associated with the Lie algebra $\mathfrak{gl}_k$ is

$$r(z) = \frac{\Omega}{z-1} + \Omega_+ = \frac{z\,\Omega_+ + \Omega_-}{z-1}. \tag{3.1}$$

Fix a nonzero complex number κ. Consider differential operators $\nabla_{z_1}, \dots, \nabla_{z_n}$ and $\widehat{\nabla}_{z_1}, \dots, \widehat{\nabla}_{z_n}$ with coefficients in $\bigl(U(\mathfrak{gl}_k)\bigr)^{\otimes n}$ depending on complex variables $z_1, \dots, z_n$ and $\lambda_1, \dots, \lambda_k$:

$$\nabla_{z_i}(z;\lambda) = \kappa\frac{\partial}{\partial z_i} - \sum_{a=1}^{k} \lambda_a (e_{aa})^{(i)} - \sum_{\substack{j=1 \\ j\neq i}}^{n} \frac{\Omega^{(ij)}}{z_i - z_j}, \tag{3.2}$$

$$\widehat{\nabla}_{z_i}(z;\lambda) = \kappa z_i\frac{\partial}{\partial z_i} - \sum_{a=1}^{k} \Bigl(\lambda_a - \frac{e_{aa}}{2}\Bigr)(e_{aa})^{(i)} - \sum_{\substack{j=1 \\ j\neq i}}^{n} r^{(ij)}(z_i/z_j). \tag{3.3}$$

The differential operators $\nabla_{z_1}, \dots, \nabla_{z_n}$ (resp. $\widehat{\nabla}_{z_1}, \dots, \widehat{\nabla}_{z_n}$) are called the *rational* (resp. *trigonometric*) *Knizhnik-Zamolodchikov* (*KZ*) *operators*. The following statements are well known.

Theorem 3.1. *The operators* $\nabla_{z_1}, \dots, \nabla_{z_n}$ *pairwise commute.*

Theorem 3.2. *The operators* $\widehat{\nabla}_{z_1}, \dots, \widehat{\nabla}_{z_n}$ *pairwise commute.*

The rational *KZ* equations associated with the Lie algebra $\mathfrak{gl}_k$ is a system of differential equations

$$\nabla_{z_i} u = 0, \qquad i = 1, \dots, n, \tag{3.4}$$

for a function $u(z_1, \dots, z_n; \lambda_1, \dots, \lambda_k)$ taking values in an n-fold tensor product of $\mathfrak{gl}_k$-modules. Similarly, the trigonometric *KZ* equations associated with the Lie algebra $\mathfrak{gl}_k$ is a system of differential equations

$$\widehat{\nabla}_{z_i} u = 0, \qquad i = 1, \dots, n. \tag{3.5}$$

for a function $u(z_1, \dots, z_n; \lambda_1, \dots, \lambda_k)$.

Introduce differential operators $D_{\lambda_1}, \ldots, D_{\lambda_k}$ and $\widehat{D}_{\lambda_1}, \ldots, \widehat{D}_{\lambda_k}$ with coefficients in $\bigl(U(\mathfrak{gl}_k)\bigr)^{\otimes n}$ depending on complex variables $z_1, \ldots, z_n$ and $\lambda_1, \ldots, \lambda_k$:

$$D_{\lambda_a}(z;\lambda) = \kappa \frac{\partial}{\partial \lambda_a} - \sum_{i=1}^{n} z_i \, (e_{aa})^{(i)} - \sum_{\substack{b=1\\ b\neq a}}^{k} \frac{e_{ab}\, e_{ba} - e_{aa}}{\lambda_a - \lambda_b} \,. \tag{3.6}$$

$$\begin{aligned} \widehat{D}_{\lambda_a}(z;\lambda) = {} & \kappa \lambda_a \frac{\partial}{\partial \lambda_a} + \frac{e_{aa}^2}{2} - \sum_{i=1}^{n} z_i \, (e_{aa})^{(i)} - {} \\ & - \sum_{b=1}^{k} \sum_{1 \leqslant i<j \leqslant n} (e_{ab})^{(i)} (e_{ba})^{(j)} - \sum_{\substack{b=1\\ b\neq a}}^{k} \frac{\lambda_b}{\lambda_a - \lambda_b} (e_{ab}\, e_{ba} - e_{aa}) \,. \end{aligned} \tag{3.7}$$

Recall that $e_{ab} = \sum_{i=1}^{n} (e_{ab})^{(i)}$. The operators $D_{\lambda_1}, \ldots, D_{\lambda_k}$ (resp. $\widehat{D}_{\lambda_1}, \ldots, \widehat{D}_{\lambda_k}$) are called the *rational* (resp. *trigonometric*) *differential dynamical* (*DD*) *operators*.

Theorem 3.3. *The operators* $\nabla_{z_1}, \ldots, \nabla_{z_n}$, $D_{\lambda_1}, \ldots, D_{\lambda_k}$ *pairwise commute.*

The theorem follows from the same result for the rational *KZ* and *DD* operators associated with the Lie algebra $\mathfrak{sl}_k$, see [FMTV].

Theorem 3.4. [TV4] *The operators* $\widehat{D}_{\lambda_1}, \ldots, \widehat{D}_{\lambda_k}$ *pairwise commute.*

The statement can be verified in a straightforward way.

Later we will formulate analogues of Theorem 3.3 for the trigonometric *KZ* operators and the trigonometric *DD* operators, see Theorems 4.1 and 5.1. They involve difference dynamical operators and difference (quantized) Knizhnik-Zamolodchikov operators which are discussed in the next two sections.

The rational *DD* equations associated with the Lie algebra $\mathfrak{gl}_k$ is a system of differential equations

$$D_{\lambda_a} u = 0 \,, \qquad a = 1, \ldots, k \,, \tag{3.8}$$

for a function $u(z_1, \ldots, z_n; \lambda_1, \ldots, \lambda_k)$ taking values in an n-fold tensor product of $\mathfrak{gl}_k$-modules. Similarly, the trigonometric *DD* equations associated with the Lie algebra $\mathfrak{gl}_k$ is a system of differential equations

$$\widehat{D}_{\lambda_a} u = 0 \,, \qquad a = 1, \ldots, k \,. \tag{3.9}$$

for a function $u(z_1, \ldots, z_n; \lambda_1, \ldots, \lambda_k)$.

Remark. Systems (3.5) and (3.8) are not precisely the same as specializations of the respective systems (1.4) and (1.3) for the Lie algebra $\mathfrak{gl}_k$. However, in both cases the difference is not quite essential and can be worked out. The form of the operators $\widehat{\nabla}_{z_i}$ and D_{λ_a} given in this section, see (3.3) and (3.6), fits the best the framework of the $(\mathfrak{gl}_k, \mathfrak{gl}_n)$ duality.

4. Rational difference dynamical equations

For any $a, b = 1, \dots, k$, $a \neq b$, introduce a series $B_{ab}(t)$ depending on a complex variable t:

$$B_{ab}(t) = 1 + \sum_{s=1}^{\infty} e_{ba}^{s} e_{ab}^{s} \prod_{j=1}^{s} \frac{1}{j\,(t - e_{aa} + e_{bb} - j)}\,.$$

The series has a well-defined action on any finite-dimensional $\mathfrak{gl}_k$-module W, giving an $\mathrm{End}(W)$-valued rational function of t. The series $B_{ab}(t)$ have zero weight:

$$\bigl[B_{ab}(t), x\bigr] = 0 \qquad \text{for any} \quad x \in \mathfrak{h}\,, \tag{4.1}$$

satisfy the inversion relation

$$B_{ab}(t)\, B_{ba}(-t) = 1 - \frac{e_{aa} - e_{bb}}{t}\,, \tag{4.2}$$

and the braid relation

$$B_{ab}(t-s)\, B_{ac}(t)\, B_{bc}(s) = B_{bc}(s)\, B_{ac}(t)\, B_{ab}(t-s)\,. \tag{4.3}$$

Relation (4.1) is clear. Relations (4.2) and (4.3) follow from [TV3], namely from the properties of functions $B_w(\lambda)$ considered there in the $\mathfrak{sl}_k$ case, see [TV1, Section 2.6]. In notation of [TV3] the series $B_{ab}(t)$ equals $p(t-1; e_{aa} - e_{bb}, e_{ab}, e_{ba})$.

Remark. The series $B_{ab}(t)$ first appeared in the definition of the extremal projectors [AST] and the extremal cocycles on the Weyl group [Zh2], [Zh3].

Consider the products $X_1, \dots, X_k$ depending on complex variables $z_1, \dots, z_n$ and $\lambda_1, \dots, \lambda_k$:

$$X_a(z;\lambda) = \bigl(B_{ak}(\lambda_{ak}) \dots B_{a,a+1}(\lambda_{a,a+1})\bigr)^{-1} \tag{4.4}$$
$$\times \prod_{i=1}^{n} \bigl(z_i^{-e_{aa}}\bigr)_{(i)}\, B_{1a}(\lambda_{1a} - \kappa) \dots B_{a-1,a}(\lambda_{a-1,a} - \kappa)\,,$$

where $\lambda_{bc} = \lambda_b - \lambda_c$. They act on any n-fold tensor product $W_1 \otimes \dots \otimes W_n$ of finite-dimensional (more generally, highest weight) $\mathfrak{gl}_k$-modules.

Let T_u be a difference operator acting on a function $f(u)$ by the rule

$$(T_u f)(u) = f(u + \kappa)\,.$$

Introduce difference operators $Q_{\lambda_1}, \dots, Q_{\lambda_k}$:

$$Q_{\lambda_a}(z;\lambda) = X_a(z;\lambda)\, T_{\lambda_a}\,.$$

They are called the *rational difference dynamical* (*qDD*) operators.

Theorem 4.1. *The operators* $\widehat{\nabla}_{z_1}, \dots, \widehat{\nabla}_{z_n}$, $Q_{\lambda_1}, \dots, Q_{\lambda_k}$ *pairwise commute.*

The theorem follows from the same result for the trigonometric KZ and rational qDD operators in the $\mathfrak{sl}_k$ case, see [TV1]. Theorem 4.1 extends Theorem 3.2, and is analogous to Theorem 3.3.

In more conventional form the equalities

$$[\widehat{\nabla}_{z_i}, Q_{\lambda_a}] = 0, \qquad [Q_{\lambda_a}, Q_{\lambda_b}] = 0,$$

respectively look like

$$\widehat{\nabla}_{z_i}(z;\lambda)\, X_a(z;\lambda) \;=\; X_a(z;\lambda)\, \widehat{\nabla}_{z_i}(z;\lambda_1,\dots,\lambda_a+\kappa,\dots,\lambda_k),$$

$$X_a(z;\lambda)\, X_b(z;\lambda_1,\dots,\lambda_a+\kappa,\dots,\lambda_k) \;=\; X_b(z;\lambda)\, X_a(z;\lambda_1,\dots,\lambda_b+\kappa,\dots,\lambda_k).$$

The *rational difference dynamical* (qDD) *equations* associated with the Lie algebra $\mathfrak{gl}_k$ is a system of difference equations

$$Q_{\lambda_a} u \;=\; u, \qquad a = 1,\dots,k, \tag{4.5}$$

for a function $u(z_1,\dots,z_n;\lambda_1,\dots,\lambda_k)$ taking values in an n-fold tensor product of $\mathfrak{gl}_k$-modules.

5. Rational difference Knizhnik-Zamolodchikov equations

For any two irreducible finite-dimensional $\mathfrak{gl}_k$-modules V, W there exists a distinguished $\mathrm{End}(V\otimes W)$-valued rational function $R_{VW}(t)$ called the *rational R-matrix* for the tensor product $V\otimes W$. The definition of $R_{VW}(t)$ comes from the representation theory of the Yangian $Y(\mathfrak{gl}_k)$.

The Yangian $Y(\mathfrak{gl}_k)$ is an infinite-dimensional Hopf algebra, which is a flat deformation of the universal enveloping algebra $U\big(\mathfrak{gl}_k[x]\big)$ of $\mathfrak{gl}_k$-valued polynomial functions. The subalgebra of constant functions in $U\big(\mathfrak{gl}_k[x]\big)$, which is isomorphic to $U(\mathfrak{gl}_k)$, is preserved under the deformation. Thus, the algebra $U(\mathfrak{gl}_k)$ is embedded in $Y(\mathfrak{gl}_k)$ as a Hopf subalgebra, and we identify $U(\mathfrak{gl}_k)$ with the image of this embedding.

There is an algebra homomorphism $ev: Y(\mathfrak{gl}_k) \to U(\mathfrak{gl}_k)$, called the *evaluation homomorphism*, which is identical on the subalgebra $U(\mathfrak{gl}_k) \subset Y(\mathfrak{gl}_k)$. It is a deformation of the homomorphism $U\big(\mathfrak{gl}_k[x]\big) \to U(\mathfrak{gl}_k)$ which sends any polynomial to its value at $x = 0$. The evaluation homomorphism is not a homomorphism of Hopf algebras.

The Yangian $Y(\mathfrak{gl}_k)$ has a distinguished one-parametric family of automorphisms ρ_u depending on a complex parameter u, which is informally called the *shift of the spectral parameter*. The automorphism ρ_u corresponds to the automorphism $p(x) \mapsto p(x+u)$ of the Lie algebra $\mathfrak{gl}_k[x]$. For any $\mathfrak{gl}_k$-module W we denote by $W(u)$ the pullback of W via the homomorphism $ev\circ\rho_u$. Yangian modules of this form are called *evaluation modules*.

For any finite-dimensional irreducible $\mathfrak{gl}_k$-modules V, W the tensor products $V(t) \otimes W(u)$ and $W(u) \otimes V(t)$ are isomorphic irreducible $Y(\mathfrak{gl}_k)$-modules, provided $t - u \notin \mathbb{Z}$. The intertwiner $V(t) \otimes W(u) \to W(u) \otimes V(t)$ can be taken of the form $P_{VW} R_{VW}(t-u)$, where $P_{VW} : V \otimes W \to W \otimes V$ is the flip map: $P_{VW} : v \otimes w \mapsto w \otimes v$, and $R_{VW}(t)$ is a rational $\mathrm{End}(V \otimes W)$-valued function, the *rational R-matrix* for the tensor product $V \otimes W$.

The R-matrix $R_{VW}(t)$ can be described in terms of the $\mathfrak{gl}_k$ actions on the spaces V and W. It is determined uniquely up to a scalar multiple by the $\mathfrak{gl}_k$ invariance,

$$\big[R_{VW}(t),\, g \otimes 1 + 1 \otimes g\big] = 0 \qquad \text{for any} \quad g \in \mathfrak{gl}_k\,, \tag{5.1}$$

and the commutation relations

$$R_{VW}(t) \Big(t\, e_{ab} \otimes 1 + \sum_{c=1}^{k} e_{ac} \otimes e_{cb}\Big) = \Big(t\, e_{ab} \otimes 1 + \sum_{c=1}^{k} e_{cb} \otimes e_{ac}\Big) R_{VW}(t)\,. \tag{5.2}$$

The standard normalization condition for $R_{VW}(t)$ is to preserve the tensor product of the respective highest weight vectors v, w:

$$R_{VW}(t)\, v \otimes w = v \otimes w\,.$$

The introduced R-matrices obey the inversion relation

$$R_{VW}(t)\, R^{(21)}_{WV}(-t) = 1\,, \tag{5.3}$$

where $R^{(21)}_{WV} = P_{WV} R_{WV} P_{VW}$, and the Yang-Baxter equation

$$R_{UV}(t-u)\, R_{UW}(t)\, R_{VW}(u) = R_{VW}(u)\, R_{UW}(t)\, R_{UV}(t-u)\,. \tag{5.4}$$

The aforementioned facts on the Yangian $Y(\mathfrak{gl}_k)$ are well known. A good introduction into the representation theory of the Yangian $Y(\mathfrak{gl}_k)$ can be found in [MNO].

Consider the $\mathfrak{gl}_k$-module $\mathbb{V}$, and let $V_l \subset \mathbb{V}$ be the irreducible component with highest weight vector x_1^l, see (2.1). We define the R-matrix $R_{\mathbb{V}\mathbb{V}}(t)$ to be a direct sum of the R-matrices $R_{V_l V_m}(t)$:

$$R_{\mathbb{V}\mathbb{V}}(t)\, v \otimes v' = R_{V_l V_m}(t)\, v \otimes v'\,, \qquad v \in V_l,\ \ v' \in V_m\,.$$

It is clear that $R_{\mathbb{V}\mathbb{V}}(t)$ obeys relations (5.1) and (5.2), as well as the inversion relation and the Yang-Baxter equation.

Consider the products $K_1, \ldots, K_n$ depending on complex variables $z_1, \ldots, z_n$ and $\lambda_1, \ldots, \lambda_k$:

$$K_i(z;\lambda) = \bigl(R_{in}(z_{in}) \ldots R_{i,i+1}(z_{i,i+1})\bigr)^{-1} \tag{5.5}$$
$$\times \prod_{a=1}^{k} \bigl(\lambda_a^{-e_{aa}}\bigr)^{(i)} R_{1i}(z_{1i} - \kappa) \ldots R_{i-1,i}(z_{i-1,i} - \kappa),$$

acting on a tensor product $W_1 \otimes \cdots \otimes W_n$ of $\mathfrak{gl}_k$-modules. Here $z_{ij} = z_i - z_j$, and $R_{ij}(t) = \bigl(R_{W_i W_j}(t)\bigr)^{(ij)}$.

Introduce difference operators $Z_{z_1}, \ldots, Z_{z_n}$:

$$Z_{z_i}(z;\lambda) = K_i(z;\lambda)\, T_{z_i}.$$

They are called the *rational quantized Knizhnik-Zamolodchikov* (*qKZ*) operators. The next theorem extends Theorem 3.4 and is analogous to Theorem 3.3.

Theorem 5.1. [FR], [TV4] *The operators* $Z_{z_1}, \ldots, Z_{z_n}$, $\widehat{D}_{\lambda_1}, \ldots, \widehat{D}_{\lambda_k}$ *pairwise commute.*

The *qKZ* operators $Z_{z_1}, \ldots, Z_{z_n}$ were introduced in [FR], and their commutativity was established therein. The fact that the *qKZ* operators commute with the operators $\widehat{D}_{\lambda_1}, \ldots, \widehat{D}_{\lambda_k}$ can be verified in a straightforward way using relations (5.1) and (5.2) for the R-matrices.

In more conventional form the equalities $[Z_{z_i}, Z_{z_j}] = 0$ and $[Z_{z_i}, \widehat{D}_{\lambda_a}] = 0$ respectively look like:

$$K_i(z;\lambda)\, K_j(z_1, \ldots, z_i + \kappa, \ldots, z_n;\lambda) = K_j(z;\lambda)\, K_i(z_1, \ldots, z_j + \kappa, \ldots, z_n;\lambda),$$

$$\widehat{D}_{\lambda_a}(z;\lambda)\, K_i(z;\lambda) = K_i(z;\lambda)\, \widehat{D}_{\lambda_a}(z_1, \ldots, z_i + \kappa, \ldots, z_n;\lambda).$$

The *rational qKZ equations* associated with the Lie algebra $\mathfrak{gl}_k$ is a system of difference equations

$$Z_{z_i} u = u, \qquad i = 1, \ldots, n, \tag{5.6}$$

for a function $u(z_1, \ldots, z_n; \lambda_1, \ldots, \lambda_k)$ taking values in an n-fold tensor product of $\mathfrak{gl}_k$-modules.

6. $(\mathfrak{gl}_k, \mathfrak{gl}_n)$ duality

In this section we are going to consider the Lie algebras $\mathfrak{gl}_k$ and $\mathfrak{gl}_n$ simultaneously. In order to distinguish generators, modules, etc., we will indicated the dependence on k and n explicitly, for example, $e_{ab}^{\langle k \rangle}$, $V_\lambda^{\langle n \rangle}$.

Consider the polynomial ring $\mathbb{P}_{kn} = \mathbb{C}[x_{11}, \ldots, x_{k1}, \ldots, x_{1n}, \ldots, x_{kn}]$ of kn variables. There are two natural isomorphisms of vector spaces:

$$\big(\mathbb{C}[x_1, \ldots, x_k]\big)^{\otimes n} \to \mathbb{P}_{kn}\,, \tag{6.1}$$

$$(p_1 \otimes \cdots \otimes p_n)(x_{11}, \ldots, x_{kn}) = \prod_{i=1}^{n} p_i(x_{1i}, \ldots, x_{ki})\,,$$

and

$$\big(\mathbb{C}[x_1, \ldots, x_n]\big)^{\otimes k} \to \mathbb{P}_{kn}\,, \tag{6.2}$$

$$(p_1 \otimes \cdots \otimes p_k)(x_{11}, \ldots, x_{kn}) = \prod_{a=1}^{k} p_a(x_{a1}, \ldots, x_{an})\,.$$

Define a $\mathfrak{gl}_k$-action on $\mathbb{P}_{kn}$ by

$$e^{\langle k\rangle}_{ab} \mapsto \sum_{i=1}^{n} x_{ai}\partial_{bi}\,, \tag{6.3}$$

where $\partial_{bi} = \partial/\partial x_{bi}$, and a $\mathfrak{gl}_n$-action by

$$e^{\langle n\rangle}_{ij} \mapsto \sum_{a=1}^{k} x_{ai}\partial_{aj}\,. \tag{6.4}$$

Proposition 6.1. *As a $\mathfrak{gl}_k$-module, $\mathbb{P}_{kn}$ is isomorphic to $\big(\mathbb{V}^{\langle k\rangle}\big)^{\otimes n}$ by (6.1). As a $\mathfrak{gl}_n$-module, $\mathbb{P}_{kn}$ is isomorphic to $\big(\mathbb{V}^{\langle n\rangle}\big)^{\otimes k}$ by (6.2).*

It is easy to see that the actions (6.3) and (6.4) commute with each other, thus making $\mathbb{P}_{kn}$ into a module over the direct sum $\mathfrak{gl}_k \oplus \mathfrak{gl}_n$. The following theorem is well known.

Theorem 6.2. *The $\mathfrak{gl}_k \oplus \mathfrak{gl}_n$ module $\mathbb{P}_{kn}$ has the decomposition*

$$\mathbb{P}_{kn} = \bigoplus_{\lambda \in \mathcal{P}_{\min(k,n)}} V^{\langle k\rangle}_{\lambda} \otimes V^{\langle n\rangle}_{\lambda}\,.$$

The module $\mathbb{P}_{kn}$ plays an important role in the representation theory and the classical invariant theory, see [Zh1], [Ho], [N].

Consider the action of *KZ*, *qKZ*, *DD* and *qDD* operators for the Lie algebras $\mathfrak{gl}_k$ and $\mathfrak{gl}_n$ on $\mathbb{P}_{kn}$-valued functions of $z_1, \ldots, z_n$ and $\lambda_1, \ldots, \lambda_k$, treating the space $\mathbb{P}_{kn}$ as a tensor product $\big(\mathbb{V}^{\langle k\rangle}\big)^{\otimes n}$ of $\mathfrak{gl}_k$-modules, and as a tensor product $\big(\mathbb{V}^{\langle n\rangle}\big)^{\otimes k}$ of $\mathfrak{gl}_n$-modules. If F and G act on the $\mathbb{P}_{kn}$-valued functions in the same way, we will write $F \simeq G$. For instance, $\big(e^{\langle k\rangle}_{aa}\big)^{(i)} \simeq \big(e^{\langle n\rangle}_{ii}\big)^{(a)}$ since both $\big(e^{\langle k\rangle}_{aa}\big)^{(i)}$ and $\big(e^{\langle k\rangle}_{aa}\big)^{(i)}$ act on $\mathbb{P}_{kn}$ as $x_{ai}\partial_{ai}$.

Introduce the following operators:

$$C_{ab}^{\langle k\rangle}(t) \;=\; \frac{\Gamma(t+1)\,\Gamma(t-e_{aa}^{\langle k\rangle}+e_{bb}^{\langle k\rangle})}{\Gamma(t-e_{aa}^{\langle k\rangle})\,\Gamma(t+e_{bb}^{\langle k\rangle}+1)}\,, \tag{6.5}$$

$$C_{ij}^{\langle n\rangle}(t) \;=\; \frac{\Gamma(t+1)\,\Gamma(t-e_{ii}^{\langle n\rangle}+e_{jj}^{\langle n\rangle})}{\Gamma(t-e_{ii}^{\langle n\rangle})\,\Gamma(t+e_{jj}^{\langle n\rangle}+1)}\,.$$

Theorem 6.3. [TV4] *For any* $i=1,\dots,n$ *and* $a=1,\dots,k$ *we have*

$$\nabla_{z_i}^{\langle k\rangle}(z;\lambda) \simeq D_{z_i}^{\langle n\rangle}(\lambda;z)\,, \qquad D_{\lambda_a}^{\langle k\rangle}(z;\lambda) \simeq \nabla_{\lambda_a}^{\langle n\rangle}(\lambda;z)\,, \tag{6.6}$$

$$\widehat{\nabla}_{z_i}^{\langle k\rangle}(z;\lambda) \simeq \widehat{D}_{z_i}^{\langle n\rangle}(\lambda;z)\,, \qquad \widehat{D}_{\lambda_a}^{\langle k\rangle}(z;\lambda) \simeq \widehat{\nabla}_{\lambda_a}^{\langle n\rangle}(\lambda;z)\,, \tag{6.7}$$

$$Z_{z_i}^{\langle k\rangle}(z;\lambda) \simeq N_i^{\langle n\rangle}(z)\,Q_{z_i}^{\langle n\rangle}(\lambda;z)\,, \qquad N_a^{\langle k\rangle}(\lambda)\,Q_{\lambda_a}^{\langle k\rangle}(z;\lambda) \simeq Z_{\lambda_a}^{\langle n\rangle}(\lambda;z)\,. \tag{6.8}$$

Here

$$N_i^{\langle n\rangle}(z) \;=\; \prod_{1\leqslant j<i} C_{ji}^{\langle n\rangle}(z_{ji}-\kappa) \prod_{i<j\leqslant n} \bigl(C_{ij}^{\langle n\rangle}(z_{ij})\bigr)^{-1} \tag{6.9}$$

and

$$N_a^{\langle k\rangle}(\lambda) \;=\; \prod_{1\leqslant b<a} C_{ba}^{\langle k\rangle}(\lambda_{ba}-\kappa) \prod_{a<b\leqslant k} \bigl(C_{ab}^{\langle k\rangle}(\lambda_{ab})\bigr)^{-1}. \tag{6.10}$$

Equalities (6.6) and (6.7) for differential operators are verified in a straightforward way. Equalities (6.8) for difference operators follow from Theorem 6.4.

Theorem 6.4. [TV4] *For any* $a,b=1,\dots,k$, $a\neq b$, *and any* $i,j=1,\dots,n$, $i\neq j$, *we have*

$$B_{ab}^{\langle k\rangle}(t)\,C_{ab}^{\langle k\rangle}(t) \simeq R_{ab}^{\langle n\rangle}(t)\,, \qquad R_{ij}^{\langle k\rangle}(t) \simeq B_{ij}^{\langle n\rangle}(t)\,C_{ij}^{\langle n\rangle}(t)\,.$$

Fix vectors $\boldsymbol{l}=(l_1,\dots,l_n)\in\mathbb{Z}^n_{\geqslant 0}$ and $\boldsymbol{m}=(m_1,\dots,m_k)\in\mathbb{Z}^k_{\geqslant 0}$ such that $\sum_{i=1}^{n} l_i = \sum_{a=1}^{k} m_a$. Let

$$\mathcal{Z}_{kn}[\boldsymbol{l},\boldsymbol{m}] \;=\; \Bigl\{\,(d_{ai})_{\substack{a=1,\dots,k\\ i=1,\dots,n}} \in \mathbb{Z}^{kn}_{\geqslant 0} \;\Bigm|\; \sum_{a=1}^{k} d_{ai}=l_i\,,\quad \sum_{i=1}^{n} d_{ai}=m_a \Bigr\}\,.$$

Denote by $\mathbb{P}_{kn}[\boldsymbol{l},\boldsymbol{m}]\subset\mathbb{P}_{kn}$ the span of all monomials $\prod_{a=1}^{k}\prod_{i=1}^{n} x_{ai}^{d_{ai}}$ such that $(d_{ai})\in\mathcal{Z}_{kn}[\boldsymbol{l},\boldsymbol{m}]$. Formulae (2.1), (6.1)–(6.4) and Proposition 6.1 imply that $\mathbb{P}_{kn}[\boldsymbol{l},\boldsymbol{m}]$ is isomorphic to each of the weight subspaces

$$(V_{l_1}^{\langle k\rangle}\otimes\dots\otimes V_{l_n}^{\langle k\rangle})[m_1,\dots,m_k] \quad\text{and}\quad (V_{m_1}^{\langle n\rangle}\otimes\dots\otimes V_{m_k}^{\langle n\rangle})[l_1,\dots,l_n]\,.$$

The isomorphisms are described in Proposition 6.6.

Let $v_i^{\langle k\rangle}$, $v_j^{\langle n\rangle}$ be highest weight vectors of the respective modules $V_{l_i}^{\langle k\rangle}$, $V_{m_j}^{\langle n\rangle}$. For an indeterminate y set $y^{[0]} = 1$ and $y^{[s]} = y^s/s!$ for $s \in \mathbb{Z}_{>0}$. For any $\mathbf{d} \in \mathcal{Z}_{kn}[\boldsymbol{l},\boldsymbol{m}]$ set $x^{[\mathbf{d}]} = \prod_{a=1}^{k}\prod_{i=1}^{n} x_{ai}^{[d_{ai}]} \in \mathbb{P}_{kn}[\boldsymbol{l},\boldsymbol{m}]$.

Lemma 6.5. *A basis of the weight subspace* $(V_{l_1}^{\langle k\rangle}\otimes\cdots\otimes V_{l_n}^{\langle k\rangle})[m_1,\dots,m_k]$ *is given by vectors*

$$v_{\mathbf{d}}^{\langle k\rangle} = \prod_{a=2}^{k}\bigl(e_{a1}^{\langle k\rangle}\bigr)^{[d_{a1}]}v_1^{\langle k\rangle}\otimes\cdots\otimes\prod_{a=2}^{k}\bigl(e_{a1}^{\langle k\rangle}\bigr)^{[d_{an}]}v_n^{\langle k\rangle}, \quad \mathbf{d} = (d_{ai}) \in \mathcal{Z}_{kn}[\boldsymbol{l},\boldsymbol{m}]. \tag{6.11}$$

A basis of the weight subspace $(V_{m_1}^{\langle n\rangle}\otimes\cdots\otimes V_{m_k}^{\langle n\rangle})[l_1,\dots,l_n]$ *is given by vectors*

$$v_{\mathbf{d}}^{\langle n\rangle} = \prod_{i=2}^{n}\bigl(e_{i1}^{\langle n\rangle}\bigr)^{[d_{1i}]}v_1^{\langle n\rangle}\otimes\cdots\otimes\prod_{i=2}^{n}\bigl(e_{i1}^{\langle n\rangle}\bigr)^{[d_{ki}]}v_k^{\langle n\rangle}, \quad \mathbf{d} = (d_{ai}) \in \mathcal{Z}_{kn}[\boldsymbol{l},\boldsymbol{m}]. \tag{6.12}$$

Proposition 6.6. *The isomorphisms* (6.1) *and* (6.2) *induce the isomorphisms*

$$(V_{l_1}^{\langle k\rangle}\otimes\cdots\otimes V_{l_n}^{\langle k\rangle})[m_1,\dots,m_k] \to \mathbb{P}_{kn}[\boldsymbol{l},\boldsymbol{m}], \qquad v_{\mathbf{d}}^{\langle k\rangle} \mapsto x^{[\mathbf{d}]},$$

$$(V_{m_1}^{\langle n\rangle}\otimes\cdots\otimes V_{m_k}^{\langle n\rangle})[l_1,\dots,l_n] \to \mathbb{P}_{kn}[\boldsymbol{l},\boldsymbol{m}], \qquad v_{\mathbf{d}}^{\langle n\rangle} \mapsto x^{[\mathbf{d}]}.$$

Since all *KZ*, *qKZ*, *DD* and *qDD* operators respect the weight decomposition of the corresponding tensor products of $\mathfrak{gl}_k$ and $\mathfrak{gl}_n$-modules, they can be restricted to functions with values in weight subspaces. Then one can read Theorem 6.3 as follows.

Theorem 6.7. *Let* ϕ *be the isomorphism of weight subspaces:*

$$\phi : (V_{l_1}^{\langle k\rangle}\otimes\cdots\otimes V_{l_n}^{\langle k\rangle})[m_1,\dots,m_k] \to (V_{m_1}^{\langle n\rangle}\otimes\cdots\otimes V_{m_k}^{\langle n\rangle})[l_1,\dots,l_n], \tag{6.13}$$

$$\phi : v_{\mathbf{d}}^{\langle k\rangle} \to v_{\mathbf{d}}^{\langle n\rangle}, \qquad \mathbf{d} \in \mathcal{Z}_{kn}[\boldsymbol{l},\boldsymbol{m}].$$

Then for any $i = 1,\dots,n$ *and* $a = 1,\dots,k$ *we have*

$$\nabla_{z_i}^{\langle k\rangle}(z;\lambda) = \phi^{-1}D_{z_i}^{\langle n\rangle}(\lambda;z)\,\phi, \qquad D_{\lambda_a}^{\langle k\rangle}(z;\lambda) = \phi^{-1}\nabla_{\lambda_a}^{\langle n\rangle}(\lambda;z)\,\phi, \tag{6.14}$$

$$\widehat{\nabla}_{z_i}^{\langle k\rangle}(z;\lambda) = \phi^{-1}\widehat{D}_{z_i}^{\langle n\rangle}(\lambda;z)\,\phi, \qquad \widehat{D}_{\lambda_a}^{\langle k\rangle}(z;\lambda) = \phi^{-1}\widehat{\nabla}_{\lambda_a}^{\langle n\rangle}(\lambda;z)\,\phi, \tag{6.15}$$

$$Z_{z_i}^{\langle k\rangle}(z;\lambda) = \phi^{-1}N_i^{\langle n\rangle}(z)\,Q_{z_i}^{\langle n\rangle}(\lambda;z)\,\phi, \tag{6.16}$$

$$N_a^{\langle k\rangle}(\lambda)\,Q_{\lambda_a}^{\langle k\rangle}(z;\lambda) = \phi^{-1}Z_{\lambda_a}^{\langle n\rangle}(\lambda;z)\,\phi. \tag{6.17}$$

Here $N_i^{\langle n\rangle}(z)$, $N_a^{\langle k\rangle}(\lambda)$ *are given by formulae* (6.9), (6.10).

Observe in addition that the restrictions of operators (6.5) to the weight subspaces are proportional to the identity operator:

$$C_{ab}^{\langle k\rangle}(t)\big|_{(V_{l_1}^{\langle k\rangle}\otimes\cdots\otimes V_{l_n}^{\langle k\rangle})[m_1,\dots,m_k]} = \prod_{s=1}^{m_b}\frac{t-m_a+s-1}{t+s}, \tag{6.18}$$

$$C_{ij}^{\langle n\rangle}(t)\big|_{(V_{m_1}^{\langle n\rangle}\otimes\cdots\otimes V_{m_k}^{\langle n\rangle})[l_1,\dots,l_n]} = \prod_{s=1}^{l_j}\frac{t-l_i+s-1}{t+s}, \tag{6.19}$$

and are rational function of t.

Theorem 6.7 can be "analytically continued" with respect to l_1, m_1. Namely, the theorem remains true if l_1, m_1 are complex numbers, while all other numbers $l_2,\dots,l_n$, $m_2,\dots,m_k$ are still integers, and $\sum_{i=1}^{n} l_i = \sum_{a=1}^{k} m_a$. In this case the modules $V_{l_1}^{\langle k\rangle}$ and $V_{m_1}^{\langle n\rangle}$ are to be irreducible highest weight modules with highest weight $(l_1,0,\dots,0)$ and $(m_1,0,\dots,0)$, respectively, and the definition of $\mathcal{Z}_{kn}[\boldsymbol{l},\boldsymbol{m}]$ remains intact except that d_{11} can be any number. Formulae (6.11) and (6.12) make sense because they do not contain d_{11}, and Lemma 6.5 holds. Formulae (6.18) and (6.19) for $a<b$ and $i<j$ make sense for complex l_1, m_1 as well, which is enough to obtain $N_i^{\langle n\rangle}(z)$ and $N_a^{\langle k\rangle}(\lambda)$ by (6.9), (6.10). The "analytic continuation" of Theorem 6.7 will be useful in application to identities of hypergeometric integrals of different dimensions.

7. Hypergeometric solutions of the Knizhnik-Zamolodchikov and dynamical equations

In the remaining part of the paper we will restrict ourselves to the case of the Lie algebra $\mathfrak{gl}_2$, which corresponds to $k=2$ in the previous sections.

Fix vectors $\boldsymbol{l}=(l_1,\dots,l_n)$ and $\boldsymbol{m}=(m_1,m_2)$ such that $\sum_{i=1}^{n} l_i = m_1+m_2$ and $m_2\in\mathbb{Z}_{\geqslant 0}$. Let

$$\mathcal{Z}[\boldsymbol{l},\boldsymbol{m}] = \Big\{(d_1,\dots,d_n)\in\mathbb{Z}_{\geqslant 0}^n \;\Big|\; \sum_{i=1}^{n} d_i = m_2\,,\quad d_i\leqslant l_i \ \text{ if } \ l_i\in\mathbb{Z}_{\geqslant 0}\Big\}.$$

Given $d_1,\dots,d_n$, set $d_{<i}=\sum_{j=1}^{i-1} d_j$, $i=1,\dots,n$.

Consider the weight subspace $(V_{l_1}\otimes\cdots\otimes V_{l_n})[m_1,m_2]$. It has a basis given by vectors

$$v_{\mathbf{d}} = e_{21}^{[d_1]}v_{l_1}\otimes\cdots\otimes e_{21}^{[d_n]}v_{l_n}\,, \qquad \mathbf{d}=(d_1,\dots,d_n)\in\mathcal{Z}[\boldsymbol{l},\boldsymbol{m}],$$

where $v_1,\dots,v_n$ are respective highest weight vectors of the modules $V_{l_1},\dots,V_{l_n}$.

Define the *master function*

$$\Phi_r(t_1,\dots,t_r;z_1,\dots,z_n;\lambda_1,\lambda_2;\boldsymbol{l}) = e^{\lambda_1\sum_{i=1}^n l_i z_i-(\lambda_1-\lambda_2)\sum_{a=1}^r t_a}\times$$

$$\times\,(\lambda_1-\lambda_2)^{-r}\prod_{1\leqslant i<j\leqslant n}(z_i-z_j)^{l_i l_j}\prod_{a=1}^r\prod_{i=1}^n(t_a-z_i)^{-l_i}\prod_{1\leqslant a<b\leqslant r}(t_a-t_b)^2\,,$$

and the *weight function*

$$w_{\mathbf{d}}(t_1,\dots,t_r;z_1,\dots,z_n) = \operatorname{Sym}\Bigg[\prod_{i=1}^n\prod_{a=1}^{d_i}\frac{1}{t_{a+d_{<i}}-z_i}\Bigg],$$

where $\mathbf{d}=(d_1,\dots,d_n)\in\mathbb{Z}^n_{\geqslant 0}\,,\ r=\sum_{i=1}^n d_i\,,$ and

$$\operatorname{Sym} f(t_1,\dots,t_r) = \sum_\sigma f(t_{\sigma_1},\dots,t_{\sigma_r})\,.$$

Fix a complex number κ. Define a $(V_{l_1}\otimes\dots\otimes V_{l_n})[m_1,m_2]$-valued function $U_\gamma(z_1,\dots,z_n;\lambda_1,\lambda_2)$ by the formula

$$U_\gamma(z_1,\dots,z_n;\lambda_1,\lambda_2;\boldsymbol{l},\boldsymbol{m}) \tag{7.1}$$

$$= \int\limits_{\gamma(z_1,\dots,z_n;\lambda_1,\lambda_2)}\big(\Phi_{m_2}(t_1,\dots,t_{m_2};z_1,\dots,z_n;\lambda_1,\lambda_2;\boldsymbol{l})\big)^{1/\kappa}$$

$$\times\sum_{\mathbf{d}\in\mathcal{Z}[\boldsymbol{l},\boldsymbol{m}]} w_{\mathbf{d}}(t_1,\dots,t_{m_2};z_1,\dots,z_n)\,v_{\mathbf{d}}\,d^{m_2}t\,.$$

The function depends on the choice of integration chains $\gamma(z_1,\dots,z_n;\lambda_1,\lambda_2)$. We assume that for each $z_1,\dots,z_n$, λ_1,λ_2 the chain lies in $\mathbb{C}^{m_2}$ with coordinates $t_1,\dots,t_{m_2}$, and the chains form a horizontal family of m_2-dimensional homology classes with respect to the multivalued function $\big(\Phi_{m_2}(t_1,\dots,t_{m_2};z_1,\dots,z_n;\lambda_1,\lambda_2;\boldsymbol{l})\big)^{1/\kappa}$, see a more precise statement below and in [FMTV].

Theorem 7.1. *For any choice of the horizontal family* γ, *the function* $U_\gamma(z_1,\dots,z_n;\lambda_1,\lambda_2;\boldsymbol{l},\boldsymbol{m})$ *is a solution of the KZ and DD equations, see* (3.4), (3.8), *with values in* $(V_{l_1}\otimes\dots\otimes V_{l_n})[m_1,m_2]$.

The theorem is a corollary of Theorem 3.1 in [FMTV]. For the *KZ* equation at $\lambda_1=\lambda_2$ the theorem follows from the results of [SV], [V].

There exist special horizontal families of integration chains in (7.1) labelled by elements of $\mathcal{Z}[\boldsymbol{l},\boldsymbol{m}]$. They are described below. To simplify exposition we will assume that $\operatorname{Re}\big((\lambda_1-\lambda_2)/\kappa\big)>0$ and $\operatorname{Im} z_1<\dots<\operatorname{Im} z_n$.

Let $\mathbf{d} = (d_1, \dots, d_n)$, $r = \sum_{i=1}^{n} d_i$. Set $\gamma_{\mathbf{d}}(z_1, \dots, z_n) = \mathcal{C}_1 \times \dots \times \mathcal{C}_r$, where $\mathcal{C}_1, \dots, \mathcal{C}_r$ is a collection of non-intersecting oriented loops in $\mathbb{C}$ such that d_i loops start at $+\infty$, go around z_i, and return to $+\infty$, see the picture for $n = 2$:

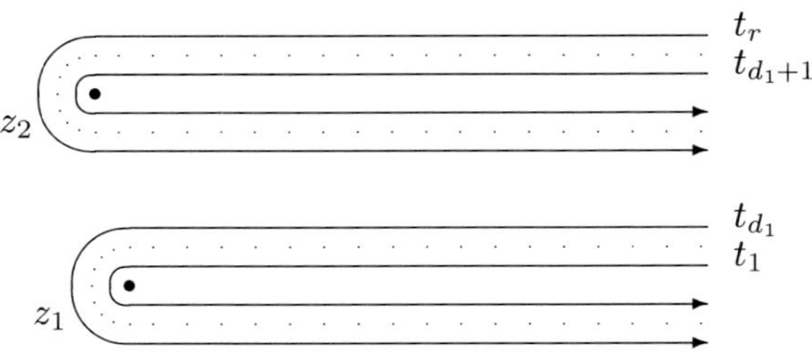

Picture 1. The contour $\gamma_{\mathbf{d}}$.

One can see that for any $\mathbf{d} \in \mathcal{Z}[\boldsymbol{l}, \boldsymbol{m}]$ the family of chains $\gamma_{\mathbf{d}}$ is horizontal. Therefore, the function $U_{\mathbf{d}}(z_1, \dots, z_n; \lambda_1, \lambda_2; \boldsymbol{l}, \boldsymbol{m}) = U_{\gamma_{\mathbf{d}}}(z_1, \dots, z_n; \lambda_1, \lambda_2; \boldsymbol{l}, \boldsymbol{m})$ is a solution of systems (3.4) and (3.8). A univalued branch of the integrand in (7.1) is fixed by assuming that at the point of $\gamma_{\mathbf{d}}$ where all numbers $t_{a+d_{<i}} - z_i$, $i = 1, \dots, n$, $a = 1, \dots, d_i$, are negative one has

$$-\pi < \arg(z_i - z_j) < 0, \qquad -2\pi < \arg(t_a - z_i) < 0, \qquad -\pi < \arg(t_a - t_b) \leqslant 0,$$

for $i = 1, \dots, n$, $j = i+1, \dots, n$, $a = 1, \dots, m_2$, $b = a+1, \dots, m_2$.

The solution $U_{\mathbf{d}}(z_1, \dots, z_n; \lambda_1, \lambda_2; \boldsymbol{l}, \boldsymbol{m})$ is distinguished by the following property.

Theorem 7.2. *Let* $\operatorname{Im}(z_i - z_{i+1}) \to -\infty$ *for all* $i = 1, \dots, n-1$. *Then for any* $\mathbf{d} \in \mathcal{Z}[\boldsymbol{l}, \boldsymbol{m}]$ *one has*

$$U_{\mathbf{d}}(z_1, \dots, z_n; \lambda_1, \lambda_2; \boldsymbol{l}, \boldsymbol{m}) \tag{7.2}$$

$$= (2\pi i)^{m_2} e^{\pi i \xi_{\mathbf{d}}(\boldsymbol{l})/\kappa} \left(\Xi_{\mathbf{d}}(z_1, \dots, z_n; \lambda_1, \lambda_2; \boldsymbol{l}, \boldsymbol{m})\right)^{1/\kappa}$$

$$\times \prod_{j=1}^{n} \prod_{s=0}^{d_j - 1} \frac{\Gamma(-1/\kappa)}{\Gamma\bigl(1 + (l_j - s)/\kappa\bigr)\,\Gamma\bigl(-(s+1)/\kappa\bigr)} \bigl(v_{\mathbf{d}} + o(1)\bigr)$$

where $\xi_{\mathbf{d}}(\boldsymbol{l}) = \sum_{1 \leqslant i \leqslant j \leqslant n} l_i d_j$ *and*

$$\Xi_{\mathbf{d}}(z_1, \dots, z_n; \lambda_1, \lambda_2; \boldsymbol{l}, \boldsymbol{m}) = \kappa^{-m_2} e^{\lambda_1 \sum_{i=1}^{n} z_i (l_i - d_i) + \lambda_2 \sum_{i=1}^{n} z_i d_i}$$

$$\times \bigl((\lambda_1 - \lambda_2)/\kappa\bigr)^{\sum_{i=1}^{n} d_i (l_i - d_i)} \prod_{1 \leqslant i < j \leqslant n} (z_i - z_j)^{(l_i - d_i)(l_j - d_j) + d_i d_j}.$$

Theorem 7.2 implies that the set of solutions $U_{\mathbf{d}}$, $\mathbf{d} \in \mathcal{Z}[\boldsymbol{l},\boldsymbol{m}]$, of systems (3.4) and (3.8) is complete, that is, any solution of those systems taking values in $(V_{l_1} \otimes \cdots \otimes V_{l_n})[m_1, m_2]$ is a linear combination of functions $U_{\mathbf{d}}$.

There is a similar statement for asymptotics of $U_{\mathbf{d}}(z_1, \dots, z_n; \lambda_1, \lambda_2; \boldsymbol{l}, \boldsymbol{m})$ with respect to λ_1, λ_2.

Theorem 7.3. *Let* $\operatorname{Re}\bigl((\lambda_1 - \lambda_2)/\kappa\bigr) \to +\infty$. *Then for any* $\mathbf{d} \in \mathcal{Z}[\boldsymbol{l},\boldsymbol{m}]$ *formula* (7.2) *holds.*

The proof of Theorems 7.2 and 7.3 uses the following Selberg-type integral

$$\int_{\gamma_m} e^{-\nu \sum_{a=1}^{m} s_a} \prod_{a=1}^{m} (-s_a)^{-1-l/\kappa} \prod_{1 \leqslant a < b \leqslant m} (s_a - s_b)^{2/\kappa} \, d^m s \tag{7.3}$$

$$= (-2\pi i)^m \, \nu^{m(l-m+1)/\kappa} \prod_{j=0}^{m-1} \frac{\Gamma(1 - 1/\kappa)}{\Gamma\bigl(1 + (l-j)/\kappa\bigr) \, \Gamma\bigl(1 - (j+1)/\kappa\bigr)},$$

where $\operatorname{Re} \nu > 0$, $\gamma_m = \bigl\{(s_1, \dots, s_m) \in \mathbb{C}^m \mid s_a \in \mathcal{C}_a, \ a = 1, \dots, m\bigr\}$, and $\mathcal{C}_1, \dots, \mathcal{C}_m$ are non-intersecting oriented loops in $\mathbb{C}$ which start at $+\infty$, go around zero, and return to $+\infty$, the loop $\mathcal{C}_a$ being inside $\mathcal{C}_b$ for $a < b$, see the picture:

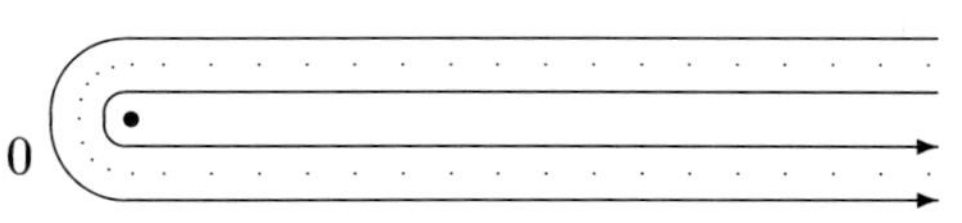

s_m
s_1
0

Picture 2. The contour γ_m.

A univalued branch of the integrand in (7.3) is fixed by assuming that at the point of γ_m where all numbers $s_1, \dots, s_m$ are negative one has $\arg(-s_1) = \dots = \arg(-s_m) = 0$ and $\arg(s_a - s_b) = 0$ for $1 \leqslant a < b \leqslant m$.

The construction of hypergeometric solutions of the trigonometric *KZ* equations (3.5) and the difference dynamical equations (4.5) is similar. We describe it below.

Define the master function

$$\Psi_r(t_1, \dots, t_r; z_1, \dots, z_n; \lambda_1, \lambda_2; \boldsymbol{l}, \boldsymbol{m}) \tag{7.4}$$

$$= \prod_{i=1}^{n} z_i^{l_i(\lambda_1 - m_1 + l_i/2)} \prod_{1 \leqslant i < j \leqslant n} (z_i - z_j)^{l_i l_j}$$

$$\times \prod_{a=1}^{r} t_a^{\lambda_2 - \lambda_1 + m_1 - m_2 + 1} \prod_{a=1}^{r} \prod_{i=1}^{n} (t_a - z_i)^{-l_i} \prod_{1 \leqslant a < b \leqslant r} (t_a - t_b)^2,$$

and a $(V_{l_1}\otimes\cdots\otimes V_{l_n})[m_1,m_2]$-valued function

$$\widetilde{U}_\delta(z_1,\dots,z_n;\lambda_1,\lambda_2;\boldsymbol{l},\boldsymbol{m}) \tag{7.5}$$
$$= \int\limits_{\delta(z_1,\dots,z_n;\lambda_1,\lambda_2)} \bigl(\Psi_{m_2}(t_1,\dots,t_{m_2};z_1,\dots,z_n;\lambda_1,\lambda_2;\boldsymbol{l},\boldsymbol{m})\bigr)^{1/\kappa} \times \sum_{\mathbf{d}\in\mathcal{Z}[\boldsymbol{l},\boldsymbol{m}]} w_{\mathbf{d}}(t_1,\dots,t_{m_2};z_1,\dots,z_n)\,v_{\mathbf{d}}\,d^{m_2}t\,.$$

The function depends on the choice of integration chains $\delta(z_1,\dots,z_n;\lambda_1,\lambda_2)$. We assume that for each $z_1,\dots,z_n$, λ_1,λ_2 the chain lies in $\mathbb{C}^{m_2}$ with coordinates $t_1,\dots,t_{m_2}$, and the chains form a horizontal family of m_2-dimensional homology classes with respect to the multivalued function $\bigl(\Psi_{m_2}(t_1,\dots,t_{m_2};z_1,\dots,z_n;\lambda_1,\lambda_2;\boldsymbol{l},\boldsymbol{m})\bigr)^{1/\kappa}$, see a more precise statement below and in [MV].

Theorem 7.4. *For any choice of the horizontal family δ, the function $U_\delta(z_1,\dots,z_n;\lambda_1,\lambda_2;\boldsymbol{l},\boldsymbol{m})$ is a solution of the trigonometric KZ and rational qDD equations, see* (3.5), (4.5), *with values in $(V_{l_1}\otimes\cdots\otimes V_{l_n})[m_1,m_2]$.*

The theorem is a direct corollary of results in [MV]. Another way of writing down hypergeometric solutions of the trigonometric *KZ* equations is given in [Ch1].

There exist special horizontal families of integration chains in (7.5) labelled by elements of $\mathcal{Z}[\boldsymbol{l},\boldsymbol{m}]$. They are described below. To simplify exposition we will assume that $\operatorname{Re}\bigl((\lambda_1-\lambda_2)/\kappa\bigr)$ is large positive and $\arg z_1<\ldots<\arg z_n<\arg z_1+2\pi$, that is, all the ratios z_i/z_j for $i\neq j$ are not real positive, and $z_1,\dots,z_n$ are ordered counterclockwise. Recall that all $z_1,\dots,z_n$ are nonzero.

Let $\mathbf{d}=(d_1,\dots,d_n)$, $r=\sum_{i=1}^{n} d_i$. Set $\delta_{\mathbf{d}}(z_1,\dots,z_n)=\mathcal{C}_1\times\cdots\times\mathcal{C}_r$, where $\mathcal{C}_1,\dots,\mathcal{C}_r$ is a collection of non-intersecting oriented loops in $\mathbb{C}$ such that d_i loops start at infinity in the direction of z_i, go around z_i, and return to infinity in the same direction, see the picture for $n=2$:

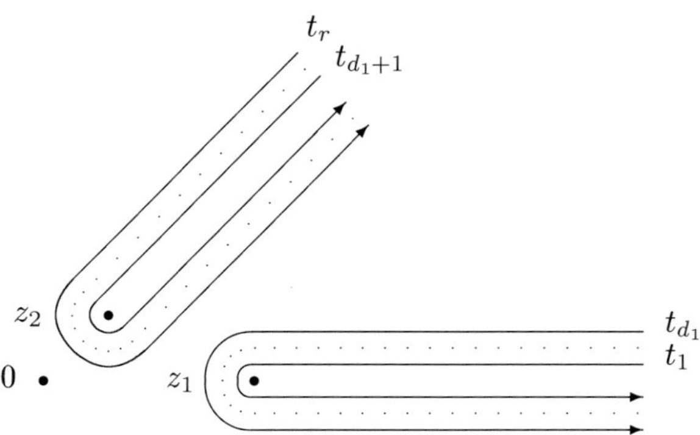

Picture 3. The contour $\delta_{\mathbf{d}}$.

One can see that for any $\mathbf{d} \in \mathcal{Z}[\boldsymbol{l},\boldsymbol{m}]$ the family of chains $\delta_{\mathbf{d}}$ is horizontal. Therefore, the function $\widetilde{U}_{\mathbf{d}}(z_1,\ldots,z_n;\lambda_1,\lambda_2;\boldsymbol{l},\boldsymbol{m}) = \widetilde{U}_{\delta_{\mathbf{d}}}(z_1,\ldots,z_n;\lambda_1,\lambda_2;\boldsymbol{l},\boldsymbol{m})$ is a solution of systems (3.5) and (4.5). A univalued branch of the integrand in (7.5) is fixed by assuming that at the point of $\delta_{\mathbf{d}}$ where all ratios $t_{a+d_{<i}}/z_i$, $i=1$, $\ldots,n$, $a=1,\ldots,d_i$, are real and belong to $(0,1)$ one has

$$\arg t_{a+d_{<i}} = \arg z_i\,, \qquad i=1,\ldots,n, \quad a=1,\ldots,d_i\,,$$

and

$$-\pi < \arg(z_i - z_j) - \arg z_i < \pi\,,$$

$$-2\pi < \arg(t_a - z_i) - \arg z_i < 0\,,$$

$$-\pi < \arg(t_a - t_b) - \arg t_a < \pi\,,$$

for $i=1,\ldots,n$, $j=i+1,\ldots,n$, $a=1,\ldots,m_2$, $b=a+1,\ldots,m_2$. Recall, it is assumed that $\arg z_1 < \ldots < \arg z_n < \arg z_1 + 2\pi$,

There is an analogue of Theorem 7.2 which describes asymptotics of the functions $\widetilde{U}_{\mathbf{d}}(z_1,\ldots,z_n;\lambda_1,\lambda_2;\boldsymbol{l},\boldsymbol{m})$ as $z_i/z_{i+1} \to 0$ for all $i=1,\ldots,n-1$. The corresponding formulae are similar to (7.2), but more involved. Asymptotics of $\widetilde{U}_{\mathbf{d}}$ with respect to λ_1,λ_2 are as follows.

Theorem 7.5. *Let* $(\lambda_1-\lambda_2)/\kappa \to +\infty$. *Then*

$$\widetilde{U}_{\mathbf{d}}(z_1,\ldots,z_n;\lambda_1,\lambda_2;\boldsymbol{l},\boldsymbol{m}) \tag{7.6}$$

$$= (2\pi i)^{m_2}\, e^{\pi i \xi_{\mathbf{d}}(\boldsymbol{l})/\kappa} \left(\widetilde{\Xi}_{\mathbf{d}}(z_1,\ldots,z_n;\lambda_1,\lambda_2;\boldsymbol{l},\boldsymbol{m})\right)^{1/\kappa}$$

$$\times \prod_{j=1}^{n}\prod_{s=0}^{d_j-1} \frac{\Gamma(-1/\kappa)}{\Gamma(1+(l_j-s)/\kappa)\,\Gamma(-(s+1)/\kappa)} \left(v_{\mathbf{d}} + o(1)\right)$$

where $\xi_{\mathbf{d}}(\boldsymbol{l}) = \sum_{1\leqslant i\leqslant j\leqslant n} l_i d_j$ *and*

$$\widetilde{\Xi}_{\mathbf{d}}(z_1,\ldots,z_n;\lambda_1,\lambda_2;\boldsymbol{l},\boldsymbol{m}) = \left((\lambda_1-\lambda_2)/\kappa\right)^{\sum_{i=1}^{n} d_i(l_i-d_i+1)}$$

$$\times \prod_{i=1}^{n} z_i^{(\lambda_1-m_1)(l_i-d_i)+(\lambda_2-m_2)d_i+((l_i-d_i)^2+d_i^2)/2} \prod_{1\leqslant i<j\leqslant n} (z_i-z_j)^{(l_i-d_i)(l_j-d_j)+d_i d_j}\,.$$

The construction of hypergeometric solutions of the *qKZ* equations (3.5) and trigonometric *DD* equations (3.9) goes along the same lines as for hypergeometric solutions of the *KZ* and rational dynamical equations, but instead of ordinary hypergeometric integrals it employs q-hypergeometric integrals of Mellin-Barnes type, see [TV2].

Define the *q-master function*

$$\hat{\Phi}_r(t_1,\dots,t_r;z_1,\dots,z_n;\lambda_1,\lambda_2;\boldsymbol{l};\kappa) \tag{7.7}$$

$$= \lambda_1^{(r/2+\sum_{i=1}^{n}(z_i l_i - l_i^2/2)-\sum_{a=1}^{r} t_a)/\kappa}\,\lambda_2^{(r/2+\sum_{a=1}^{r} t_a)/\kappa}\,(\lambda_1-\lambda_2)^{-r/\kappa}$$

$$\times \prod_{a=1}^{r}\prod_{i=1}^{n}\frac{\Gamma\bigl((t_a-z_i)/\kappa\bigr)}{\Gamma\bigl((t_a-z_i+l_i)/\kappa\bigr)}\prod_{1\leqslant a<b\leqslant r}\frac{\Gamma\bigl((t_a-t_b+1)/\kappa\bigr)}{\Gamma\bigl((t_a-t_b-1)/\kappa\bigr)},$$

the *rational weight function*

$$\hat{w}_{\mathbf{d}}(t_1,\dots,t_r;z_1,\dots,z_n;\boldsymbol{l}) = \prod_{1\leqslant a<b\leqslant r}\frac{t_a-t_b}{t_a-t_b-1}$$

$$\times\ \mathrm{Sym}\Bigl[\prod_{j=1}^{n}\prod_{a=1}^{d_j}\Bigl(\frac{1}{t_{a+d_{<j}}-z_j+l_j}\prod_{1\leqslant p<j}\frac{t_{a+d_{<j}}-z_p}{t_{a+d_{<j}}-z_p+l_p}\Bigr)\prod_{1\leqslant a<b\leqslant r}\frac{t_a-t_b-1}{t_a-t_b}\Bigr],$$

where $\mathbf{d}=(d_1,\dots,d_n)\in\mathbb{Z}^n_{\geqslant 0}$, $r=\sum_{j=1}^{n} d_j$, and the *trigonometric weight function*

$$W_{\mathbf{d}}(t_1,\dots,t_r;z_1,\dots,z_n;\boldsymbol{l}) = \prod_{1\leqslant a<b\leqslant r}\frac{\sin\bigl(\pi(t_a-t_b)/\kappa\bigr)}{\sin\bigl(\pi(t_a-t_b-1)/\kappa\bigr)}$$

$$\times\ \mathrm{Sym}\Bigl[\prod_{j=1}^{n}\prod_{a=1}^{d_j}\Bigl(\frac{e^{\pi i(z_j-t_{a+d_{<j}})/\kappa}}{\sin\bigl(\pi(t_{a+d_{<j}}-z_j+l_j)/\kappa\bigr)}\prod_{1\leqslant p<j}\frac{\sin\bigl(\pi(t_{a+d_{<j}}-z_p)/\kappa\bigr)}{\sin\bigl(\pi(t_{a+d_{<j}}-z_p+l_p)/\kappa\bigr)}\Bigr)$$

$$\times\prod_{1\leqslant a<b\leqslant r}\frac{\sin\bigl(\pi(t_a-t_b-1)/\kappa\bigr)}{\sin\bigl(\pi(t_a-t_b)/\kappa\bigr)}\Bigr].$$

For simplicity of exposition from now on we assume that κ is a real positive number and the ratio λ_2/λ_1 is not real positive. For any $\mathbf{d}\in\mathcal{Z}[\boldsymbol{l},\boldsymbol{m}]$ define a $(V_{l_1}\otimes\dots\otimes V_{l_n})[m_1,m_2]$-valued function

$$\widehat{U}_{\mathbf{d}}(z_1,\dots,z_n;\lambda_1,\lambda_2;\boldsymbol{l},\boldsymbol{m}) \tag{7.8}$$

$$= \int\limits_{\mathbb{I}(z_1,\dots,z_n;\boldsymbol{l})}\hat{\Phi}_{m_2}(t_1,\dots,t_{m_2};z_1,\dots,z_n;\lambda_1,\lambda_2;\boldsymbol{l})\,W_{\mathbf{d}}(t_1,\dots,t_{m_2};z_1,\dots,z_n;\boldsymbol{l})$$

$$\times\sum_{\mathbf{p}\in\mathcal{Z}[\boldsymbol{l},\boldsymbol{m}]}\hat{w}_{\mathbf{p}}(t_1,\dots,t_{m_2};z_1,\dots,z_n;\boldsymbol{l})\,v_{\mathbf{p}}\,d^{m_2}t,$$

the integration contour $\mathbb{I}(z_1,\dots,z_n;\boldsymbol{l})$ being described below. For the factors $(\lambda_2/\lambda_1)^{t_a/\kappa}$ in the integrand it is assumed that $0<\arg(\lambda_2/\lambda_1)<2\pi$.

The integral in (7.8) is defined by analytic continuation with respect to z_1, $\ldots, z_n$ and $\boldsymbol{l} = (l_1, \ldots, l_n)$ from the region where $\operatorname{Re} z_1 = \cdots = \operatorname{Re} z_n = 0$ and $\operatorname{Re} l_i < 0$ for all $i = 1, \ldots, n$. In that case

$$\mathbb{I}(z_1, \ldots, z_n; \boldsymbol{l}) = \{(t_1, \ldots, t_{m_2}) \in \mathbb{C}^{m_2} \mid \operatorname{Re} t_1 = \cdots = \operatorname{Re} t_{m_2} = \varepsilon\}$$

where ε is a positive number less then $\min(-\operatorname{Re} l_1, \ldots, -\operatorname{Re} l_n)$. In the considered region of parameters the integrand is well defined on $\mathbb{I}(z_1, \ldots, z_n; \boldsymbol{l})$ and the integral is convergent. It is also known that $\widehat{U}_{\mathbf{d}}(z_1, \ldots, z_n; \lambda_1, \lambda_2; \boldsymbol{l}, \boldsymbol{m})$ can be analytically continued to a value of $\boldsymbol{l}$ in $\mathbb{Z}^n_{\geqslant 0}$ and generic values of $z_1, \ldots, z_n$, if $\mathbf{d} \in \mathcal{Z}[\boldsymbol{l}, \boldsymbol{m}]$ at that point, and the analytic continuation is given by the integral over a suitable deformation of the imaginary plane $\{(t_1, \ldots, t_{m_2}) \in \mathbb{C}^{m_2} \mid \operatorname{Re} t_1 = \cdots = \operatorname{Re} t_{m_2} = 0\}$, see [MuV].

Theorem 7.6. *For any* $\mathbf{d} \in \mathcal{Z}[\boldsymbol{l}, \boldsymbol{m}]$ *the function* $\widehat{U}_{\mathbf{d}}(z_1, \ldots, z_n; \lambda_1, \lambda_2; \boldsymbol{l}, \boldsymbol{m})$ *is a solution of the rational qKZ and trigonometric DD equations, see* (5.6), (3.9), *with values in* $(V_{l_1} \otimes \cdots \otimes V_{l_n})[m_1, m_2]$.

The part of the theorem concerning the *qKZ* equations is a direct corollary of the construction of q-hypergeometric solutions of the *qKZ* equations given in [TV2], [MuV]. The part of the theorem on the trigonometric *DD* equations is obtained in [TV8].

The solution $\widehat{U}_{\mathbf{d}}(z_1, \ldots, z_n; \lambda_1, \lambda_2; \boldsymbol{l}, \boldsymbol{m})$ of systems (5.6) and (3.9) is distinguished by the following property.

Theorem 7.7. *Let* $\operatorname{Re}(z_i - z_{i+1}) \to +\infty$ *for all* $i = 1, \ldots, n-1$. *Then for any* $\mathbf{d} \in \mathcal{Z}[\boldsymbol{l}, \boldsymbol{m}]$ *one has*

$$\widehat{U}_{\mathbf{d}}(z_1, \ldots, z_n; \lambda_1, \lambda_2; \boldsymbol{l}, \boldsymbol{m}) \tag{7.9}$$

$$= (-2i)^{m_2}\, m_2!\, e^{\pi i \zeta_{\mathbf{d}}(\boldsymbol{l})/\kappa} \left(\widehat{\Xi}_{\mathbf{d}}(z_1, \ldots, z_n; \lambda_1, \lambda_2; \boldsymbol{l}, \boldsymbol{m})\right)^{1/\kappa}$$

$$\times \prod_{j=1}^{n} \left(d_j! \prod_{s=0}^{d_j - 1} \frac{\Gamma((s - l_j)/\kappa)\,\Gamma(1 + (s+1)/\kappa)}{\Gamma(1 + 1/\kappa)} \right) (v_{\mathbf{d}} + o(1))\,.$$

where $\zeta_{\mathbf{d}}(\boldsymbol{l}) = \sum_{i=1}^{n} d_j(2l_j - d_j + 1)/2$, *and*

$$\widehat{\Xi}_{\mathbf{d}}(z_1, \ldots, z_n; \lambda_1, \lambda_2; \boldsymbol{l}, \boldsymbol{m}) = \lambda_1^{\sum_{i=1}^{n}(z_i(l_i - d_i) - l_i^2/2 + d_i^2/2)}\, \lambda_2^{\sum_{i=1}^{n} d_i(z_i - l_i + d_i/2)}$$

$$\times (\lambda_1 - \lambda_2)^{\sum_{i=1}^{n} d_i(l_i - d_i)} \prod_{1 \leqslant i < j \leqslant n} ((z_i - z_j)/\kappa)^{(l_i - d_i)(l_j - d_j) + d_i d_j - l_i l_j}$$

Recall that κ *is assumed to be a real positive number.*

Theorem 7.2 implies that the set of solutions $\widehat{U}_{\mathbf{d}}$, $\mathbf{d} \in \mathcal{Z}[\boldsymbol{l},\boldsymbol{m}]$, of systems (5.6) and (3.9) is complete, that is, any solution of those systems taking values in $(V_{l_1} \otimes \cdots \otimes V_{l_n})[m_1, m_2]$ is a linear combination of functions $U_{\mathbf{d}}$.

The proof of Theorem 7.2 uses the following Selberg-type integral

$$\int\limits_{\mathbb{I}_m(l)} (-x)^{\sum_{a=1}^m s_a} \prod_{a=1}^m \Gamma(s_a)\,\Gamma(-s_a - l/\kappa) \prod_{\substack{a,b=1\\ a\neq b}}^m \frac{\Gamma(s_a - s_b + 1/\kappa)}{\Gamma(s_a - s_b)}\, d^m s \tag{7.10}$$

$$= (2\pi i)^m\, (-x)^{(m-1-2l)m/2\kappa}\, (1-x)^{m(l-m+1)/\kappa}$$

$$\times \prod_{j=0}^{m-1} \frac{\Gamma((j-l)/\kappa)\,\Gamma(1+(j+1)/\kappa)}{\Gamma(1+1/\kappa)},$$

where $-\pi < \arg(-x) < \pi$ and $-\pi < \arg(1-x) < \pi$. The integral is defined by analytic continuation from the region where κ is real positive and $\operatorname{Re} l$ is negative. In that case

$$\mathbb{I}_m(l) = \bigl\{(s_1, \ldots, s_m) \in \mathbb{C}^m \mid \operatorname{Re} s_1 = \cdots = \operatorname{Re} s_m = -\operatorname{Re} l/2\bigr\}.$$

In the considered region of parameters the integrand in (7.10) is well defined on $\mathbb{I}_m(l)$ and the integral is convergent, see [TV1].

8. Duality for hypergeometric and q-hypergeometric integrals

In this section we consider the $(\mathfrak{gl}_k, \mathfrak{gl}_n)$ duality for the case of $k = n = 2$, and apply the results of the previous sections to obtain identities for hypergeometric and q-hypergeometric integrals of different dimensions.

Further on we fix complex numbers l_1, m_1 and nonnegative integers l_2, m_2 such that $l_1 + l_2 = m_1 + m_2$. Set $\boldsymbol{l} = (l_1, l_2)$ and $\boldsymbol{m} = (m_1, m_2)$.

Let V_l be the irreducible highest weight $\mathfrak{gl}_2$-module with highest weight $(l, 0)$ and highest weight vector v_l. The weight subspace $(V_{l_1} \otimes V_{l_2})[m_1, m_2]$ has a basis given by vectors

$$v_b(\boldsymbol{l},\boldsymbol{m}) = \frac{1}{(m_2-b)!\, b!}\, e_{21}^{(m_2-b)} v_{l_1} \otimes e_{21}^{b} v_{l_2}, \qquad b = 0, \ldots, \min(l_2, m_2),$$

provided that l_1 is not a nonnegative integer or $m_2 \leqslant l_1$. Otherwise, the vectors $v_0(\boldsymbol{l},\boldsymbol{m}), \ldots, v_{m_2-l_1-1}(\boldsymbol{l},\boldsymbol{m})$ equal zero and the basis is given by the rest of the vectors $v_b(\boldsymbol{l},\boldsymbol{m})$. Say that b is admissible if $v_b(\boldsymbol{l},\boldsymbol{m}) \neq 0$.

The weight subspaces $(V_{l_1} \otimes V_{l_2})[m_1, m_2]$ and $(V_{m_1} \otimes V_{m_2})[l_1, l_2]$ are isomorphic. The isomorphism $\phi : (V_{l_1} \otimes V_{l_2})[m_1, m_2] \to (V_{m_1} \otimes V_{m_2})[l_1, l_2]$, cf. (6.13), sends the vector $v_b(\boldsymbol{l},\boldsymbol{m})$ to $v_b(\boldsymbol{m},\boldsymbol{l})$.

Given an admissible integer b let $\mathbf{d} = (m_2 - b, b)$ and $\mathbf{d}' = (l_2 - b, b)$. Consider $(V_{l_1} \otimes V_{l_2})[m_1, m_2]$-valued functions

$$U_b(z_1, z_2; \lambda_1, \lambda_2; \boldsymbol{l}, \boldsymbol{m}) = U_{\mathbf{d}}(z_1, z_2; \lambda_1, \lambda_2; \boldsymbol{l}, \boldsymbol{m})$$

and

$$U'_b(\lambda_1, \lambda_2; z_1, z_2; \boldsymbol{m}, \boldsymbol{l}) = \phi^{-1}\big(U_{\mathbf{d}'}(\lambda_1, \lambda_2; z_1, z_2; \boldsymbol{m}, \boldsymbol{l})\big),$$

where the functions $U_{\mathbf{d}}$ and $U_{\mathbf{d}'}$ are defined in Section 7, cf. (7.1) and Picture 1.

Theorem 8.1. [TV6] *For any* $b = 0, \dots, \min(l_2, m_2)$ *one has*

$$A_b(\boldsymbol{l}, \boldsymbol{m}) U_b(z_1, z_2; \lambda_1, \lambda_2; \boldsymbol{l}, \boldsymbol{m}) = A_b(\boldsymbol{m}, \boldsymbol{l}) U'_b(\lambda_1, \lambda_2; z_1, z_2; \boldsymbol{m}, \boldsymbol{l}) \tag{8.1}$$

where

$$A_b(\boldsymbol{l}, \boldsymbol{m}) = (-2i)^{-m_2} \kappa^{(m_1+1)m_2/\kappa} e^{-\pi i (m_1+m_2-b)m_2/\kappa} \times \prod_{s=0}^{m_2-b-1} \frac{1}{\sin\big(\pi(s+1)/\kappa\big)} \prod_{s=0}^{m_2-1} \frac{\Gamma\big(1+(l_1-s)/\kappa\big)}{\Gamma(-1/\kappa)\,\Gamma\big(1+(s+1)/\kappa\big)}.$$

The idea of the proof of the statement is as follows. By Theorems 7.1 and 6.7 the functions $U_b(z_1, z_2; \lambda_1, \lambda_2; \boldsymbol{l}, \boldsymbol{m})$ and $U'_b(\lambda_1, \lambda_2; z_1, z_2; \boldsymbol{m}, \boldsymbol{l})$ are solutions of the rational differential *KZ* and dynamical equations (3.4) and (3.8). Theorem 7.2 implies that the functions $U_b(z_1, z_2; \lambda_1, \lambda_2; \boldsymbol{l}, \boldsymbol{m})$ with admissible b's form a complete set of solutions, which means that the functions $U'_b(\lambda_1, \lambda_2; z_1, z_2; \boldsymbol{m}, \boldsymbol{l})$ are their linear combinations. The transition coefficients can be found by comparing asymptotics of U_b and U'_b as $z_1 - z_2$ goes to infinity, see Theorems 7.2 and 7.3.

Remark. Equality (8.1) holds for vector-valued functions. That is, it contains several identities of the form: a hypergeometric integral of dimension m_2 (a coordinate of U_b) equals a hypergeometric integral of dimension l_2 (the corresponding coordinate of U'_b).

Consider $(V_{l_1} \otimes V_{l_2})[m_1, m_2]$-valued functions

$$\widehat{U}_b(z_1, z_2; \lambda_1, \lambda_2; \boldsymbol{l}, \boldsymbol{m}) = \widehat{U}_{\mathbf{d}}(z_1, z_2; \lambda_1, \lambda_2; \boldsymbol{l}, \boldsymbol{m})$$

and

$$\widetilde{U}'_b(\lambda_1, \lambda_2; z_1, z_2; \boldsymbol{m}, \boldsymbol{l}) = \prod_{s=0}^{l_2-1} \frac{\Gamma\big((z_1 - z_2 + s - l_1)/\kappa\big)}{\Gamma\big((z_1 - z_2 + s + 1)/\kappa\big)} \; \phi^{-1}\big(\widetilde{U}_{\mathbf{d}'}(\lambda_1, \lambda_2; z_1, z_2; \boldsymbol{m}, \boldsymbol{l})\big),$$

where $\mathbf{d} = (m_2 - b, b)$, $\mathbf{d}' = (l_2 - b, b)$, and the functions $\widehat{U}_{\mathbf{d}}$, $\widetilde{U}_{\mathbf{d}'}$ are defined in Section 7, cf. (7.8), (7.5) and Picture 3.

Theorem 8.2. [TV6] *For any* $b = 0, \dots, \min(l_2, m_2)$ *one has*

$$\widehat{A}_b(\boldsymbol{l}, \boldsymbol{m}) \widehat{U}_b(z_1, z_2; \lambda_1, \lambda_2; \boldsymbol{l}, \boldsymbol{m}) = \widetilde{A}_b(\boldsymbol{m}, \boldsymbol{l}) \widetilde{U}'_b(\lambda_1, \lambda_2; z_1, z_2; \boldsymbol{m}, \boldsymbol{l}) \tag{8.2}$$

where

$$\widehat{A}_b(\boldsymbol{l},\boldsymbol{m}) = (2\pi i)^{-m_2} \prod_{s=0}^{m_2-b-1} \sin\bigl(\pi(l_1-s)/\kappa\bigr) \prod_{s=0}^{m_2-1} \frac{\Gamma\bigl(1+(l_1-s)/\kappa\bigr)\,\Gamma(1+1/\kappa)}{\Gamma\bigl(1+(s+1)/\kappa\bigr)},$$

and

$$\widetilde{A}_b(\boldsymbol{l},\boldsymbol{m}) = (2\pi i)^{-l_2}\, e^{\pi i(-b^2+b(l_2-l_1)+l_1m_2-l_2m_1-m_2(m_2-1)/2)/\kappa}$$
$$\times \prod_{s=0}^{b-1} \frac{1}{\sin\bigl(\pi(s+1)/\kappa\bigr)} \prod_{s=0}^{l_2-1} \frac{\Gamma\bigl(1+(m_1-s)/\kappa\bigr)\,\Gamma\bigl(-(s+1)/\kappa\bigr)}{\Gamma(-1/\kappa)}.$$

The idea of the proof is similar to that of Theorem 8.1. By Theorems 7.6 and 7.4, 6.7 the functions $\widehat{U}_b(z_1,z_2;\lambda_1,\lambda_2;\boldsymbol{l},\boldsymbol{m})$ and $\widetilde{U}'_b(\lambda_1,\lambda_2;z_1,z_2;\boldsymbol{m},\boldsymbol{l})$ are solutions of the rational *qKZ* equations (5.6). Theorem 7.7 implies that the functions $U_b(z_1,z_2;\lambda_1,\lambda_2;\boldsymbol{l},\boldsymbol{m})$ with admissible b's form a complete set of solutions over the field of κ-periodic functions of z_1,z_2 (λ_1,λ_2 are treated as parameters in the present consideration). Therefore, the functions $U'_b(\lambda_1,\lambda_2;z_1,z_2;\boldsymbol{m},\boldsymbol{l})$ as functions of z_1,z_2 are linear combinations of $U_b(z_1,z_2;\lambda_1,\lambda_2;\boldsymbol{l},\boldsymbol{m})$ with periodic coefficients. The coefficients can be found by comparing asymptotics of U_b and U'_b as z_1-z_2 goes to infinity, see Theorems 7.7 and 7.5.

Though one does not need the fact that the functions $\widehat{U}_b(z_1,z_2;\lambda_1,\lambda_2;\boldsymbol{l},\boldsymbol{m})$ and $\widetilde{U}'_b(\lambda_1,\lambda_2;z_1,z_2;\boldsymbol{m},\boldsymbol{l})$ solve the trigonometric *DD* equations (3.9) in the proof of Theorem 8.2, this fact is reflected in formula (8.2) – the coefficients $\widehat{A}_b(\boldsymbol{l},\boldsymbol{m})$ and $\widetilde{A}_b(\boldsymbol{l},\boldsymbol{m})$ do not depend on λ_1,λ_2.

Remark. Similar to (8.1), equality (8.2) contains several identities of the form: a q-hypergeometric integral of dimension m_2 (a coordinate of $\widehat{U}_b$) equals a hypergeometric integral of dimension l_2 (the corresponding coordinate of $\widetilde{U}'_b$).

For $l_2=m_2=1$ formula (8.2) yields the classical equality of integral representations of the Gauss hypergeometric function ${}_2F_1$. For instance, taking $b=0$ and the coordinate at $v_0(\boldsymbol{l},\boldsymbol{m})$, one gets after simple transformations:

$$\frac{1}{2\pi i}\,\frac{\Gamma(\gamma)}{\Gamma(\alpha)\Gamma(\beta)} \int_{-i\infty-\varepsilon}^{+i\infty-\varepsilon} (-x)^s\, \frac{\Gamma(-s)\Gamma(s+\alpha)\Gamma(s+\beta)}{\Gamma(s+\gamma)}\, ds$$
$$= (1-x)^{\gamma-\alpha-\beta}\, \frac{\Gamma(\gamma)}{\Gamma(\alpha)\Gamma(\gamma-\alpha)} \int_1^{+\infty} t^\beta\,(t-1)^{\alpha-1}\,(t-x)^{\beta-\gamma}\, dt$$
$$= \frac{\Gamma(\gamma)}{\Gamma(\alpha)\Gamma(\gamma-\alpha)} \int_0^1 u^{\alpha-1}\,(1-u)^{\gamma-\alpha-1}\,(1-ux)^{-\beta}\, du = {}_2F_1(\alpha,\beta;\gamma;x),$$

where

$$\alpha=-l_1/\kappa, \qquad \beta=(z_1-z_2-l_1)/\kappa, \qquad \gamma=(z_1-z_2-l_1+1)/\kappa.$$

Here it is assumed that $\operatorname{Re}\gamma > \operatorname{Re}\alpha > 0$, $\operatorname{Re}\beta > 0$, $0 < \varepsilon < \min(\operatorname{Re}\alpha, \operatorname{Re}\beta)$, and $-\pi < \arg(-x) < \pi$, $-\pi < \arg(1-x) < \pi$. The second equality is obtained by the change of integration variable $u = (t-1)/(t-x)$.

Theorems 8.1 and 8.2 exhibit the $(\mathfrak{gl}_k, \mathfrak{gl}_n)$ duality for hypergeometric integrals for $k = n = 2$. The proofs of the theorems essentially involve explicit formulae for Selberg-type integrals (7.3) and (7.10). Those integrals are associated with the Lie algebra $\mathfrak{sl}_2$. To extend the duality of hypergeometric integrals to the case of arbitary k, n one needs to know suitable generalizations of the Selberg integral associated with the Lie algebras $\mathfrak{sl}_k$ for $k > 2$. For $k = 3$ the required generalizations were obtained in [TV5], and similar ideas can be used to construct the required Selberg-type integrals associated with the Lie algebras $\mathfrak{sl}_k$ for $k > 3$.

Acknowledgements

The author was supported in part by RFFI grant 02–01–00085a and CRDF grant RM1–2334–MO–02.

References

[AST] R.M. Asherova, Yu.F. Smirnov and V.N. Tolstoy, *Projection operators for simple Lie groups*, Theor. Math. Phys. **8** (1971), no. 2, 813–825.

[B] P. Baumann, *q-Weyl group of a q-Schur algebra*, Preprint (1999).

[Ch1] I. Cherednik, *Integral solutions of trigonometric Knizhnik-Zamolodchikov equations and Kac-Moody algebras*, Publ. RIMS, Kyoto Univ. **27** (1991), no. 5, 727–744.

[Ch2] I. Cherednik, *Lectures on Knizhnik-Zamolodchikov equations and Hecke algebras*, Math. Society of Japan Memoirs **1** (1998), 1–96.

[EFK] P.I. Etingof, I.B. Frenkel and A.A. Kirillov Jr., *Lectures on representation theory and Knizhnik-Zamolodchikov equations*, Math. Surveys and Monographs, vol. 58, AMS, Providence RI, 1998.

[EV] P. Etingof and A. Varchenko, *Dynamical Weyl groups and applications*, Adv. Math. **167** (2002), no. 1, 74–127.

[FMTV] G. Felder, Ya. Markov, V. Tarasov and A. Varchenko, *Differential equations compatible with KZ equations*, Math. Phys. Anal. Geom. **3** (2000), no. 2, 139–177.

[FR] I.B. Frenkel. and N.Yu. Reshetikhin, *Quantum affine algebras and holonomic difference equations*, Comm. Math. Phys. **146** (1992), 1–60.

[H1] J. Harnad, *Dual isomonodromic deformations and moment maps to loop algebras*, Comm. Math. Phys. **166** (1994), no. 2, 337–366.

[H2] J. Harnad, *Quantum isomonodromic deformations and the Knizhnik-Zamolodchikov equations*, CRM Proc. Lecture Notes **9** (1996), 155–161.

[Ho] R. Howe, *Perspectives on invariant theory: Schur duality, multiplicity-free actions and beyond*, Israel Math. Conf. Proc. **8** (1995), 1–182.

[KZ] V. Knizhnik and A. Zamolodchikov, *Current algebra and Wess-Zumino model in two dimensions*, Nucl. Phys. B **247** (1984), 83–103.

[MNO] A. Molev, M. Nazarov and G. Olshanski, *Yangians and classical Lie algebras*, Russian Math. Surveys **51** (1996) 205–282.

[MTV] Ya. Markov, V. Tarasov and A. Varchenko, *The determinant of a hypergeometric period matrix*, Houston J. Math. **24** (1998), no. 2, 197–220.

[MV] Ya. Markov and A. Varchenko, *Hypergeometric solutions of trigonometric KZ equations satisfy dynamical difference equations*, Adv. Math. **166** (2002), no. 1, 100–147.

[MuV] E. Mukhin and A. Varchenko, *The quantized Knizhnik-Zamolodchikov equation in tensor products of irreducible sl(2)-modules*, Calogero-Moser-Sutherland models (Montreal, QC, 1997), CRM Series Math. Phys., Springer, New York, 2000.

[N] M. Nazarov, *Yangians and Capelli identities*, Amer. Math. Society Transl. Ser. 2 **181** (1998), 139–163.

[R] N. Reshetikhin, *The Knizhnik-Zamolodchikov system as a deformation of the isomonodromy problem*, Lett. Math. Phys. **26** (1992) no. 3, 167–177.

[ST] Yu.F. Smirnov and V.N. Tolstoy, *Extremal projectors for usual, super and quantum algebras and their use for solving Yang-Baxter problem*, in Selected topics in QFT and Mathematical Physics, (Proc. of the V-th Intern. Conf., Liblice, Czechoslovakia, June 25–30, 1989), World Scientific, Singapore, 1990.

[SV] V.V. Schechtman and A.N. Varchenko, *Arrangements of hyperplanes and Lie algebras homology*, Invent. Math. **106** (1991), 139–194.

[TL] V. Toledano-Laredo, *A Kohno-Drinfeld theorem for quantum Weyl groups*, Duke Math. J. **112** (2002), no. 3, 421–451.

[TV1] V. Tarasov and A. Varchenko, *Jackson integral representations of solutions of the quantized Knizhnik-Zamolodchikov equation*, Leningrad Math. J. **6** (1995), no. 2, 275–313.

[TV2] V. Tarasov and A. Varchenko, *Geometry of q-hypergeometric functions as a bridge between Yangians and quantum affine algebras*, Invent. Math. **128** (1997), 501–588.

[TV3] V. Tarasov and A. Varchenko, *Difference equations compatible with trigonometric KZ differential equations*, Int. Math. Res. Notices (2000), no. 15, 801–829.

[TV4] V. Tarasov and A. Varchenko, *Duality for Knizhnik-Zamolodchikov and dynamical equations*, Acta Appl. Math. **73** (2002), no. 1-2, 141–154.

[TV5] V. Tarasov and A. Varchenko, *Selberg integrals associated with* $\mathfrak{sl}_3$, Lett. Math. Phys. **65** (2003), no. 2, 173–185.

[TV6] V. Tarasov and A. Varchenko, *Identities for hypergeometric integrals of different dimensions*, Lett. Math. Phys. (2005), to appear; `math.QA/0305224`.

[TV7] V. Tarasov and A. Varchenko, *Identities between q-hypergeometric integrals and hypergeometric integrals of different dimensions*, Adv. in Math. **191** (2005), no. 1, 29–45.

[TV8] V. Tarasov and A. Varchenko, *Differential equations compatible with rational qKZ difference equations*, Lett. Math. Phys. (2005), to appear; `math.QA/0403416`.

[V] A. Varchenko, *Multidimensional hypergeometric functions and representation theory of Lie algebras and quantum groups*, Advanced Series in Math. Phys., vol. 21, World Scientific, Singapore, 1995.

[Zh1] D.P. Zhelobenko, *Compact Lie groups and their representations*, Transl. Math. Mono., vol. 40 AMS, Providence RI, 1983.

[Zh2] D.P. Zhelobenko, *Extremal cocycles on Weyl groups* Func. Analys. Appl. **21** (1987), no. 3, 183–192.

[Zh3] D.P. Zhelobenko, *Extremal projectors and generalized Mickelsson algebras on reductive Lie algebras* Math. USSR, Izvestiya **33** (1989), no. 1, 85–100.

V. Tarasov
St. Petersburg Branch of Steklov Mathematical Institute
Fontanka 27
St. Petersburg 191023, Russia
Department of Mathematical Sciences
Indiana University Purdue University at Indianapolis
Indianapolis, IN, 46202-3216, USA
e-mail: vt@pdmi.ras.ru, vt@math.iupui.edu